Quols.

anthrop.

HUMAN ORIGINS

VOLUME I

TO
MY WIFE

PREFACE

IN scope, this work covers not only the origin and development of the genus *Homo,* but also the origin and development of human mentality as reflected in man's discoveries, inventions, and all the activities which enter into the warp and woof of human culture. Since it has to do with man's organic evolution as well as his cultural evolution, the name *Human Origins* has been chosen as the title best designed to describe the field to be covered. The theme of cultural evolution has received fuller treatment than that of organic, or physical, evolution; hence the sub-title, "A Manual of Prehistory."

The growth of our knowledge in the field of human origins has been slow and laborious, aided on the one hand by the growth of a number of cognate sciences, and on the other hand hindered by the various difficulties that are of necessity inherent in self-investigation. Among these difficulties prejudice and tradition stand out prominently. The cognate sciences that have been most helpful in solving the problems of human origins are geology and paleontology, zoölogy and comparative anatomy, psychology, ethnology and ethnography, the social sciences, archeology and comparative philology. The origin of things human is so closely linked with the general field of paleontology that the institute recently founded in Paris by the late Prince of Monaco for their special study was called the Institute of Human Paleontology. History does not take cognizance of man's physical evolution, but it does have a definite bearing on cultural evolution; it bears the same relation to prehistory as biology does to paleontology.

The terms "prehistoric" and "prehistory," so much in vogue now, have been in use but a comparatively short time. It will be of interest to our American readers to recall that the first appearance of one of these words in the title of a book was on March 12,

vii

1862, and that the author was the Canadian (Sir) Daniel Wilson of Toronto. His manuscript of *Prehistoric Man: Researches into the Origin of Civilization in the Old and the New World* was sent to the publishers (Macmillan, London) in January, 1861. Many are the authors who have since made titular use of one or the other of these two terms.

The gap between history and prehistory is filled by archeology, which has well earned the right to be called the retriever of history, for through it the historic period is in a constant process of being not only rounded out, but also pushed farther back into the remote past. There are thus two arms to the service of archeology as the retriever of history—a prehistoric and an historic arm. As examples of the latter we have Mesopotamian archeology, Egyptology, and Greek, Etruscan, and Roman archeology. These, which deal with the period of recorded history, are subjects beyond the scope of the present work.

The term "history" is applied to the record of man's doings as revealed through written documents; the term "prehistory" is applied to the human period antedating the historic. Back of prehistory lies the great field of the geologic past. Prehistory therefore is a middle term, a link connecting history with geology; it is a field the boundary lines of which are still in process of determination. Through it new ground has been gained for history; on the other hand, its own field has been found to overlie more and more the field of geology. This recognition of common ground is reflected in the extent to which the geologic method is employed by the prehistorian as well as in prehistoric terminology.

The adoption by prehistorians of the stratigraphic method as a guiding principle has placed prehistoric research on a scientific basis and given a new and sustained impetus to archeologic endeavor. It is in recognition of the importance of this method that the author has devoted a long appendix to culture sequence as revealed by practically all the known Paleolithic stations. This feature is the result of a long period of painstaking labor, including the checking up of the data by a personal visit to a majority of the sites listed; and in order to make this section of the greatest possible use to the student, the summary of each station is accompanied by its own carefully selected bibliography. This portion

of the book is unique, and the author hopes it may be as useful
to his readers as it has been to him in the preparation of the rest
of the work.

The studies that have culminated in these two volumes have
covered a period of nearly thirty years and are based chiefly on the
author's personal observations in the field supplemented by studies
in the principal museums. The volumes therefore represent what
the author has learned about human origins as revealed through
researches in the European field.

Many of the data gathered come from first-hand knowledge,
which only actual personal excavation can give. Some of the work
was done alone, some in collaboration with other prehistorians.
Part of it could not have been so successfully accomplished without
the local guidance of a host of the author's European colleagues;
and to these, one and all, he wishes to acknowledge his deep
indebtedness. Space for a complete list of these has been reserved
on the pages of the author's memory, since space in this preface
is of necessity denied to all save a few names. To mention only
those now living, special indebtedness must be acknowledged to the
Abbé H. Breuil, professor in the Institut de Paléontologie Humaine,
Paris; Professors M. Boule of the Muséum d'Histoire Naturelle
and R. Verneau of the Laboratory of Anthropology, Paris; Count
Begouen, professor in the University of Toulouse; Count R. de
Saint-Périer, Morigny (Seine-et-Oise); Professors L. Capitan, L.
Manouvrier, and A. de Mortillet of the École d'Anthropologie de
Paris; D. Peyrony, inspector of prehistoric monuments, Les Eyzies
(Dordogne); Dr. Henri-Martin, explorer of the Mousterian station
of La Quina (Charente); Professor H. Obermaier, University of
Madrid; Dr. A. Rutot, Brussels; Dr. O. Tschumi, Berne; Dr. D.
Viollier, Zurich; Dr. Paul Vouga, Neuchâtel; Emil Bächler, St.
Gallen; Professor R. R. Schmidt, founder and present director
of the Urgeschichtliches Forschungsinstitut, Tübingen; Drs. Josef
Bayer and Adolph Mahr, Vienna; Drs. K. Absolon and Josef
Schranil, Czechoslovakia; Dr. A. Smith Woodward, Natural
History Museum, South Kensington, London; Reginald A. Smith,
British Museum; Professor G. Elliot Smith, University College,
London; J. Reid Moir, Ipswich (Suffolk); and Dr. Ernst Antevs,
Sweden.

In a different way, but no less essentially helpful, has been the assistance rendered by Miss Clara M. LeVene, research assistant in the Peabody Museum of Yale University. And above all the author is indebted to his wife, who has constantly shared his labors both in the field and in the office, and who should receive much of the credit for whatever of merit the work may possess.

G. G. MacC.

CONTENTS

CHAPTER III

CHAPTER IV

CHAPTER V

CHAPTER VI

CHAPTER VII

CHAPTER VIII

GLOSSARY

ABIES PECTINATA. European silver fir.

ACHEN. The name of a lake in the Tyrol, also of a stage of ice retreat during the Würm Glacial Epoch.

ACHEULIAN. Third epoch of the Lower Paleolithic Period, named from Saint-Acheul (Somme), France.

ADAPIS. Extinct (Eocene of Europe) genus of the Lemuroid suborder of Primates.

ÆSCULUS HIPPOCASTANUM. Common horse-chestnut.

ALCES LATIFRONS. Moose, a species of the deer family.

ALCES PALMATUS. Elk, a species of the deer family.

ALINEMENT. A row of menhirs, q.v.

ALLUVIUM. The most recent sedimentary deposits, especially such as occur in the valleys of large rivers.

AMORPHOUS. Without definite shape; formless.

AMPHORA. Tall, two-handled earthenware jar with slender neck and usually pointed base, used to hold wine, oil, etc.

AMYGDALOID. Almond-shaped.

ANAPTOMORPHUS HOMUNCULUS. Extinct Primate species belonging to the Lemuroid group.

ANCYLUS FLUVIATILIS. River limpit, distinguished by its patelliform shell; a species of gastropod or snail.

ANDROMEDA POLYFOLIA. Species of the heath family with showy white flowers and evergreen leaves.

ANEPIGRAPHIC. Without inscription.

ANTENNÆ. Paired, lateral, movable, jointed appendages to the head of an insect; also applied to a certain type of sword-handle decoration belonging to the Hallstatt Epoch.

ANTHROPODUS. Extinct (Upper Miocene of Europe) Anthropoid ape.

ANTHROPOMORPH (same as Simiidae). Tailless ape.

ANTICYCLONE. The reverse of cyclone: an atmospheric condition of high central pressure, with currents flowing outward.

ANUBIS. An Egyptian god represented as having the head of a jackal or dog; conductor or guardian of the dead.

ARCHEOLEMUR. Extinct (Pleistocene) ape-like lemur.

ARCTOMYS MARMOTTA. Marmot of Europe and Asia, the corresponding form in America being the woodchuck.

ARTHRITIS. Inflammation of a joint, producing exostosis.

ARTIFACT. An object produced by human art.

ARVICOLA AMPHIBIUS. Water rat.

ARVICOLA NIVALIS. Alpine rat (snow mouse).

ARVICOLA RATTICEPS. Northern vole; a species of rodent, dark grayish brown dorsally and grayish white ventrally.

ARVICOLA TERRESTRIS. Field vole (mouse).

ASTRAGALUS. Ankle bone, articulating with the tibia.

ATYPIC. Differing from the type; irregular.

AURIGNACIAN. First epoch of the Upper Paleolithic Period, named from Aurignac (Haute-Garonne), France.

AZILIAN. Epoch of transition from the Paleolithic to the Neolithic Period, named from Mas d'Azil (Ariège), France.

BASALT. A volcanic rock.

BATON. Prepared horn of reindeer or stag with one or more perforations, and generally ornamented by means of engravings or carvings in the round.

BELGRANDIA MARGINATA. Species of shell sometimes associated with Acheulian cultural remains.

BISON PRISCUS. Extinct variety of European bison.

BORDS-DROITS. Descriptive of bronze axes with straight borders dating from Bronze Age II.

BOS PRIMIGENIUS. Wild ox; the *urus* of Caesar's text; collateral ancestor of the large long-horned cattle of western Europe.

BOS TAURUS AKERATOS. Hornless ox.

BOS TAURUS BRACHYCEROS. Short-horned domestic ox.

BOS TAURUS URUS. Wild ox.

BOULDER CLAY. Unstratified gritty clay laden with rock débris of various shapes and sizes.

BRACHYCEPHALIC. Short-headed.

BRADYLEMUR. African Pleistocene Lemuroid Primate.

BRECCIA. Rock made up of angular fragments or pebbles embedded in a matrix which may or may not be of the same nature and origin.

BÜHL. A name derived from the hillocks formed by ice above Kufstein (Tyrol), also the name of a stage of ice advance during the Würm Glacial Epoch.

BUNDLE BURIAL. Inhumation burial with arms and legs flexed.

BURIN. Flint tool with beveled point, same as graver; characteristic of the Upper Paleolithic Period.

BUSKED. Shaped like the nose of an animal (horse, for example) in profile; term also applied to a type of Upper Paleolithic flint graver.

CAENOPITHECUS. Extinct (Upper Eocene of Europe) Primate of the Lemuroid suborder.

CAIRN. Artificial heap of earth and stones, usually covering burials; a barrow.

CALCANEUM. Heel bone.

CALLITHRIX. Titi monkey (broad-nose or Platyrrhine type).

CALOTTE. The upper or top portion of a human cranium.

CAMPIGNIAN. Early epoch of the Neolithic Period, named from Campigny (Seine-Inférieure), France.

CANIS FAMILIARIS PALUSTRIS. Domestic swamp dog.

CANIS LAGOPUS. Arctic fox.

CANIS LUPUS. Wolf.

CANIS VULPES. Fox.

CAPELLA RUPICAPRA. Chamois.

CAPRA HIRCUS RÜTIMEYERI. Domestic goat.

CAPRA IBEX. Ibex.

CAPSIAN. Equivalent in northern Africa of the Upper Paleolithic Period; named from Capsa, the Latin equivalent of Gafsa (Tunis).

CARDIUM EDULE. Cockle; common edible bivalve mollusk.

CARIES. A disease of the teeth.

CARINATE. Having a keel; keel-shaped.

CARNACIAN. Late epoch of the Neolithic Period, named from Carnac in Brittany.

CARNIVORA. An order of mammals which feeds on flesh.

CASTOR FIBER. Beaver.

CATARRHINIAN. Having the nose directed downward.

CEBUS. Platyrrhine (broad-nose) monkey (Pleistocene of South America).

CELT. A prehistoric implement or weapon of stone or bronze, somewhat resembling a chisel or ax.

CERVUS ALCES. Moose.

CERVUS CANADENSIS. Wapiti, or elk of America.

CERVUS CAPREOLUS. Roe deer.

CERVUS DAMA. Fallow deer.

CERVUS ELAPHUS. Red deer.

CERVUS MEGACEROS. Giant deer.

CERVUS TARANDUS. Reindeer; same as *Rangifer tarandus*.

CHAMAEDONT. Term applied to teeth with relatively low crowns.

CHAMPLEVÉ. Having the ground cut out, or lowered.

CHELLEAN. Second epoch of the Lower Paleolithic Period, named from Chelles (Seine-et-Marne), France.

CICATRIZATION. Healing of a wound in flesh or bone; state of being healed with a cicatrix.

CIST. A type of Neolithic burial; casket, chest, prototype of the coffin.

CLAVICLE. Collar bone.

CLAVIFORM. Club-shaped.

CLINKER-BUILT. Made of boards which overlap one another; used in connection with boats.

COLEOPTER. Beetle.

COMPRESSOR. Bone anvil or chopping block dating especially from the Mousterian Epoch.

CONCHOID OF PERCUSSION. Shell-shaped concavity left on a flint nucleus after the removal of a chip or flake.

CONCRETION. An aggregation of mineral matter in concentric layers, often around an organic nucleus.

CORBICULA FLUMINALIS. Siphonate bivalve river mollusk.

COUP DE POING (*same as* hand ax or cleaver). Stone implement chipped on both faces to an almond shape.

CRANNOG. Name given to lake dwellings of the special type found in Ireland and Scotland.

CRENELATED. Notched, channeled, indented.

CROMLECH. *See* Dolmen.

CRYPTOPITHECUS. An Eocene Lemuroid.

CUPULES. Cup-shaped pits or concavities.

CURSIVE WRITING. Writing in which the letters or characters are joined together; flowing.

CYNOCEPHALUS. Dog-faced baboon.

CYNOMORPH (*same as* Cercopithecidae). Tailed monkeys.

CYPRAEA LURIDA. Cowry; a species of univalve mollusk.

CYPRAEA PYRUM. Cowry; a species of univalve mollusk.

CYRENA FLUMINALIS. Siphonate bivalve river mollusk.

DAUN. Name given to a group of mountain peaks southwest of Innsbruck; also name of last stage of ice advance during the Würm Glacial Epoch.

DECKENSCHOTTER. *See* Outwash.

DENTALIUM. Genus of mollusk with tooth-shaped shell.

DETRITUS. Loose, uncompacted fragments of rock, either waterworn or angular.

DIAPHYSIS. The shaft of a long bone.

DIASTEMA. Interval between any two consecutive teeth, especially between the canines and incisors in the upper jaw, and the canines and premolars in the lower jaw of the apes.

DINOTHERIUM GIGANTEUM. Extinct (Miocene of Europe) species related to the elephants and mastodons.

DIPLOË. Spongy structure of bone between the hard dense inner and outer tables of the cranial bones.

DOLICHOCEPHALIC. Long-headed.

DOLICHOPITHECUS. Extinct (Pliocene of Europe) Catarrhine or narrow-nosed monkey.

DOLMEN. An ancient sepulchral monument of unhewn stones set on end or on edge, so as to form a chamber, and covered with a single huge stone or several stones.

DRYAS. Small genus of rosaceous plants, found in alpine and arctic regions of the northern hemisphere.

DRYOPITHECUS. Extinct (Miocene of Europe) Anthropoid ape.

ELEPHAS ANTIQUUS. Straight-tusked elephant.

ELEPHAS MERIDIONALIS. Southern elephant, first appeared in the Upper Pliocene.

ELEPHAS PRIMIGENIUS. True mammoth.

ELEPHAS TROGONTHERII. Regarded by Pohlig as the direct successor of *E. meridionalis* (southern elephant).

ENEOLITHIC. Transition stage between the Neolithic Period and the Bronze Age.

EOANTHROPUS DAWSONI. Species of early man as represented by the cranium and lower jaw found at Piltdown (Sussex), England; the dawn man.

EOCENE. First epoch of the Tertiary Period.

EOLITH. Intentionally chipped or utilized stone, shaped or employed during the Tertiary Epoch.

EOLITHIC. Dawn stage of the Stone Age; that part of the Stone Age which falls within the Tertiary Period.

EPIPHYSIS. A subordinate part of a bone formed by a separate center of ossification and remaining for some time distinct from the main portion; the end of a long bone as distinguished from the diaphysis or shaft.

EPISTROPHEUS. Second cervical or neck vertebra.

EQUUS CABALLUS. Domestic horse.

EQUUS HEMIONUS. Wild ass.

EQUUS PRZEWALSKII. Wild horse of the Asiatic steppe type.

EQUUS STENONIS. Large early type of true horse.

ERIODES. South American spider monkey.

ERRATIC. A boulder transported from its original site by natural agencies.

EUGENIA JAMBOLANA. Black plum; Java plum.

FAGUS. Beech.

FELIS CATUS. Wildcat.

FELIS LEO. Lion.

FELIS LYNX. Lynx.

FELIS PARDUS. Leopard, panther.

FELIS SPELAEA. Cave lion.

FENNO-SCANDIA. Region comprising Finland, Sweden, Norway, and Denmark.

FIBROLITE. Mineral of a white or gray color and fibrous to columnar structure.

FIBULA. Ornamental brooch of the type of a safety-pin.

FICUS. Fig.

FLUVIATILE. Pertaining to rivers.

FONT-ROBERT POINT. Flint point with base chipped on both sides to form a peduncle or stem; named from the cave of Font-Robert (Corrèze), France.

FORAMEN MAGNUM. Great foramen at the base of the cranium through which the spinal cord passes.

FOSSA CANINA. Depression in the outer surface of the upper jaw bone serving as an attachment for the muscle which lifts the corners of the upper lip, exposing the canine teeth.

FRET. Ornament formed of bands or fillets variously combined; first used in prehistoric times and later elaborated by the Greeks.

GAMMARACANTHUS LORICATUS. Species of crustacean.

GAZELLA DEPERDITA. A Pliocene gazelle.

GENIAL TUBERCLES. Median spinous processes on the inner face of the lower jaw; spina mentalis.

GENIOGLOSSAL (muscle). Muscle attached to the genial tubercles of the lower jaw, the tongue and hyoid bone.

GENIOHYOID (muscle). Muscle attached to the genial tubercles of the lower jaw and the hyoid bone.

GLENOID CAVITIES (or fossae). Cavities in the base of the cranium into which the condyles (articular extremities) of the lower jaw fit.

GONION. Craniometric median point on the lower jaw below the chin.

GRATTOIR. Scratcher, flint flake retouched at one or both ends.

GRAVER. Flint implement with beveled point for working in bone, ivory, and reindeer horn as well as for the execution of Paleolithic objects of art.

GRAVETTE BLADE. Small, slender, pointed flint blade, the penknife of the Upper Aurignacians; named from the station of La Gravette (Dordogne), France.

GRECQUE. Same as Fret.

GSCHNITZ. Name of a small stream near Innsbruck; also of a stage of ice advance during the Würm Glacial Epoch.

GULO BOREALIS. Wolverine.

GÜNZ. Name of a small river in the Alpine foothills; also of the First Glacial Epoch.

HADROPITHECUS. African Pleistocene ape.

HALBERT (same as halberd). Broad blade with sharp edges ending in a point, mounted on a long handle.

HAPALE. Short-tusked marmoset.

HELIX HISPIDA. Species of snail.

HELVE. Handle of a tool.

HIPPARION GRACILE. Fossil (Tertiary) species of the horse family, small in size.

HIPPOPOTAMUS AMPHIBIUS. Species of hippopotamus from the Nile River.

HIPPOPOTAMUS MAJOR. Species of extinct hippopotamus.

HOMO AURIGNACENSIS. Species of fossil man, closely akin to existing races.

HOMO HEIDELBERGENSIS. Extinct species of man, based on a lower jaw found near Heidelberg, Germany.

HOMO NEANDERTALENSIS. Extinct species of man, dating from the Mousterian Epoch.

HOMO SAPIENS. Species to which existing races of men belong.

HUMERUS. Bone of the upper arm.

HYAENA SPELAEA. Cave hyena.

HYPSILOID. Curved or arched like U; shaped like the Greek letter upsilon.

HYPSODONT. Applied to teeth having long crowns.

IDEOGRAPH. A picture, symbol, or sign of an object.

IDOTHEA ENTOMON. Crustacean found along the Baltic littoral and in certain Scandinavian lakes.

ILIUM (*pl.* ilia). Upper portion of the hip bone.

INCISURA MANDIBULAE. Notch in the ascending branch of the lower jaw, between the condyle and the coronoid process.

INDURATED. Hardened.

INION. Craniometric median point on the occipital bone of the cranium at the upper limit of the surface to which the neck muscles are attached.

INNOMINATE. A large irregular bone resulting from the growing together of three of the pelvic bones.

INSECTIVORES. A group of small mammals, most of which feed on insects.

ISCHIUM. Lower portion of the hip bone on which the weight of the body rests when the individual is seated.

JUGLANS BERGAMENSIS. Species of the walnut tree.

KITCHEN MIDDEN. A mound composed of kitchen refuse from ancient dwellings.

LACUSTRINE. Pertaining to lakes.

LAGOMYS PUSILLUS. Pika, or tailless hare.

LAGOPUS ALBUS. Ptarmigan, white grouse.

LAGOPUS ALPINUS. Alpine ptarmigan.

LAPIDARIAN. Pertaining to stones.

LAUFEN. Name of a village near Salzach; also of a stage of ice retreat during the Würm Glacial Epoch.

LEPUS CUNICULUS. Rabbit.

LEPUS TIMIDUS. European hare.

LEPUS VARIABILIS. Arctic hare.

LEVALLOIS FLAKE. Flint implement made from a large oval flake, occurring in certain late Acheulian and early Mousterian deposits.

LIMANDE. French for flounder (flat fish); term applied to a special type of Acheulian stone implement.

LIMNAEA PEREGRA. Species of snail.

LINEA ASPERA. Longitudinal crest or ridge on the back of the shaft of the femur or thigh bone.

LITTORINA LITTOREA. Common periwinkle of Europe and the Atlantic coast of America; sea snail.

LITTORINA OBTUSA. Species of sea snail.

LITTORINA RUDIS. Species of sea snail common to both Europe and America.

LOESS. Term applied to extensive deposits of fine-grained, even-textured materials yellowish brown in color and for the most part of eolian origin; loam.

LUSUS NATURAE. Freak of nature.

LUTRA VULGARIS. Otter.

MACACUS. Macaque, one of the Primates.

MACHAERODUS. Saber-toothed tiger.

MACTRA SUBTRUNCATA. Species of bivalve mollusk.

MAGDALENIAN. Closing epoch of the Upper Paleolithic Period, named from the ruins of La Madeleine (Dordogne), France.

MAGLEMOSEAN. Scandinavian equivalent of Azilian, or epoch of transition from the Paleolithic to the Neolithic Period.

MAGMA. Crude mixture, especially of organic matter.

MARTES SYLVATICA. Marten, a species of carnivorous quadruped (*Martes* same as *Mustela*).

MASTODON ANGUSTIDENS. Species of an extinct proboscidean quadruped of the family Elephantidae.

MASTODON ARVERNENSIS. Species of an extinct proboscidean quadruped of the family Elephantidae.

MEGALITHIC. Pertaining to Neolithic and Bronze-Age monuments composed of large stones (dolmens, menhirs, etc.).

MELES TAXUS. Badger.

MENHIR. A prehistoric monument consisting of a single tall stone.

MESOPITHECUS. A Pliocene ape.

MESORRHINIAN. Having a nose of medium width.

METACARPAL. One of the bones in the hand, between the fingers and wrist.

METATARSAL. One of the bones in the foot between the toe bones and ankle.

METOPIC SUTURE. Median suture in the frontal bone of the cranium; this suture usually disappears in childhood, but occasionally persists to old age.

MICHELIA. Genus of the Magnolia family.

MICROLITH. Diminutive stone tool, occurring in various culture levels from the Aurignacian to the Tardenoisian inclusive.

MINDEL. Name of a small river in the Alpine foothills, also of the Second Glacial Epoch.

MIOCENE. Third epoch of the Tertiary Period.

MOERIPITHECUS. An African Tertiary ape.

MORAINE. Accumulations of rock and detrital material along the lateral and terminal margins of a glacier.

MOUSTERIAN. Epoch between the Lower and Upper Paleolithic Periods, named from Le Moustier (Dordogne), France.

MUS MUSCULUS. Common house mouse.

MUS SYLVATICUS. Field mouse.

MUSTELA MARTES. Pine marten.

Mustela vulgaris. Common weasel.

Mya arenaria. Species of clam.

Mycetes. Howling monkey.

Mylo-hyoid groove. Oblique groove leading from the mandibular or inferior dental foramen on the inner face of each half of the lower jaw, which lodges the mylo-hyoid vessels and nerve.

Mylo-hyoid muscle. Muscle attached to the inner face of the lower jaw in the region of the molar teeth and to the hyoid bone.

Mylo-hyoid ridge. Slightly oblique ridge on the inner face of each half of the lower jaw, serving for the attachment of the mylo-hyoid muscle.

Myodes obensis. Siberian lemming, a species of rodent.

Myodes torquatus. Banded lemming.

Mytilus edulis. Common mussel found on most coasts.

Nassa neritea. Species of gastropod.

Nassa reticulata. Species of gastropod.

Navicella. Small boat or skiff; a type of fibula used during the Hallstatt Epoch.

Necrolemur. Eocene lemur.

Necrosis. Death of a circumscribed piece of tissue, bony or otherwise.

Neolithic. Last period of the Stone Age.

Nephritis. Inflammation of the kidneys.

Nevé. That part of a glacier which is above the snow line.

Non-remanié. Not derived from an older deposit.

Norma basilaris (or inferior). Basal aspect of the skull.

Norma frontalis. Frontal aspect of the skull.

Norma lateralis. Profile of the skull.

Norma occipitalis. Posterior aspect of the skull.

Norma verticalis. Top view of the skull; all these aspects of the skull are taken with the skull oriented in a given horizontal plane or (in 1 and 5) perpendicular to that plane.

Notharctus. An Eocene Lemuroid.

Nuchal. Pertaining to the nape of the neck.

Obsidian. A glassy volcanic rock.

Œnochoë (or oinochoë). Pottery vase with single handle and three-lobed rim, used for dipping wine from the crater (basin or vase), and filling drinking cups.

Oligocene. Second epoch of the Tertiary Period.

Oppidum. Provincial town.

Orbits. The cavities in the skull containing the eyes.

Oreopithecus. A Miocene ape.

Osteonia. Tumor composed of bony tissue.

Ostrea edulis. Oyster.

Outwash. Deposits of fluvio-glacial origin, or formed by the action of water and ice.

Ovibos moschatus. Musk-ox.

Ovis aries palustris. Sheep of a variety common to Swiss pile villages.

Paleolith. Stone implement dating from the Pleistocene Epoch.

PALEOLITHIC. Old Stone Age, corresponding roughly to the Pleistocene Epoch.

PALMETTE. Carved or painted ornament resembling a palm leaf.

PARIETAL. Pertaining to walls, as parietal frescoes; also one of the paired bones of the cranium.

PATINA. Incrustation which forms on stone or bronze after exposure to the weather, or after burial; also applied to the surface texture gained by other art objects through the action of time.

PECTEN JACOBAEUS. Species of mollusk; scallop.

PEDUNCULATE. Having a peduncle, or stalk.

PERCUSSION. The sharp striking of one body against another.

PERIOSTITIS. Inflammation of the periosteum.

PETROGLYPH. A figure or legend cut on rock.

PICEA BALSAMI. Species of spruce.

PICEA SERIANA. Species of spruce.

PIRIFORM APERTURE. Anterior nasal opening in the human skull.

PISTILIFORM. Shaped like the pistil or seed-bearing organ of a flower.

PITHECANTHROPUS ERECTUS. Ape-man, species based on a cranial cap, teeth, and femur found at Trinil, Java.

PITHECOID. Resembling or pertaining to the higher apes, as typified by the genus *Pithecus*.

PLANERA. Genus of ulmaceous trees; elm.

PLATYCEPHALIC. Flat-skulled.

PLATYRRHINIAN. Broad-nosed.

PLEISTOCENE (*same as* Quaternary). Fourth grand division of geologic time.

PLEURONECTES LIMANDA. Species of flounder; flat fish.

PLIOCENE. Last epoch of the Tertiary Period.

PLIOPITHECUS. A Pliocene gibbon.

POITREL. Armor used to protect the breast of a warhorse.

PORTLANDIA ARCTICA. Species of bivalve mollusk; same as *Yoldia arctica*.

POTSHERD. A fragment of earthenware or pottery.

PRESLE. A chalk talus or scree.

PROGNATHISM. The condition of having protrusive jaws.

PRONATION. Position of the forearm, in which the two bones are more or less crossed, and the palm of the hand turned downward.

PUPA MUSCORUM. Species of pulmonate land snails.

PYORRHEA ALVEOLARIS. Abscess of the bone surrounding the roots of the teeth.

QUATERNARY. *See* Pleistocene.

QUERCUS. Oak.

QUERN. Stone handmill for grinding grain.

RACLOIR. French for scraper; a type of stone implement much in use during the Mousterian Epoch.

RADIUS. Outer one of the two bones of the forearm.

RANGIFER TARANDUS. *See Cervus tarandus.*

RHINOCEROS ETRUSCUS. Etruscan rhinoceros.

RHINOCEROS MERCKII. Broad-nosed rhinoceros.

RISS. Name of a small river in the Alpine foothills, also of the Third Glacial Epoch.

RISSOA ULVAE. Species of shell found in clay deposits of interglacial age.

RIVER DRIFTS. Deposits along river courses including sands, silts, and loess.

ROBENHAUSIAN. Epoch of the Neolithic Period during which polished flint implements first appeared; named from Robenhausen, Switzerland.

RONDELLE. Any round object; term especially applied to disks of bone cut from the human cranium.

RUBBLE. *See* Outwash.

SACRUM. Fused portion of the vertebral column to which are attached the ilia of the pelvic girdle.

SAGAIE (*same as* assagai). Spear, lance, or javelin.

SALIX POLARIS. Species of willow.

SARSEN STONES. Blocks of silicious sandstone from a deposit in the valleys between Salisbury and Swindon.

SCHOTTERFELDER. Outwash or rubble beds; a class of deposits resulting from glacial action.

SCIURUS VULGARIS. Squirrel.

SCREE (*same as* Talus). Pile of débris at the base of a cliff.

SHERD. A fragment of pottery.

SIGMOID. Shaped like the letter S; sinuous.

SIMIAN GUTTERS. Paired depressions in the upper jaw bone at the level of the lower margin of the anterior (outer) nasal opening.

SIMIIDAE. *Same as* Anthropomorph.

SITULA. Bucket; urn.

SNOW LINE. The line above which snow fields last throughout the year.

SOLUTREAN. Next to the last epoch of the Paleolithic Period; named from Solutré (Saône-et-Loire), France.

SOMATOLOGICAL. Pertaining to the physical nature of man.

SOREX VULGARIS. Common shrew.

SPELEOLOGY. The study of caves.

SPERMOPHILUS. Genus of rodent related to the marmots and true squirrels.

SPERMOPHILUS RUFESCENS. Steppe suslik or pouched marmot.

SPIENNEAN. Equivalent of Robenhausian, *q.v.*

SPOKESHAVE. Notched scraper; tool that could be used in shaping spokes.

STEATOPYGY. Accumulation of fat on the buttocks of certain Africans, especially Hottentot women.

STEGODON. A late Pliocene elephant.

STELA. Upright slab of stone, either sepulchral or for the inscription of laws, decrees, etc.

STEREOSCOPIC VISION. The blending into one image of two pictures of an object seen from slightly different points of view.

STRANGLED BLADE. Flint blade with a lateral notch at the same level on each margin.

STRATIGRAPHY. The study of strata or beds of rock.

SUCCINEA OBLONGA. Species of land shell occurring in loess.

SUPINATION. Position of the forearm in which the radius and ulna are parallel, and the palm of the hand turned upward.

SUS SCROFA DOMESTICUS. Domestic pig.

SUS SCROFA FERUS. Wild boar.

SUS SCROFA PALUSTRIS. Swamp pig.

TALPA EUROPAEA. English mole.

TALUS. *See* Scree.

TAPES DECUSSATUS. Species of bivalve.

TARDENOISIAN. Final episode in the transition from the Paleolithic to the Neolithic.

TECTIFORM. Roof-shaped.

TELLINA BALTICA. Species of marine bivalve mollusk.

TERRAMARA (*pl.* terremare). Marl earth; also applied to pile dwellings of the type found in Italy.

TERTIARY. Third grand division of geologic time.

THAUMATURGIC. Of, or pertaining to, miracles; magical.

THUNDER STONES. Polished stone implements, which until recently were quite generally believed to have fallen from the sky during thunder storms.

TIBIA. Shin bone.

TILL. *See* Boulder clay.

TOGGLE. Species of button, or pin tapering at both ends and attachable at the center.

TORQUE. An ornament of twisted wire.

TRAGOCEROS. A Pliocene antelope.

TRAPA HEERI. Species of Old World aquatic plant.

TRAUMATIC. Of, or pertaining to, wounds.

TRAVERTINE. A porous rock formed by a limy deposit from springs.

TREPANATION. An early form of Trephination, *q.v.*

TREPHINATION. A surgical operation which consists in removing a disk of bone from the skull.

TRILITHON. An ancient monument resembling a gateway, with two upright stones supporting a horizontal one.

TRISKELE. Symbol consisting of three human legs bent at the knee and joined at the thigh.

TROCHANTER (great). Irregular eminence at the upper end of the femur or thigh bone, rising above the neck of the femur but lower than the femoral head.

TROCHANTER (small). Eminence just below and back of the neck of the femur.

TROGLODYTE. Cave-dweller.

TROGLODYTES. Name of the genus to which the chimpanzee belongs.

TUFA. A rock having a rough or cellular texture, sometimes a fragmental volcanic material, and sometimes a calcareous deposit from springs.

TUMULUS. An artificial mound, usually sepulchral.

TUNDRA. A rolling treeless plain.

TURBINATED. Top-shaped.

TYPOLOGY. Doctrine of types or symbols.

ULNA. Inner one of the two bones of the forearm.

UNIO LITTORALIS. Species of mussel.

URSUS ARCTOS. Brown bear.

URSUS ARVERNENSIS. Auvergne bear.

URSUS SPELAEUS. Cave bear.

VITIS. Large genus of climbing shrubs including grapevines.

VITIS VINIFERA. European grape.

VULTUR FULVUS. Species of the vulture family.

WIND VALLEY (*same as* wind gap). A valley formed by stream action but in which a stream no longer exists.

WÜRM. Name of a small river in the Alpine foothills, also of the Fourth Glacial Epoch.

YOLDIA. *See Portlandia arctica.*

KEY TO ABBREVIATIONS

AA   *Archiv für Anthropologie*, Brunswick, 1866–

AAE   *Archivio per l'antropologia e la etnologia*, Florence, 1871–

ABB   *Archives belges de biologie*, Ghent, 1880–

AFAS   *Association française pour l'avancement des sciences*, Paris, 1872–

AIB   *Académie des inscriptions et belles-lettres (Comptes rendus)*, Paris, 1857–

Anat. A.   *Anatomischer Anzeiger*, Jena, 1886–

Anthr.   *L'anthropologie*, Paris, 1890–

AP   *Annales de paléontologie*, Paris, 1906–

APAR   PIETTE, *L'Art pendant l'âge du Renne* (100 plates), Paris, 1907

Arch.   *Archaeologia*, London, 1749–

AS   *Académie des sciences (Comptes rendus)*, Paris, 1835–

ASNZ   *Annales des sciences naturelles zoölogiques*, Paris, 1854–

BA   *Bulletin archéologique*, Paris, 1883–

BARB   *Bulletin de l'Académie royale de Belgique (Classe des Sciences)*, Brussels, 1836–

BAUB   *Beiträge zur Anthropologie und Urgeschichte Bayerns*, Munich, 1877–

BMSA   *Bulletins et mémoires de la Société d'anthropologie de Paris*, 1900–

BSA   *Bulletins de la Société d'anthropologie de Paris*, 1859–

BSAB   *Bulletin de la Société d'anthropologie de Bruxelles*, 1882–

BSAHC   *Bulletin de la Société archéologique et historique de la Charente*, Angoulême, 1845–

BSBG   *Bulletin de la Société belge de géologie, de paléontologie et d'hydrologie*, Brussels, 1887–

BSGF   *Bulletin de la Société géologique de France*, Paris, 1830–

BSPF   *Bulletin de la Société préhistorique française*, Paris, 1904–

CA   CARTAILHAC and BREUIL, *La Caverne d'Altamira près Santander (Espagne)* (287 pp., 37 plates), Monaco, 1906 (see *PGMCP*)

CAF   *Congrès archéologique de France (Comptes rendus)*, Caen, 1843–

CFG   CAPITAN, BREUIL, and PEYRONY, *La Caverne de Font-de-Gaume* (279 pp. and 65 pls.), Monaco, 1910 (see *PGMCP*)

CIA   *Congrès international d'anthropologie et d'archéologie préhistoriques*, Neuchâtel, Switzerland, 1866–

CIPP   *Comision de investigaciones paleontologicas y prehistoricas*, Madrid, 1915–

CPF   *Congrès préhistorique de France (Comptes rendus)*, Périgueux, 1905–

CRC   ALCALDE DEL RIO, BREUIL, and SIERRA, *Les Cavernes de la Région cantabrique (Espagne)*, Monaco, 1911 (see *PGMCP*)

DASGN   *Denkschriften der allgemeinen schweizerischen Gesellschaft für die gesammten Naturwissenschaften*, Zurich, 1829–

DME   HOERNES, *Der diluviale Mensch in Europa*, Brunswick, 1903

DVD   SCHMIDT, KOKEN, and SCHLIZ, *Die diluviale Vorzeit Deutschlands*, Stuttgart, 1912

HF   Obermaier, *El Hombre Fosil*, Madrid, 1916
HP   *L'homme préhistorique*, Paris, 1903–

JAI   *Journal of the Royal Anthropological Institute of Great Britain and Ireland*,
           London, 1871–
JAP   *Journal of Anatomy and Physiology*, London, 1867–

KB   *Korrespondenz Blatt der deutschen Gesellschaft für Anthropologie, Ethnologie
           und Urgeschichte*, Brunswick and Munich, 1870–

MAGW   *Mittheilungen der anthropologischen Gesellschaft in Wien*, Vienna, 1871–
MAGZ   *Mittheilungen der antiquarischen Gesellschaft*, Zurich, 1841–
MARB   *Mémoires de l'Académie royale de Belgique (Classe des sciences)*, Brussels,
           1820–
Mat.   *Matériaux pour l'histoire . . . . de l'homme*, Paris, 1864–1888
MSA   *Mémoires de la Société d'anthropologie de Paris*, Paris, 1860–1899
MSAB   *Mémoires de la Société d'anthropologie de Bruxelles*, Brussels, 1882–
MV   Obermaier, *Der Mensch der Vorzeit*, Berlin, Munich, and Vienna, 1912

NDSNG   *Neue Denkschriften der schweizerischen Naturforschenden Gesellschaft*
           (continuation of *DASGN*), Zurich, 1837
NF   *Nordiske Fortidsminder* (publication of the Kgl. Nordiske Oldskriftsel-
           skab, with résumés in French), Copenhagen, 1890–

PB   Breuil and Obermaier, *La Pileta à Benoajan (Malaga), Espagne*,
           Monaco, 1915 (see *PGMCP*)
PGMCP   *Peintures et Gravures Murales des Cavernes Paléolithiques* (publications of
           the Institut de Paléontologie Humaine), Monaco, 1906–
PM   *Palæontological Memoirs*, London, 1868–
PP   Breuil, Obermaier, and Alcalde del Rio, *La Pasiega à Puente-Viesgo
           (Santander), Espagne* (64 pp., 29 plates), Monaco, 1913 (see *PGMCP*)
Preh.   G. and A. de Mortillet, *Le Préhistorique*, 3rd edition, Paris, 1900
PRS   *Proceedings of the Royal Society*, London, 1800–
PT   *Philosophical Transactions*, London, 1665–
PZ   *Præhistorische Zeitschrift*, Berlin, 1909–

QJGS   *Quarterly Journal of the Geological Society*, London, 1845–

RA   *Revue anthropologique* (continuation of *REA*), Paris
R. Arch. *Revue archéologique*, Paris, 1844–
RD   Buckland, *Reliquiae Diluvianae*, London, 1823
Rd'A   *Revue d'anthropologie*, Paris, 1872–1889
REA   *Revue mensuelle de l'École d'anthropologie de Paris*, Paris, 1891–
RP   *Revue préhistorique*, Paris, 1906–

VBGA   *Verhandlung der Berliner Gesellschaft für Anthropologie, Ethnologie und
           Urgeschichte* (appendix for *ZE*), Berlin, 1869–

WPZ   *Wiener Præhistorische Zeitschrift*, Vienna, 1914–

ZE   *Zeitschrift für Ethnologie* (see *VBGA*), Berlin, 1869–

HUMAN ORIGINS

HUMAN ORIGINS

INTRODUCTION

THE STAGES IN HUMAN EVOLUTION

The quest of origins has always been fascinating to mankind; it is the natural outgrowth of man's endeavor to link cause with effect. The oldest historic records testify to man's interest in the beginnings of all things, the human race included. No other myths are so nearly universal as creation myths; no other problem in all the realm of things in their initial stages is at once so stimulating to the imagination as the problem of the origin of things human—the physical man, the human intellect, the human culture complex. In all three categories (physical, intellectual, and cultural), there is a gap between man and the other primates, the gap being least marked in the physical domain.

A difficulty inherent in the quest of human origins is that man must perforce be his own examiner and classifier, and sit in judgment upon the results of self-investigation. To perform these functions without bias is a crucial test even for the human intellect. Another obstacle inherent in self-scrutiny is the determination of and agreement upon the sum total of the physical, mental, and cultural characters that enter into the making of *Homo*. Granting that these characters can be listed and agreed upon, there still remains the problem as to which of them are primary and which are of secondary importance. There are, however, certain outstanding features which at once arrest the attention, such as the power of speech, brain elaboration, and erect posture. The latter can be traced back to a pre-human stage, and therefore has only an indirect bearing on the passage from a simian stage to man's estate—a passage which was obviously coincident with an

1

active period of brain elaboration, one of the fruits of which might well have been the acquisition of articulate speech.

It is no part of the purpose of this book to trace in detail the organic evolution of man from more lowly forms of life. That is the province of the zoölogist, and our concern is with the development of *Homo* as disclosed by the material records of prehistory. Nevertheless, it may be well at the outset of a study of human origins to indicate in a very general way the main stages in human evolution as the process is viewed by modern science.

To discover the roots of the human stem one must dig in Eocene soil, the earliest epoch of the geologic age of mammals. There one finds among mammals a group of insectivores, small of body but relatively large of brain and closely allied to the primate stock, also a primitive lemuroid (*Tarsius*). A study of these forms reveals the influence of an arboreal mode of life on the development of the mammalian brain. G. Elliot Smith points out that in the forerunners of the Mammalia the cerebral hemisphere was predominantly olfactory in function; that even when the true mammal emerged and all the other senses received due representation in the cerebral cortex, the animal's behavior was still influenced to a much greater extent by smell impressions than by those of the other senses. Arboreal life would suffice to change all this; for it would limit the usefulness of the olfactory organs and at the same time favor the development of hearing, touch, and especially vision. It was the special cultivation of vision that differentiated the lemurs from their ancestors and opened the way for a period of brain elaboration that eventually culminated in man.

The fullest possible reliance upon these senses, particularly upon vision and touch, would be of special advantage to an arboreal primate. The development of the sense of sight is well illustrated in the living representative of a group of lower Eocene primates, namely, *Tarsius,* in which sight completely overshadows the sense of smell. The reduction of the snout automatically increased the range of usefulness of vision, because it removed the chief physical obstruction to the fullest use of conjugate movements of the eyes.

The overlapping of the visual fields was the first step toward the acquisition of true stereoscopic vision; it was stereoscopic vision that immeasurably increased the appreciation of form, color, and details of objects and transformed "some Eocene Tarsioid into a monkey" with machinery for bringing about conjugate movements of the eyes. The making of such movements involves an elaboration of that part of the cerebral cortex which is concerned with eye-movement control and the function of attention. Visual concentration whets curiosity to examine; and to satisfy curiosity the hand is called upon to adjust movable objects for inspection. Thus the tactile sense becomes intimately linked up with vision.

There came a time when the call of the brain for the freedom of the hand outweighed the needs of the hand for support; and in winning the freedom of the hand, the brain won its own freedom to a field of almost unlimited possibilities for expansion, for hand freedom means erect posture and a brain case poised where it may best expand, at the top of a practically vertical spinal column rather than suspended from the end of a horizontal one. Given a human hand and a nascent human brain, only time and a suitable natural environment were necessary to produce the fabric of our civilization.

The researches of Elliot Smith, Hedd, and Hunter show that the areas of the brain that have to do with sensory discrimination are precisely those which were the first to expand, namely, the parietal, prefrontal and inferior temporal areas of the cerebrum. *Pithecanthropus, Eoanthropus, Homo rhodesiensis, Homo neandertalensis,* and *Homo sapiens* represent so many stages in an ascending series of cerebral expansion.

Improvement of the vision and of tactile discrimination lessened the chances of error of judgment and led to increased confidence in the validity of mental processes as a guide to action. When the mind of man became capable of solving a problem by means of simple analysis, it found a new world, unattainable to its nearest of animal kin, a world which man has been exploring for hundreds of thousands of years and the possibilities of which will continue to stimulate intellectual achievement without apparent limit.

In this new world, where one can arrive at results by indirection, one is apt to ignore the importance of the human hand, that marvelously adaptable tool of the human brain. The clenched fist can deal a telling blow, the grip can strangle, and the nails can rend only when in relatively close proximity to the object attacked or attacking. Since he had neither fleetness of foot nor unusual strength of arm, primitive man had to use artificial substitutes for these qualities. The use of stick or club makes possible a heavier blow and at a longer range. Moreover, a club or stone may be used as a projectile, thus immeasurably increasing the zone of safety without appreciably lessening the power of attack or defense.

The problem was, therefore, one of utilization of materials at hand. The most utilizable of all stones is flint, because of its hardness and mode of fracture, which leaves a comparatively straight edge. Moreover, flint flakes are produced by purely natural means. Sometimes they may have been produced accidentally, and the mind that observed the connection between the accidental product and itself as the producer is entitled to be called human.

Given exceptional minds capable of arriving at a goal by indirection, by intermediate stages, we have the prerequisite for the origin and development of the whole fabric of our material and intellectual culture. The translation of individual experience into racial experience has been accelerated by the development of rudimentary speech. The brain of even the most primitive monkeys reveals a distinct expansion of the cerebral areas concerned with hearing and the control of sound production.

For untold eons the common stock of knowledge was passed on from generation to generation and from age to age by word of mouth. In such circumstances it is easy to see how myth, legend, and tradition would flourish. The exceptional individual, the discoverer, of one age became the hero or god of the next. In all ages the course of progress has been thought out by the gifted few, while the rate of progress has depended upon the ability of the many to profit by the achievements of the few.

The anatomist, the psychologist, the anthropologist have all contributed to a better understanding of human origins. The field of each has its special limitations. Much can be learned by a study of living races, both civilized and primitive, especially the latter; but the traditional background of even the most primitive living races is exceedingly complex in comparison with that of the earliest known fossil races.   In the last analysis there is little left on which to base one's conclusions as to how fossil man reacted to mental stimuli save that which may be deduced from a comparative study of fossil human skeletal remains and associated cultural remains.   The skeletal remains are rare; cultural remains are relatively plentiful; but in both categories the record is only fragmentary.   In order to make the most of the available records, a knowledge of prehistoric sequence is of the utmost importance. Chronology is the essential framework of prehistory in the same sense that it is of history.   The difference is one of degree rather than of kind: history requires absolute chronology, prehistory only relative chronology.

PART I

THE OLD STONE AGE
AND THE DAWN OF
MAN AND HIS ARTS

CHAPTER I

THE DEVELOPMENT OF PREHISTORIC CHRONOLOGY

The history of the development of a relative chronology for that part of the age of man which antedates history is a good example of how far ahead of their times great thinkers sometimes are, and how long it takes the great mass of their fellows to profit by their achievements. Prehistoric chronology is based on a triple division —the Ages of Stone, Bronze, and Iron. Able thinkers long ago declared such a sequence to have existed, but their conclusions failed to become a part of the common stock of knowledge chiefly because of the tendency of the human mind *en masse* to resist the intrusion of knowledge beyond that which is carried by the traditional current of thought.

Early Discoveries

Hesiod (about 850 B.C.) speaks of a period when bronze had not been superseded by iron. Pausanias (about 468 B.C.) relates that in Homeric times all weapons were of bronze. Virgil gives bronze arms to the heroes of the Æneid, also to certain peoples of Italy. In his poem *De Rerum Natura* Lucretius (about 98-55 B.C.) says: "The earliest weapons were the hands, nails, and teeth; then came stones and clubs. These were followed by iron and bronze, but bronze came first, the use of iron not being known until later."

Coming down to the seventeenth century, we find that Sir William Dugdale speaks of stone celts, in his *Antiquities of Warwickshire* (1656), as "weapons used by the Britons before the art of making arms of brass or iron was known."

In Père Montfaucon's *Antiquité Expliquée* there is an interesting account of a discovery made in 1685 at Evreux, Normandy, by M. de Cocherel. After removing two suspicious-looking surface stones and the earth beneath them, Cocherel found a sepulture composed of five enormous stones. Within this sepulture were two skulls and under each a "hard stone cut in the form of an ax." One of these axes was of jade, and perforated. The author adds that this stone is "good against the *Epilepsy* and the *Nephritis.*" Beneath these remains was a great stone which, upon being lifted, disclosed two other skeletons with a hatchet of stone under the head of each. "In the same place there were three urns filled with coals."

By enlarging the pit, the workmen encountered "from sixteen to eighteen other bodies stretched out side by side on the same line," their heads[1] to the south, and under each head a stone and a hatchet. The stone hatchets were all of the same shape but of different colors. "There were found three bones pointed like the head of a halbert, which formerly had been fixed to long staffs for to make lances and pikes of them. One of these was the bone of a horse's leg. There were also points found, some of ivory and others of stone, which had served for heads of arrows. From whence it appears that these barbarians had not any use either of iron or copper or of any other metal. A little piece of harts horn that was found in the same place had served for to fasten one of these axes in; this horn had a hole in one of its extremities for to fix an helve therein."[2]

By the side of these bodies, but where the ground was eight inches higher, was found a large quantity of burnt bones, including portions of a human skull, and also an earthen pot, broken and filled with charcoal and a thick deposit of ashes.

Montfaucon concludes that this was the sepulture of two nations of great antiquity. "Those whole bodies laid on the same line were of some barbarous nation that knew not yet either the use of iron, or of any other metal"; the two in the sepulture built of great stones he supposed to be chiefs of that nation because of the jade ax. The burnt bones were thought to be of Gauls, who were known to have incinerated their dead.

As early as 1734 it was pointed out by Mahudel that man did not begin to make use of bronze and iron until a long time after

---

[1] One of these heads had the skull pierced in two places; but it seemed as if the wounds had been cured, and the skull closed up again" (the first notice of what probably represents a case of prehistoric trephining).

[2] From translation by David Humphreys (London, 1722), Vol. V, p. 132.

the birth of the world. He argued that implements were essential
in acquiring the necessities of life by means of both agriculture and
the chase; that stone was not only appropriate, but also readily ob-
tainable for such purposes; furthermore, that only certain stones
are adapted to such uses, depending upon their hardness and mode
of fracture. The so-called "thunderstones" were found to belong
in this class; hence they did not fall from heaven, but were the work
of man; an added proof of this was that the later tools of bronze
were modeled after the stone tools.

A similar view was taken by another early French writer,
Goguet (1758), who says: "Toute l'antiquité s'accorde à dire qu'il
a été un temps où le monde était privé de l'usage des métaux"; and,
"L'usage du cuivre a precédé celui du fer"; again, "les pierres, les
cailloux, les os, les cornes d'animaux, les arêtes de poisson, les
couquilles, les roseaux, les épines, servoient à tous les usages où les
nations policées employent aujourd'hui les métaux. Les sauvages
nous retracent une peinture fidelle de ces anciens peuples et de
l'ignorance des premiers tems." Goguet likewise draws a lesson
from the fact that in his time the Japanese still employed copper and
bronze in the manufacture of implements that in Europe would
have been made of iron.[3]

A passage from Bishop Littleton is of special interest in this
connection:[4] "There is not the least doubt of these stone implements having been fabricated in the earliest times, and by barbarous people, before the use of iron or other metals was known;
and from the same cause spears and arrows were headed with flint
and other hard stones, abundance of which, especially of the latter,
are found in Scotland, where they are, by the vulgar, called *Elfs
arrows.*"

But the statements of these pioneers left no permanent impression on the thought of their times. Polished stone hatchets were
still regarded by the majority of people as "thunderstones" which
had fallen from the sky during storms. Olaf Worm of Copenhagen, the recognized authority on northern archeology until the
beginning of the nineteenth century, writes in the *Museum Wor-*

---

[3] Kaempfer, *Histoire du Japon.*
[4] *Archæologia*, Vol. II (1766), p. 118.

mianum (published in 1655, after his death) of the common belief in thunderstones as follows: "These are called *ceraunicæ* because they are supposed to fall with the lightning from the sky. They are of various forms, often spindle-shaped or like a hammer or hatchet with a hole in the middle. Opinions differ as to their origin, since some believe they are not thunderstones but petrified iron implements, seeing they resemble the latter in shape so closely."

"This explanation does not hold, however," he continues, "because people whose word can be trusted have picked them up on spots where lightning has struck. . . . But there are drawbacks to this view also, especially on account of the various forms of these objects, including perforations; and yet there is much to be said in its favor. Others again believe that these stones were formed in the earth and were later caught up by the storm, carried high into the heavens, and then descended with the lightning to the earth."

As late as 1802, Thorlacius expressed his views, in a paper on "Thor and his Hammer," as follows: "The objects found in the mounds are nothing else than symbols of the weapons employed by the god of thunder in chasing and destroying evil spirits and dangerous giants. These stone objects are emblems of the three-fold action of thunder, namely; the hammer represents its power of grinding to pieces; the pointed implement its boring or piercing action; while the chisel and wedge-shaped objects symbolize its splitting action. They could not be ordinary tools and weapons as these have been made of metal since the earliest times." The tumuli were still looked upon as altars and temple sites, hence artifacts found in them must be offerings to the gods. According to Dr. J. Walter Fewkes, the native inhabitants of Porto Rico still call a stone ax *"piedra del rayo"* (thunderstone).

The first step in the process of placing prehistoric archeology on a scientific basis was made in Scandinavia. In 1806 an action of far-reaching consequence was taken by Denmark in the appointment of a scientific commission to direct investigation in the history and natural history (including geology) of the country. The attention of this commission was attracted from the first to the numerous dolmens and shell heaps. No connection, however, could be found between these and the history and sagas of the country. Meanwhile, important archeological collections were made, especially by Professor R. Nyerup. In 1816 these collections became the nucleus for the Royal Danish Museum of Antiquities in Copenhagen, now the National Museum.

C. J. Thomsen, director of this museum from 1816 to 1865, was the first paleo-ethnologist to make use of the methods of geology and paleontology. Seconded by a number of Danish and Swedish scientists (including Forchhammer the geologist, Worsaae and Steenstrup, subsequent directors of the Copenhagen museum, and Nilsson, professor of zoölogy at the University of Lund, Sweden), Thomsen succeeded in establishing a relative chronology for prehistoric times, based upon the development of human industry, a chronology which has served as a foundation for all later classifications. After having worked over the problem for twenty years, he divided the prehistoric into the Ages of Stone, Bronze, and Iron, and published his results in 1836.

Although, as we have seen, the idea was not original with the Scandinavians, the prevailing views held by his predecessors tend to add to the glory of Thomsen, whose classification, applied to a great museum, was immediately accepted in Scandinavia and later throughout the world. Whereas Thomsen mentioned only one Stone Age, subsequent discoveries in other parts of Europe, especially in the valley deposits and caves, made it necessary to subdivide this age; and the ages of Bronze and Iron have also undergone subdivisions with the progress of the science. Worsaae, and later Gabriel de Mortillet, Montelius, and Breuil have had much to do with giving relative chronology its present form.

TERREMARE, CRANNOGS, AND LAKE DWELLINGS

During the second half of the eighteenth century numerous artificial deposits in the form of large flat mounds, especially frequent in the Italian provinces of Modena, Parma, and Reggio, began to be exploited as a fertilizer. These deposits were called *marna,* which in its scientific guise became *mara* or *terramara.* In the removal of this fertilizing soil many antiquities came to light. Both the antiquities and the mounds were assigned partly to the Boii, a Celtic (Bronze Age) race, and partly to the Romans by the celebrated naturalist Venturi in his *Storia di Scandiano,* published in 1822.

As early as 1839 Sir W. R. Wilde began the exploration of the Irish crannogs or small, partly artificial lake islands. His at-

FIG. I. REMAINS OF A NEOLITHIC LAKE DWELLING AT AUVERNIER, ON LAKE NEUCHÂTEL, SWITZERLAND.

The decayed ends of the piles on which the village was built show as stumps above the water. Excavation at such sites has supplied data for the reconstruction of the lake dwelling shown in Fig. 284. Photograph by the author (1908).

tention had been called to them through an accidental discovery: workmen engaged in clearing a stream at Lagore, near Dunshaughlin, County Meath, had encountered large quantities of antiquities and bones, including human skeletal remains. Many other crannogs came to light, forty-six in all being recorded by 1857. These discoveries in Italy and Ireland were only partially understood, and failed, as pioneer work so often does, to make a permanent impression on a nascent science. This was reserved for workers in the Swiss field.

The Swiss were the first to follow in the lead of Scandinavia, Ferdinand Keller's discoveries of the Swiss lake dwellings in the

winter of 1853–54 confirming the exactness of Thomsen's classification. During that winter an extraordinary combination of
drought and long continued cold produced a rare phenomenon in
the region of the Alps. The lakes reached a level one foot lower
than had ever been recorded before; the shores receded and islands
never before seen appeared. The first lake dwelling that came to
light, in the bay between Obermeilen and Dollikon on the Lake of
Zurich, was discovered by workmen who were taking advantage
of the low water to excavate in order to make a fill with earth
thus rescued from the lake bed. As soon as they began to dig,
they found not only the heads of piles, but also many staghorns,
as well as pottery and implements of various kinds. Long before
this, fishermen had been troubled by "stumps" of a supposed "submerged forest that had caught their nets." These now proved to
be piles of prehistoric dwellings. It was recalled that as early as
1829, piles and various antiquities had been uncovered in an excavation made in front of Obermeilen for the purpose of deepening
the harbor; but the matter does not seem to have been followed
up and the incident was all but forgotten.

Swiss prehistorians took advantage of the opportunity afforded
by the prolonged drought of 1920–21 to collect new data bearing
on pile villages. Some of these villages were found to be much
greater in size than had hitherto been supposed. Moreover, a
definite pile-village culture sequence was established by means of
Vouga's excavations at several important sites on Lake Neuchâtel
(Fig. 1).

Recent excavations of lake and moor villages in southern Germany by Schmidt, Reinerth, and Kraft have thrown new light on
prehistoric house construction.

River Drift

The stone implements referred to by the early writers already
mentioned were of polished stone. With two exceptions, rude
chipped implements had attracted no attention until 1838, when
Boucher de Perthes submitted his first *"haches diluviennes"* to
the Société d'Émulation d'Abbeville. The following year he exhibited some of these implements in Paris.

Boucher de Perthes was a firm believer in the Deluge, proofs of which he saw in the valley terraces of the Somme. In his *De la Création, Essai sur l'Origine et la Progression des Êtres* (Paris, 1838), he stated that sooner or later traces of antediluvian man would be found. His belief was based upon: (1) the tradition of a race destroyed by the Flood; (2) the geological proofs of that deluge; (3) the existence during that epoch of mammals akin to man; (4) the certainty that the globe was inhabitable by man during the epoch in question. Curiously enough, his discoveries served to render uneasy the believers in the inviolability of the Mosaic account, while geologists were simply amused by the seriousness with which he took the story of the Deluge.

The significance of Boucher de Perthes' discoveries dawned even upon himself very gradually. His first convert was made at Amiens in 1855 in the person of Rigollot, who surrendered to the evidence afforded by the discoveries made at Saint-Acheul, a suburb of Amiens. French geologists and antiquarians in general, however, held aloof until after three British scientists, Falconer, Prestwich, and Evans, had visited the Somme valley.

In 1858, Hugh Falconer, the distinguished English paleontologist, crossed the Channel for the express purpose of investigating Boucher de Perthes' claims, and became, to use his own words, "satisfied that there was a great deal of fair presumptive evidence in favor of many of his speculations regarding the remote antiquity of these industrial objects and their association with animals now extinct."

Acting upon Falconer's suggestion, Sir Joseph Prestwich, one of England's foremost geologists, visited Abbeville and Amiens in 1859, where he was joined by Sir John Evans, whose description of what was seen and done is here quoted: "We examined the local collections of flint implements and the beds in which they were said to have been found; and in addition to being perfectly satisfied with the evidence adduced as to the nature of the discoveries, we had the crowning satisfaction of seeing one of the worked flints still *in situ,* in its undisturbed matrix of gravel, at a depth of 17 feet from the original surface of the ground. . . . From the day on which Sir Joseph Prestwich gave an account to the Royal Society (*Proceedings,* p. 50, 1859) of the results of his visit to the valley of the Somme, the authenticity of the discoveries of Boucher de Perthes was established."

During all this time neither Boucher de Perthes nor his English champions seem to have been aware of the fact that John Frere had made an earlier discovery of similar implements in England. In his "Account of Flint Weapons Discovered at Hoxne in Suffolk," read before the Society of Antiquaries of London, June 22, 1797,[5] Frere expresses the belief that "the situation in which these weapons were found may tempt us to refer them to a very remote period indeed; even beyond that of the present world." They were found in great numbers at a depth of about 3.66 meters (12 feet), in a stratified deposit that had been "dug into for the purpose of raising clay for bricks." Their mode of occurrence led Frere to suppose that the spot was the place of their manufacture and not of their accidental deposit. The implements were found at the rate of five or six to the square yard. The exploiter of the brick works is said to have emptied basketfuls of them into the ruts of the adjoining road before learning that they were "objects of curiosity."

FIG. 2. THE FIRST PALEOLITHIC FLINT IM-
PLEMENT EVER RECOGNIZED AND PRE-
SERVED AS A RELIC OF ANCIENT MAN.

This Chellean implement of the cleaver type was discovered about 1690 in Pleistocene gravels at Gray's Inn Lane, London. A century and a half later it was recognized in the collections of the British Museum as identical with implements recently discovered in the Somme valley in France. Scale, $\frac{2}{3}$. Photograph from the British Museum.

Nor was John Frere the first discoverer of these rude implements. He too was ignorant of a still earlier discovery, the ear-

<hr>

[5] *Archæologia*, Vol. XIII (1800), p. 204.

liest recorded of all.   About the year 1690, a roughly chipped,
pointed implement of flint (Fig. 2), in association with the remains of elephant, was dug from the Quaternary gravels at Gray's
Inn Lane, London, which after being preserved for more than
150 years in the Sloane Collection and in the British Museum,
was ultimately recognized by Sir Wollaston Franks as identical
with those discovered in the river gravels of the Somme valley
during the second quarter of the nineteenth century.

According to Sir John Evans, a rude engraving of this Gray's
Inn Lane specimen illustrates a letter by Mr. Bagford on the antiquities of London printed under date of 1715 in Hearne's edition
of Leland's *Collectanea* (Vol. I, p. 64).   From his account it
seems to have been found near the skeleton of an elephant.   Evans
also says that excavations at Gray's Inn Lane in 1883–84 brought
to light several implements of various forms, but none quite so
fine as the one discovered near the close of the seventeenth century.

It is not generally known that river-drift implements were
found in Italy soon after Boucher de Perthes' first discovery at
Abbeville and before that of Rigollot at Saint-Acheul.   As early
as 1850, Scarabelli announced the discovery of chipped implements
in the foothills of the Apennines near Imola and in Quaternary
deposits of the Santerno River in Emilia.   In 1846, similar discoveries were made at a number of localities in the environs of
Rome.   In 1869, Nicollucci and Gastaldi called attention to stone
implements collected near Sora (Caserta), and two years later
Bellucci and Rosa found splendid examples in Umbria and in the
valley of the Vibrata.   The Umbrian finds came from Quaternary
deposits of the Tiber and its tributaries, and occasionally from the
surface.

Another early discovery that has almost been lost sight of was
that of Noulet, in 1851, who found rude implements of quartzite
in gravel beneath a loam at an altitude of 15 to 30 meters (49.2
to 98.5 feet) above the Garonne at L'Infernet, near Toulouse.
Ridiculed by members of the Toulouse Academy of Sciences,
Noulet held back his manuscript on this station for eleven years.

In the exploration of valley deposits of Belgium and of the
Somme, Rutot and Commont have proved themselves to be worthy
successors of Boucher de Perthes and Rigollot.   For many years

and until his death in 1918, Commont was indefatigable in his studies of the four terraces of the Somme valley, especially in the vicinity of Amiens. His researches have thrown much new light upon the physiographic features of the terraces, as well as on the faunal and cultural remains found therein.

FIG. 3. THE VALLEY TERRACE OF SAN ISIDRO, ON THE MANZANARES RIVER, NEAR MADRID, SPAIN.

This hillock shows a typical section of the thick alluvial (river-drift) deposit of Pleistocene age covering the floor of the ancient valley above the present bed of the Manzanares River. In ascending order the strata are gravels, sands, and clays (see also Fig. 43). In exploiting these and other similar deposits for road and building materials many Paleolithic implements and fossil animal bones have been discovered. Such remains may have been deposited *in situ* during occupation of the valley floor in dry periods, or carried into the valley from higher levels by the action of streams. At San Isidro the gravels yield remains of the Chellean Epoch; the sands and clays, remains of Acheulian age. Photograph by the author.

Modern house and road construction require much building material of a kind furnished by valley terraces—coarse gravels especially for road metal, sands for numerous purposes, and the finer, more recent deposits for brick making (Fig. 3). In some localities deposits acres in extent have been completely removed. Had it not been for these requirements of present-day civilization, we might have remained in ignorance concerning some of the earliest cultural stages of our ancestors. Curiously enough, the development of the automobile industry, bringing to the fore rubber-tired vehicles, seems destined to have a bearing on future river-drift discoveries. Flint as a road metal tends more than some other materials to puncture rubber tires. The demands on the gravel beds are thus not so great now as they once were; hence, fewer flint implements come to light than formerly.

The culture-bearing valley deposits are still well-nigh inexhaustible. The one serious drawback is that finds made in them must ever be from the very nature of the case in a large measure fortuitous. Untutored workmen are constantly digging in hundreds of sand, gravel, and clay pits over wide areas. Without an international subsidy on a large scale, continuous expert control is out of the question. The result is that important data are being overlooked constantly, and valuable specimens are smashed by pick and shovel and irretrievably lost to view.

Caves and Rock Shelters

Data for a very important chapter on the antiquity of man have been furnished by the caves, and rock shelters (Fig. 4). Early exploration was for the sake of collecting fossil ivory, which was used in those days for medicinal purposes; later, attention was directed to fossil remains of all sorts from the viewpoint of paleontology.

During the seventeenth and eighteenth centuries, the caverns of the Harz Mountains, of Hungary, and of Franconia (Bavaria) were explored. Beginning with the nineteenth century, there was increased interest in cavern exploration, and corresponding progress was made concerning the nature and meaning of the remains thus brought to light. But the human bones and artifacts occasionally met with in the caverns were still looked upon as more recent than the remains of extinct animals.

The Rev. J. MacEnery, the Roman Catholic chaplain at Tor Abbey, Devonshire, was probably the first to note the association in caverns of flint implements with fossil animal remains. His researches in Kent's Cavern near Torquay from 1825 to 1841 convinced him of the coexistence of man and the fossil animal bones in question, but the prevailing belief of the time was against his views. MacEnery died in 1841 without publishing his manuscripts, which were lost for nearly twenty years. Fortunately they were recovered; an abstract of them was printed in 1859, and about ten years later they were published in full by W. Pengelly.

FIG. 4.   THE CHÂTEAU ROCK SHELTER AT LES EYZIES, FRANCE.

Natural rock shelters may be weathered out of escarpments wherever a bed of soft rock is overlain by a layer of harder rock, of either the same or a different kind. Such shelters were natural abodes of primitive man, who inhabited them sometimes intermittently and sometimes for long periods of time. The floor deposits of many of them are rich in prehistoric remains. At this site all but a small part of the Paleolithic deposits were destroyed when the château was built on the floor of the shelter during the Middle Ages, but enough remained to show that it was an important center of Magdalenian culture. A partial restoration of the château by the French Government for museum purposes is distinguishable at the extreme right of the shelter. Photograph by the author.

While MacEnery's explorations were in progress at Kent's Cavern, similar discoveries were being made in France and Belgium. In 1828 Tournal announced the discovery of human bones and potsherds associated with faunal remains, in part extinct, in the cave of Bize (Aude). The following year a paper was published by de Christol announcing the discovery of human remains in cave deposits near Pondres (Gard), which also contained the bones of hyena and rhinoceros. The pottery found there and at Bize is intrusive and of later age. The period to which the human remains belong is still in doubt. In 1833–34 Schmerling published a large volume on numerous caverns about Liège, in two of which

FIG. 5. ENTRANCE TO THE CAVERN OF CASTILLO, NEAR PUENTE VIESGO, SPAIN.

Caverns formed by subterranean streams were another natural abode of primitive man, and in them the remains of his occupation are preserved. Floor deposits near the mouth yield his kitchen refuse, implements, and, rarely, his fossil bones; the walls and ceilings bear examples of his art in painting and sculpture. Castillo is one of the most important Spanish caverns for Paleolithic mural art and the sequence of cultures exhibited by the floor deposits. Photograph by the author.

he had found human bones associated with the remains of elephant, rhinoceros, and extinct species of carnivora.

Sometime between the years 1834 and 1845, Brouillet discovered the first example of the art of fossil man—an engraving on bone from the cave of Chaffaud (Vienne). This specimen, supposed to be of Celtic age, was deposited in the Musée de Cluny, Paris, in 1851; not until 1861 was its great antiquity recognized (by Edouard Lartet), after which it was transferred to the Museum of National Antiquities at Saint-Germain (Fig. 6).

In 1848 a primitive type of human cranium was found in a small cave at Gibraltar. Not much importance was attached to this find until after parts of a skeleton, including a cranium similar in type, were discovered in 1856 in a cave of the Neander valley near Düsseldorf—a type since known as *Homo neandertalensis*.

Following the publication of Darwin's *Origin of Species,* there was a revival of speleologic exploration. In 1860–61 Edouard Lartet explored the caves of Massat (Ariège) and Aurignac (Haute-Garonne). In 1862–63, in collaboration with Henry Christy, he explored Bruniquel (Tarn-et-Garonne) and various

FIG. 6. THE FIRST EXAMPLE OF PALEOLITHIC ART EVER RECOGNIZED AND PRESERVED
AS A RELIC OF ANCIENT MAN.

The cannon (lower leg) bone of a reindeer bearing this engraving was discovered by Brouillet about 1840 in the cave of Chaffaud (Vienne), France. The engraving represents two hinds, both wounded by darts. Now in the museum at Saint-Germain. Scale, ⅘. After Breuil.

caves and rock shelters in the valley of the Vézère (Dordogne). Work in the caves of southern France was continued with brilliant success by Edouard Piette and Emile Cartailhac. In 1870, Emile Rivière began his excavations in the Grimaldi caves, east of Mentone.

One of the most notable events in the annals of cavern exploration was the discovery by Sautuola in 1879 of frescoes on the ceiling of the cavern of Altamira near Santander, Spain. These frescoes consisted of a remarkable group of animal figures representing, for the most part, extinct species. From the beginning, Sautuola was convinced of their authenticity and consequent great antiquity. In 1880 he published a paper on the subject privately, but his fate was that which so often befalls those whose discoveries lead to conclusions that run counter to the prevailing views of the

time. Sautuola died, and the event was in a fair way to be forgotten when, nearly twenty years later, similar discoveries in
France finally brought to him a posthumous but nevertheless full
vindication.

The period from 1895 to the present has witnessed a rapid
succession of discoveries of prehistoric mural figures in the caves
and rock shelters of France and Spain. The more or less

Fig. 7. Ceiling of the Principal Gallery in the Cavern of Altamira, Spain.

Here in 1879 Sautuola made the first discovery of Paleolithic mural art in the frescoes
of animals with which the ceiling is covered (see Fig. 110). The natural embossments
on the limestone no doubt account for the contorted positions in which many of the
animal figures are represented. After Cartailhac and Breuil.

prosaic exploration of floor deposits in caves and rock shelters has
continued with gratifying results: the caves in the region of Mentone under the patronage of the Prince of Monaco; those of France
by Cartailhac, Breuil, Capitan, Peyrony, *et al.*; those of Spain by
the Marquis of Cerralbo and several local Spanish savants, as well
as by Cartailhac, Breuil, and Obermaier; while much effective
work has recently been done in the caves of southern Germany
by R. R. Schmidt and others.

DIVISIONS OF THE STONE AGE

When Thomsen published his relative chronology for prehistoric times, the oldest known cultural epoch was called simply the Stone Age. Boucher de Perthes' discovery of a stone industry in the deposits of the Somme valley led to a realization that Thomsen's Stone Age represented only a later phase of the period. On the suggestion of Sir John Lubbock (Lord Avebury), the term *Neolithic* was accepted for this later phase, and *Paleolithic* for the older. Broadly speaking, the Neolithic Period was characterized by implements in which polishing was employed as a final shaping process; the Paleolithic was characterized by the complete absence of polishing as a shaping process. One of the most frequent Paleolithic implements, the flint cleaver (French *coup de poing,* "hand-ax"), crudely chipped on both sides to an amygdaloid or almond-shaped form, was supposed to represent the oldest phase of the Paleolithic Period. These implements occur in undoubted Quaternary deposits, while Neolithic implements occur only in later deposits.

Recognition of the authenticity of paleoliths was scarcely more than achieved when a new struggle broke forth over the question of the nature of certain flints found *in situ* in Tertiary deposits. To the period represented by these pre-Paleolithic, pre-Pleistocene flints, Gabriel de Mortillet gave the name *Eolithic,* a term eminently in keeping with the other two grand divisions of the Stone Age. The struggle of eoliths for recognition has been so long and bitter, and the issue so important, that a comprehensive review of the evidence will be presented in a subsequent chapter (Chapter III).

CHRONOLOGICAL TABLES

It is apparent from what precedes that pioneer work in three important branches of prehistoric archeology was being carried on during the nineteenth century almost simultaneously: the beginnings of a system of relative chronology by Thomsen; the successful search in Quaternary valley deposits for industrial remains of ancient man by Boucher de Perthes; and the discovery by Mac-Enery and others that human occupation of the caverns dates back to Quaternary times. At the same time Louis Agassiz was laying the foundation for a proper understanding of glacial phenomena (discussed in Chapter II), a subject of fundamental importance in estimating the age of human skeletal and cultural remains.

TABLE I. GEOLOGIC CHRONOLOGY

HOLOCENE	*Geologic Present*	Era of man
QUATERNARY (Pleistocene)	*Ice Age*	
TERTIARY (Cenozoic)	*Pliocene* *Miocene* *Oligocene* *Eocene*	Era of mammals (First traces of man)
MESOZOIC	*Cretaceous* *Jurassic* *Triassic*	Age of reptiles (First mammals)
PALEOZOIC	*Permian* *Carboniferous* *Devonian* *Ordovician* *Cambrian*	First reptiles First amphibians Age of fishes Cryptogam plants
ARCHEOZOIC		First traces of life

The present system of classifying prehistory is the outgrowth, therefore, of nearly a century of extensive and intensive study, especially in the European field, and is primarily applicable to that field. Whether the system can be extended geographically so as to apply without serious changes to other parts of the world occupied by Paleolithic man remains to be seen. Prehistory has been found to cover such a long period of time that it can be measured in geologic and paleontologic units. Moreover, a thorough study of cultural evolution under the control of paleontologic and stratigraphic data has yielded means of determining finer divisions of the time scale than are afforded by fossils and geologic deposits alone. Thus, while dependent upon and linked up with geologic

TABLE II. THE CHRONOLOGY OF PREHISTORY

	Iron Age		La Tène	III II I	100–Christian era 300–100 B.C. 500–300 B.C.
			Hallstatt	II I	Antennae swords and poniards Long swords of both bronze and iron
RECENT	*Bronze Age*		IV Winged and end-socket axes III Axes with transverse ridges II Plain-border axes I Flat axes		
		Neolithic	Carnacian		Stone cist Many-chambered dolmen Small dolmen
			Robenhausian Campignian Maglemosean Azilian- Tardenoisian		Polished flint implements Pottery Painted pebbles, harpoons of staghorn

	Stone Age	*Paleolithic*	Magdalenian	VI
				V
				IV
				III
				II
				I
					Evolution of Poly-
					  harpoon of chrome
					  reindeer horn frescoes
					
					Evolution of the javelin

			Solutrean	Upper Middle Lower	Point with lateral notch at base Laurel-leaf point Font-Robert point
QUATERNARY (*Pleistocene*)			Aurignacian	Upper Middle Lower	Art Graver (*burin*); Scratcher (*grattoir*) Transition
			Mousterian	Upper Lower	
			Acheulian	Upper Lower	Evolution of the cleaver (*coup de poing*) and scraper (*racloir*)
			Chellean	Upper Lower	
			Pre-Chellean		Transition
TERTIARY		*Eolithic*			Utilization of natural flint flakes; first attempts at artificial production

chronology, prehistory has developed its own more elaborate system of nomenclature in which the cultural note is dominant. The prehistorian thinks in terms of *Paleolithic* instead of *Pleistocene,* and of *Magdalenian Epoch* in place of *Reindeer Epoch.*

The conception of the length of time that man has existed upon the earth is very different now from what it was even less than one hundred years ago. As our knowledge of the record increases, it is found that instead of having begun only a few thousand years ago, the prehistoric period had its beginnings hundreds of thousands of years ago. This is not inconsistent with the views of the geologist concerning the age of the earth or the length of time that has elapsed since the most lowly forms of life appeared upon it.

Lord Kelvin calculated the age of the earth by the time that it would take its crust to cool, arriving at a maximum of 100,000,000 years. Others have attacked the problem by estimating the time it would take the land to furnish the sea with its present degree of saltness, or rather, its percentage of sodium. Their figures are not very different from the foregoing. Professor B. B. Boltwood's calculation of the earth's age, based on the time it takes radium to disintegrate, gives a maximum of 1,600,000,000 years. According to Professor Jeans, some 500,000,000 years ago the near approach of a star to our sun caused the latter to throw off the planets.

Sir Archibald Geikie employed as a time measure the thickness of the sedimentary formations, and reached a result of some 100,000,000 years. The latest researches of Professor Charles Schuchert lead to the conclusion that this figure is much too small, so that the tendency now is to lengthen rather than to shorten geologic time. Professor John Joly estimates (1922) that some 150,000,000 to 200,000,000 years have elapsed since the beginning (about the close of the Archeozic Period) of sedimentation on the earth.

A geologic table (Table I) of classification in its simplest outlines will suffice for a work of this nature, and a second generalized table (Table II) will introduce the reader at once to prehistoric nomenclature, leaving the detail on which it is based to be dealt with in subsequent chapters.

CHAPTER II

Geologic research has established the fact that the Pleistocene, in which man's Paleolithic history falls, was marked by the most widespread glaciation in the earth's history. This epoch, therefore, well deserves to be known as the Ice Age. Northwestern Europe, northern North America, the polar regions, and the principal mountain ranges in middle latitudes were the centers of dispersal from which ice spread in all directions, advancing and retreating with climatic oscillations. The northern hemisphere alone will come within the scope of the present survey of Ice Age phenomena.

GEOGRAPHIC EXTENT AND CAUSES OF PLEISTOCENE GLACIATION

The southern limit of the maximum ice extension in North America is traceable from the Atlantic coast across New Jersey, Pennsylvania, and Ohio to the Ohio river at Cincinnati; westerly to St. Louis and northeastern Kansas; thence the direction shifts to the northwest through Nebraska, South Dakota, and southwestern North Dakota; westerly again to a point on the Pacific coast a short distance south of the Strait of Juan de Fuca, with two projecting tongues, one in western Montana, the other in Washington and Oregon.

In Europe the continental glacier, during its maximum extension (Mindel), covered all of Fenno-Scandia, Ireland, Great Britain as far south as the Thames, Holland, nearly all of northern Germany, and two-thirds of Russia. The Alpine glaciers of to-day are but the merest isolated remnants of the great field of ice that covered the Alpine region during the last of the glacial epochs.

The causes that have been invoked to account for the Ice Age may be classed as terrestrial, solar, and cosmic. The exact reason

29

for the climatic change leading to glaciation is still a matter of dispute, though a great number of theories have been advanced to explain it. The chief general factor seems to have been a drop in temperature, but other factors, especially precipitation, locally played a great, perhaps a dominating, rôle.

The Work of Glaciers

In order to understand the glacial action of the Pleistocene, the student's best guide is present-day glaciation, which may be inspected in practically all its phases by a trip to the frozen north or to mountainous regions of sufficient height to insure perpetual snow. Glacial action may also be studied during the winter months in regions where glaciation does not exist, since it includes the work of frost and snow as well as of glaciers proper.

The physical and chemical properties of water, and its abundance over the earth, make of it nature's most powerful agent of construction as well as of destruction. Its property of freezing and thawing serves to increase its capacity for accomplishment. It attacks rocks at their weakest points, taking advantage of their porosity and planes of fracture. In the change from freezing to thawing, the fracture planes are not only widened until the mass is separated into blocks, but the blocks themselves are by degrees pulverized. Heavy rains, and especially the melting of snow and ice in the spring, cause the débris, or *talus,* to creep slowly to lower levels. Thus steep slopes are by degrees clothed with talus which itself becomes an even easier prey to further action of the frost than the rock mass that gave it birth.

Such movements of earthy débris, but on a large scale, have been noted in the Arctic regions and on mountain heights. Sven Hedin, the Swedish explorer, found his way almost completely blocked by water-laden moving sheets of detritus on the Thibetan highlands. The so-called "earth glaciers" of certain valleys in the Rockies are saturated and set in motion by the melting of snows in the spring. This phenomenon of "flowing soil" throws much light on the origin of certain widely distributed Pleistocene deposits.

Wherever snow accumulates more rapidly than it can be melted or evaporated, glacial snows and ice result, the size of the resultant glacier depending not only upon latitude, but also upon climate, altitude, and geographic conditions. The line above which snow fields last throughout the year is called the *snow line*. It is not fixed, but oscillates with the seasons and the years. It is lowest at the poles, and highest, not at the equator as might be expected, but near the tropics (30°–23°), where it is about 5,000 meters (16,417 feet) above sea level; it drops from the tropics toward the equator, and from inland mountain passes toward the sea. In the Alps of to-day, the snow line is on an average about 2,700 meters (8,865 feet) above sea level.

A glacier is composed of two parts—the *nevé* above the snow line, and the *tongue* below the snow line. It may be defined as

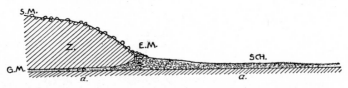

FIG. 8. LONGITUDINAL SECTION OF THE END OF A GLACIER.

Z., tongue; S.M., lateral moraine; E.M., end moraine; G.M., ground moraine; SCH., outwash or rubble field; a., ground level. After Penck.

a stream of ice which flows from higher to lower levels. In its course it often ascends slight elevations. Since ice is a crystalline body, it cannot be plastic in the strict sense, its apparent plasticity being due to the phenomenon of regelation. Glacial ice is granular; hence, when it is subjected to pressure, melting is caused at certain portions of the contact between grains, and the film of water slips away to a position of less pressure and there refreezes. Thus, through regelation, a pseudo-plastic motion is produced in masses of ice. Owing to the latent heat of water, the property that ice has of melting under pressure does not impose any limit to the thickness that an ice mass may attain.

The most rapidly moving part of a glacier is central and superficial, its sides and bottom being retarded by friction. The rate of movement depends upon three factors—temperature, slope of ground, and volume of ice; hence the motion is greater by day than by night, and greater in summer than in winter. In small glaciers it may be only a fraction of a meter a day; in some of the

Fig. 9. The Valley of Lauterbrunnen, Switzerland, an example of a valley over-deepened by glacial action.

large ice flows of the Arctic region the daily movement may exceed 15 meters (49.2 feet).

An ice particle that falls or forms above the snow line enters the body of the glacier and follows the longitudinal stream lines until it finally reappears on the lower terminal border of the glacier. There is also a lateral movement of the ice particles which causes them eventually to reappear on the marginal borders of the glacier. Rock particles follow the same lines of movement as the ice particles.

A stream of ice is capable of carrying a great load of débris, much of which is picked up from its own channel. The tongues descend to lower levels, or retreat, depending upon factors of temperature and precipitation. The ice melts only upon reaching the required altitude or latitude; when this point is reached, the load is dropped, forming *moraines*. A stationary tongue tends to accumulate a *terminal moraine;* a retreating tongue lays bare a *ground moraine,* sometimes capped by a *surface moraine;* an advancing tongue may level a terminal moraine representing some preceding stationary stage of glaciation (Fig. 8).

Streams flowing from glaciers carry quantities of powdered rock in suspension, testifying to the amount of work that a glacier in action is capable of doing. It is estimated that the stream issuing from the Aar glacier transports 208,617 kilograms (280 short tons) of mud daily; that from the Vatnajökull in Iceland, draining an ice field 20,992 square kilometers (8,200 square miles) in extent, about 13,605,442,177 kilograms (15,000,000 short tons).

CHRONOLOGY AS REVEALED BY REMAINS OF ANCIENT GLACIERS

Conspicuous among the débris left by the retreat of ancient glaciers are erratic blocks, terminal moraines, and scourings on rock outcrops. Other features that would not escape notice are: the unstratified gritty clay laden with rock débris of various shapes and sizes, known as *till* or *boulder clay;* outwash or rubble beds (*Schotterfelder*) ; over-deepened valleys (Figs. 9 and 10) ; and the profusion of lakes to be found in regions that were once glaciated. Especially important because of their bearing on the antiquity of man are the so-called *river drifts,* including gravels and sands, the

silts, and loess.   These formations have been especially explored
in regions that were never actually covered by the ice, for example,
in southern England, Belgium, and northern France.   Their precise
meaning and relation to the various phases of the Ice Age have
not as yet been fully determined, although recent researches have
thrown much needed light on this obscure and complicated problem.

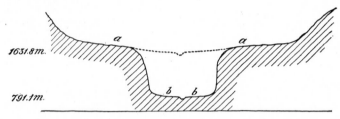

FIG. 10.   CROSS-SECTION OF THE OVER-DEEPENED VALLEY OF LAUTERBRUNNEN
SHOWING THE AMOUNT OF WORK DONE BY THE GLACIER.

The present bottom of the valley at *b, b,* is 840.7 meters (2,757.5 feet) below the
former bottom at *a, a.*   Adapted from Geikie.

A study of the remains of mountain glaciers reveals the fact
that in past times glaciation must have been much more pronounced
than at present (Fig. 11).   The work of Albrecht Penck and E.
Brückner in the Alpine field has done much to clarify the problem
of Ice Age chronology, including the number and character of its
phases.   On the plains north of the Alps they found spreads of
rubble or outwash, an older (*älterer Deckenschotter*) and a younger
(*jüngerer Deckenschotter*); they were also able to distinguish two
outwash formations in the valleys of rivers that flow from the
Alpine highlands, one in the upper terrace (*Hochterrassenschotter*)
and one in the lower (*Niederterrassenschotter*).   South of Memmingen it was possible to trace the outwash of the lower terrace
to a point where it overlies the outwash of the upper terrace.   A
like superposition also occurs in the Salzach, Traun, and Enns
valleys, as well as at Haidhausen, south of Munich.   Here Penck
found the outwash of the upper terrace superposed on the younger
of the two *Deckenschotter.*   That the foregoing is the true order
of succession is likewise seen in the greater weathering of the older
deposits; their materials are often cemented into a firmer mass.

All four of the outwash deposits belong to the Quaternary or
Pleistocene Epoch, because in the region of the Traun and Enns,

characteristic Pleistocene shells are found in the oldest of the four. Unfortunately, little is as yet known concerning other fossil remains from any of the four deposits in question.

That all four outwash deposits are of fluvio-glacial origin there is no longer any doubt. Each can be traced to its respective moraine in the direction of the Alps. The outwash of the lower terraces passes into the younger moraines and that of the upper terraces into the older moraines. The younger outwash spread of the plains is connected with the maximum morainic extension. As yet this connection has been noted only in Brandholz, south of

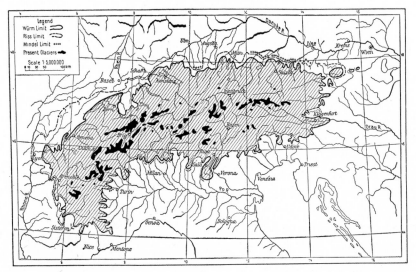

FIG. 11. MAP SHOWING THE EXTENT OF THE FOUR GLACIATIONS OF THE ALPS.

Memmingen, and near Wolfsrathshausen. The older outwash spread of the plains has also been traced to its moraines at Obergünzburg.

The presence of these four successive outwash formations can be explained only on the supposition that they correspond to four successive epochs of glaciation in the Alpine foothills. This is of fundamental importance in its bearing on our conception of the Ice Age. Instead of being one epoch, it is a concatenation of four epochs. Penck has given the name *Günz* to the First Glacial Epoch because in the region of the Günz River he was able to connect the older outwash with its moraine; the glaciation corresponding

to the younger outwash as developed in the district of the Mindel
River, he called the *Mindel*. The upper terrace outwash of the
Riss valley (Rhine glacier) being in distinct connection with its
moraines, Penck named the Third Glacial Epoch the *Riss*. Finally,
since in the Würm valley Penck found a connection between the
fresh lower-terrace outwash and young moraines, the corresponding glaciation is called the *Würm*. For the interglacial epochs
which alternated with these glacial epochs, Penck proposes the
names *Günz-Mindel, Mindel-Riss,* and *Riss-Würm*. Thus the
names of the four small rivers become in a sense generic and can
be applied, not only to epochs, but to their corresponding deposits,
so that one can speak of Günz, Mindel, Riss, and Würm outwash.

Outwash formations in the foothills of the Alps are generally
covered by loess and silt which Penck believes to be one and the
same formation. As one approaches the upper Danube, the cleft
or pulverulent loess, composed of dust and fine sand and containing shells, becomes thicker and covers the country like a uniform
mantle; while farther south, toward the Alpine foothills, it covers
for the most part only the outwash fields. It would seem here
that each of the first three outwash formations is connected with
its loess, like river gravel with valley silt. Along the Danube,
on the contrary, the loess far surpasses the Pleistocene outwash
deposits in both its vertical and horizontal limits; hence, it was
brought about through agencies quite different from those which
produced the outwash fields. One sees, for example, unweathered
loess capping weathered outwash, making it impossible to assume
that the loess immediately succeeded the outwash. Penck mentions
two different kinds of loess, the pulverulent and the tough brown
variety.

Nowhere did Penck and Brückner find typical loess overlying
the low-terrace outwash beds, although it covered near-by upper-
terrace outwash beds; hence the conclusion by them that the lower
terrace is younger than the latest loess. They found an old loess
near Höllriegelsgereuth between the younger outwash spread and
the upper-terrace outwash. Since the lower of these outwashes
belongs to the Mindel Glacial Epoch, and the upper to the Riss
Glacial Epoch, this old loess would date from the Mindel-Riss
Interglacial Epoch. The loess covering the upper-terrace outwash

would also correspond to an interglacial epoch, the Riss-Würm. The terminal moraines of the Mindel and Riss Glacial Epochs reach farther out into the plains than do those of the Günz and Würm Epochs, which indicates a greater development and possibly a longer duration of the Second and Third Glacial Epochs.

That warm phases occurred during each of the four glacial epochs is highly probable, but it is difficult to trace them except in the Fourth Glacial Epoch. In the Salzach region near Laufen, Penck and Brückner found a lower-terrace outwash bed between two moraines. The conditions were such that its presence could be explained by a temporary retreat of the Würm glacier rather than by a true interglacial epoch. This retreat, which happened at a time when the Fourth Glacial Epoch was at its maximum, is referred to as the *Laufen* stage. There was still another advance before the final retreat of the Würm glaciation.

Following the final maximum advance of the Würm ice, there was another marked retreat, during which the great terraces of the Inn valley in the Tyrol were laid down. Since these deposits include the dam back of which Achen lake was formed, Penck has called this retreat of the Würm glacier, the *Achen* stage. The Achen oscillation represents the retreat of the Würm ice up the Inn valley to Imst, a distance of 180 kilometers (112.5 miles) from its maximum advance.

Then came an advance of the ice once more down the valley some 120 kilometers (75 miles), where it remained long enough to form a typical complex of outwash and moraines in the Inn valley above Kufstein. The numerous hillocks formed here by the ice are locally called *Bühle*. Hence Penck has chosen the name, *Bühl* stage, for this advance of the Würm ice. During this stage the snow line was from 200 to 300 meters (657 to 985 feet) higher than that of the maximum Würm glaciation.

After the Bühl stage, two other distinct advances of the Würm glaciation occurred during its final retreat. In the little valley of the Gschnitz, a tributary of the Wipp, south of Innsbruck, at a height of 1,200 meters (3,940 feet), there is evidence of a temporary advance of ice to which has been given the name *Gschnitz* stage. Glacial accumulations in many other small valleys of the region, at about the same level, also mark this pause in the temporary advance of the ice. During the Gschnitz stage the snow line was at a height of some 2,000 to 2,100 meters (6,567 to 6,895 feet).

Within the terminal moraines of the Gschnitz glacier, and at a height of 1,900 meters (6,238 feet), are numerous old terminal moraines indicating the level at which the gradually retreating Würm ice made its last determined stand. At this time the snow line must have been at a height of more than 2,200 meters (7,223 feet), perhaps 2,500 meters (8,208 feet), and thus some 300 to 400 meters (985 to 1,313 feet) below the present snow line. Because of the typical morainic embankments in the valley of the upper Stubai southwest of Innsbruck, Penck called this the *Daun* stage, employing the term given to several mountain peaks of the locality (Daunkopf, Daunkogl, Daunbühl).

Dr. Josef Bayer, a serious student of glacial phenomena in Austria and the Alps, has attempted to shorten the latter part of the Ice Age by cutting out the whole of Penck's Riss-Würm Interglacial Epoch, substituting therefor Penck's Achen retreat and minimizing the phases that follow. This would mean a linking together of the Riss and the Würm into practically one glacial epoch, broken only by the Achen retreat. This retreat is represented in Austria by the *Göttweiger Verlehmungszone,* a clay deposit which separates the lower from the upper recent loess.

INTERGLACIAL DEPOSITS

Changes from glacial to interglacial epochs, as essentially caused by amelioration of the climate, were accompanied by changes in the fauna and flora. During the glacial epochs central Europe was covered by tundras; phases midway between glacial and interglacial produced a steppe climate similar to that of southeastern Russia at the present time; and during the interglacial epochs the climate was so warm that the hippopotamus and southern elephant migrated to northern France and even as far as England.

For a long time the identity of interglacial deposits remained in doubt, but the evidence accumulated slowly and surely until it could no longer be ignored. Fossiliferous deposits indicative of a warm climate, overlying boulder clay and covered by other deposits undeniably of glacial origin, have been found in almost every European country that has been subjected to glaciation, as well as in regions beyond the limits of the maximum extent of the ice sheet.

In Scotland James Geikie found an intercalation of interglacial deposits between bottom moraines; he found also a succession of flora in most of the Scotch peat bogs that indicates climatic oscillations. Long ago he called attention to the presence of two "buried forests" in the peat mosses of the British Isles and northwestern Europe, the lower separated from the upper by a bed of peat varying in thickness. The buried forests imply a genial climate, but the layer of peat between them contains arcto-alpine plants.

That the well-known shell-bearing clay of Kirmington (Lincolnshire), south of the estuary of the Humber, is an interglacial deposit is evident from the report (1904) of a committee appointed by the British Association for the Advancement of Science. The evenly stratified clay deposit, occurring at a height of about 30.5 meters (100 feet) above sea level, contains shells of *Cardium edule, Mactra subtruncata, Scrobicularia piperata, Rissoa ulvae, Tellina baltica,* and *Mytilus edulis.* This clay rests on an earlier, and is capped by a later, boulder clay.

One of the best known interglacial deposits, the Cromer Forest Bed (Norfolk), is a good illustration of Pleistocene climatic oscillation. The lowest deposits are for the most part of marine origin, containing fossils which indicate a warm climate (Pliocene). Passing upward through the series, northern types gradually increase and southern types diminish, until by the time the Weybourn beds are reached, the marine molluscan fauna is thoroughly arctic in character. Overlying the Arctic shell beds of the Weybourn Crag (Günz Glacial) are the fresh-water and estuarine "Forest-Bed" series (Günz-Mindel Interglacial) in which driftwood and stumps of trees occur. With rare exceptions these and other plant remains belong to living species found in Norfolk—alder, beech, birch, elm, hawthorne, maple, pine, and spruce. The associated mammalian remains include *Elephas meridionalis* and *E. antiquus, Rhinoceros etruscus,* bear, bison, boar, deer, hyena, wolf, saber-toothed tiger, and others. Arctic conditions are reflected in the overlying marine deposits, above which comes a fresh-water bed containing remains of an arctic flora including the polar willow (*Salix polaris*).

Of special interest from the historical viewpoint is the deposit at Hoxne (Suffolk), already mentioned, in which John Frere found rude flint implements which he reported as early as 1797. Subsequent to 1859 Prestwich and Evans ascertained that these implements came from one of a series of fresh-water deposits occupying a hollow eroded in the old boulder clay. In 1896 a committee from the British Association investigated the Hoxne site and determined the exact conditions governing the deposition of the fresh-water beds, which lie in a rather narrow channel eroded in the boulder clay. The lacustrine clays at the bottom of the fresh-water series contain many plant remains and grade upward into a bed of lignite 30.5 centimeters (1 foot) thick; the flora is temperate and for the most part indigenous. On the lignite rests a dark carbonaceous bed of silt and sand 3.96 meters (13 feet) thick, the flora of which is arctic. This bed contains no paleolithic flint implements, but they are found in the succeeding brick earth (presumably that of the old loess).

In the vicinity of Hull there is a thick fresh-water bed of sand and gravel intercalated between beds of older and younger boulder clay, characterized by the presence of a fresh-water shell, *Corbicula fluminalis*. Although now extinct in Europe, this species still lives in the Nile and certain rivers of Asia. Its fossil remains, associated with extinct and exotic Mammalia, are of common occurrence in the Quaternary deposits of the Thames, Seine, Somme, and their tributaries.

Glacial deposits are not found in the Seine valley; nevertheless one is perfectly justified in characterizing as interglacial the tufa[1] at La Celle-sous-Moret (Fig. 12), resting as it does on Pleistocene river gravels and covered by a loess deposit. Artifacts of the Acheulian type have been found in the upper part of this tufa. In the lower part are found plant remains more southern than those in the upper part. They include Canary laurel, fig, box, Judas tree, and other specimens pointing to a climate like that now met with on the Dalmatian coast. The fig is no longer indigenous north of Provence, and one must go to the southern part of the department of Var (Mediterranean coast) to find the Canary laurel in a wild

---

[1] A soft granular, or porous, rock, either calcareous or sandy.

state. The laurel and fig require a mild winter rather than a warm summer; since the laurel blooms in winter, the winters of northern France must have been mild. On the other hand, from the presence in the Pleistocene tufas of several species that cannot endure a hot, arid climate, it is evident that the summers were more humid than at present, although not so hot.

The best known Alpine record of interglacial climate is afforded by the Hötting breccia, on the left bank of the Inn near Innsbruck. The breccia, at an elevation of 1,150 meters (3,776 feet) above the sea, rests on moraines of the Mindel or the Riss Epoch and is covered by moraines and outwash of the Würm Epoch. During the

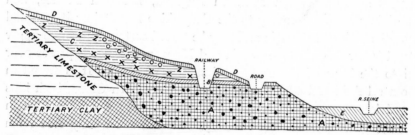

FIG. 12. GEOLOGICAL SECTION OF THE SEINE VALLEY AT LA CELLE-SOUS-MORET, SEINE-ET-MARNE, FRANCE.

A, Pleistocene river gravels with remains of the straight-tusked elephant (*Elephas antiquus*); *B*, thin layer of clay; *C*, calcareous tufa (interglacial) with Canary laurel, fig, Judas tree, etc., below (*x, x, x*) and Acheulian implements (*o, o, o*) associated with a colder type of flora (*z, z, z*) above; *D*, loess; *E*, recent river alluvia. After Geikie.

cold epoch which preceded its formation, the snow line was at least 1,200 meters (3,940 feet) lower than it is to-day. The deposit is from 10 to 20 meters (32.8 to 65.6 feet) thick, whitish to reddish in color, and consists in part of erratic material from the heart of the Alps.

The Hötting flora has been the subject of a special study by R. von Wettstein, who has determined more than forty species, twenty-nine of which still occur in Alpine valleys. Six species are no longer to be found growing at a height of 1,200 meters (3,940 feet) above sea level. Six of the forms are quite foreign to the Alps; the most common of these, the Pontic alpine rose (*Rhododendron ponticum*), to-day grows wild in the Caucasus, along the southern shores of the Black Sea, and in southwestern Spain. Geikie points out that in the regions which are now its habitat, the height of the snow line is 2,900 meters (nearly 9,800 feet), whereas the snow line in the region of Hötting does not exceed 1,118 meters (3,700 feet).

The Loess

Loess has played an important rôle in the history of mankind, especially in the art of agriculture and in building operations. It is composed of fine-grained materials, even-textured, and yellowish-brown in color. Its principal constituents are powdered calcium carbonate, quartz, and feldspar. It is spread over great stretches of Europe, in some places thick and in others thin. A great loess mantle extends from southern England, northern France, and Belgium across Germany to the Carpathian mountains; it is found on the plains of Bavaria and may be traced eastward through upper and lower Austria to Moravia and Hungary; thence it passes over into the loess of the steppes of southern Russia. It is found in central Asia and covers a great part of northern China, both hills and lowlands. The Balkans, the southern slopes of the Alps in Italy, and the foothills of the Pyrenees are not wholly free from loess. For characteristic deposits of loess, the region about Paris, the upper Rhine valley, and lower Austria take first rank. The more important loess deposits of the United States are spread over the valley of the Mississippi River and its tributaries.

There has been much discussion concerning the origin of loess. Because of its position, structure, and content, the general consensus of opinion seems to be that loess is, for the most part, an eolian (wind-borne) formation. It is not confined to positions suitable for water deposition, but clothes slopes and rises to considerable heights; in the Riesengebirge, for example, it reaches an elevation of 400 meters (1,313 feet) above sea level. The valley slopes protected from the prevailing winds are generally covered by the thickest loess mantles. Moreover, the lines of stratification are not such as would be formed by water; and the animal remains found in the loess are in a large measure land shells, fresh-water shells being rare and fishes entirely wanting. The three most abundant species of land shells are *Helix hispida, Pupa muscorum,* and *Succinea oblonga.* Mammalian remains found in the loess are not those of forms that live principally in the water.

There is the added testimony of Paleolithic hearths. At Willendorf in the Danube valley, for example, one finds evidences of human occupation at nine successive levels in the loess deposit, proof that a group of Paleolithic hunters returned as many times to the spot after leaving it for one reason or another. There was nothing in the mode of occurrence of the relics to suggest the action of water.

Eolian loess is not being formed at the present time in Europe; however, one may still see it in the process of formation in the steppes of Asia. It is probable, therefore, that at one time the climate of western Europe was similar to the present climate of the Asiatic steppes. While the great loess mantle is eolian, there are restricted loess deposits connected with valley terraces that owe their formation to the agency of water.

The loess of central and western Europe is supposed to be exclusively of Quaternary age, but it was deposited during various epochs. Loess was formed coincident with certain phases of the Ice Age. According to Penck, an interglacial epoch is composed of three successive phases, in one of which conditions favoring loess formations obtain:

1. A phase of erosion and deposition with the silting up of stream beds formed during the preceding glacial epoch. Climate, as well as fauna and flora, are not yet definitely determined.

2. Erosion, deposition, and weathering. Forest conditions; epoch of the Hötting breccia, *Schieferkohlen* (slate coal), and interglacial tufas. Temperate fauna including *Elephas antiquus, Rhinoceros merckii,* and red deer.

3. A phase of loess formation. Steppe climate, colder fauna including *E. primigenius, Rh. tichorhinus,* and horse. Prelude to the succeeding glacial epoch.

Some investigators believe that loess was formed mainly during the retreat of the ice. The outblowing glacial winds would have helped to create ideal conditions for the formation of loess. Shimek and others note that there is often an accumulation of stones and pebbles at the base of the loess. The wind alone is competent to produce such a gravel pavement without destroying the old topography. According to Hobbs, out-blowing glacial winds often attain a velocity of 128 kilometers (80 miles) an hour in Greenland and Antarctica. Such winds usually lose rapidly in velocity only a few score miles from the ice. This decrease in wind velocity might explain the occasional vertical gradation from gravel through sand to coarse, and then to normal, fine, loess.

The loess of northern France has been the subject of special study by Ladrière and later by Commont. In this region it is largely a sort of calcareous mud covering the gravel beds of the river terraces.

The classification of Quaternary deposits by Ladrière, with its equivalent by Commont, is given in the following table:

	LADRIERE	COMMONT
UPPER QUATERNARY (Recent loess)	Brick earth (*limon supérieur*) Recent loess (*ergeron*) Upper gravels	Recent loess with gray loess (*limon gris à Succinées*) at its base
MIDDLE QUATERNARY (Middle-Ancient loess)	Gray loess (*limon gris à Succinées*) Red loess (*limon rouge fendillé*) Soft loess (with black points) Streaked loess (*limon panaché*) Middle gravels	Ancient loess of the third terrace
. . . .	*Presle*	*Presle* . . . .
LOWER QUATERNARY (Ancient loess)	Peat Potter's earth Lower gravels	Alluvia of various ages

The so-called *brick earth* is simply the altered superficial zone of the recent loess or *ergeron*. Being exposed, it has become decalcified and therefore suitable for brickmaking, hence its name. The older loess likewise has its superficial altered zone, the red loess (*limon rouge fendillé*), corresponding to the brick earth of the recent loess, because it marks an old land surface. The old loess, being quite generally decalcified, is employed also in brickmaking; but that portion of the recent loess called *ergeron*, not having been as yet sufficiently decalcified, is not exploited by brickmakers. A loess that effervesces when tested with hydrochloric acid is recent; one that does not respond to this test is ancient.

The size of the concretions (*loess-männchen* or *poupées*) is also a criterion of age, the concretions of the ancient being much larger than those of the recent loess; in the latter they average about the size of a walnut.

In the valley of the Somme at Amiens the recent loess (with its brick earth) caps the two terraces nearest to the present bed of the river, that is, the fourth and third terraces in point of age. On the third terrace it overlies the ancient loess (with its *limon rouge*). There is no loess on the second terrace at Amiens; but on the oldest or first terrace (at Montières-les-Amiens), Commont found a very old loess, older than the ancient loess of the third, or next to the youngest, terrace.

The deposit of ash-gray, brown, or reddish-brown color containing shells of *Succinea* was classed by Ladrière, from the standpoint of stratigraphy, as being the uppermost in the middle loess series. But if judged from its faunal and cultural content, Commont would place it at the bottom of the upper series, because he found in it Mousterian implements associated with a northern fauna—mammoth, woolly rhinoceros, and reindeer. He found this gray loess also at the base of the recent loess in Alsace, at Achenheim, and Hochfelden. This loess, which might possibly date from the initial stages of the last glacial epoch, is found not only beneath the recent loess of the fourth terrace, but also at high levels on the borders of plateaus.

The middle ancient loess, like the recent, was formed by successive stages. At the top is (*d*) the red loess (*limon rouge fendillé*), an altered product which in places attains a thickness of several meters, thus attesting to the great age of the loess of which it is the protecting member. Below the red loess is (*c*) the soft loess with black particles (iron and manganese), and large concretions similar to those in the old loess of the Rhine valley. Next come (*b*) the streaked clay loess (*panaché*), and (*a*) a flinty layer which in places has the character of a veritable fluviatile gravel cutting into the older gravels. Below the flinty layer comes the chalk talus (*éboulis*) called *presle*.

Little is left of the first or oldest terrace at Amiens. Remnants are found at an elevation of 70 to 75 meters (230 to 246 feet) above sea level and at some distance back from the river at Mon-

tières and Saint-Acheul. The second terrace is about 55 meters
(180.5 feet) above the sea. The Pre-Chellean gravels of the Route
de Boves belong in this terrace. The best known gravel pits of
Saint-Acheul are in the third terrace, which has an elevation of
about 42 meters (138 feet). The fourth terrace at Amiens is not
much above the present bed of the Somme and only 20 to 28 meters
(65.7 to 92 feet) above sea level. The fourth terrace sinks below
the level of the Somme at Abbeville owing to the subsequent land
depression that gave rise to the English Channel; its position is in
accord with the banks of the fossil Channel river of which the
Somme was a tributary.

What is the age of the loess deposits of the Somme valley and
what was their relation to the Ice Age? The great ice sheet, even
during its maximum extension, fell far short of the Somme valley.
Nevertheless one is justified in assuming that these valley terraces
are in some way connected with the various phases of the Ice Age.

The third terrace at Saint-Acheul corresponds to that of
Menchecourt at Abbeville in which Prestwich found marine shells
and crystalline rocks carried by floating ice. The Menchecourt
terrace corresponds to the raised beaches of Étaples and Sangatte;
the latter is covered by two chalky formations similar to the recent
and middle-ancient loess at Saint-Acheul. The many fragments
of crystalline rocks transported from Brittany and Jersey by floating ice were deposited on these raised beaches during an invasion
by the sea, which, according to Commont, dates from the close of
the Chellean Epoch. It was the coming on of this period of cold
which produced the chalk talus or *presle,* whose stratigraphic position is at the base of the middle-ancient loess. In this deposit, or
its equivalent, both at Abbeville and at Saint-Acheul, Commont
has found flints broken up by frost, also frost-cracked Chellean
implements.

The loess deposits of the Somme and of the Rhine valleys are
identical. The recent loess corresponds to the *jüngerer loess* with
its three zones and, at the base, the gray loess; also the same fauna
and industry. The middle ancient loess is the same as the *älterer
loess* with its thick altered red zone at the top and large concretions in sands at its base; warm fauna (*Elephas antiquus* and
Rhinoceros merckii). The middle-ancient loess of the Somme

valley is characterized throughout by Acheulian culture; the same Acheulian industry has been found in the *älterer loess* at Achenheim (Alsace) by R. R. Schmidt and P. Wernert.

Commont, quoting Haug, says that the recent loess of the Rhine valley rests on the lowest terrace, which, toward Basle, may be connected with the Würmian fluvio-glacial deposits. He concludes, therefore, that all the recent loess is either Würmian or post-Würmian. Commont likewise quotes Haug for his statement that the middle-ancient loess of the Rhine valley rests on alluvia of the third (next to the lowest) terrace, itself in relation with the Rissian fluvio-glacial deposits. The ancient loess would thus belong to the Riss-Würm Interglacial Epoch and the Acheulian industry would be interglacial, as would the Pleistocene dunes of Ghyvelde near Dunkirk which contain shells of *Cyrena fluminalis*. Commont believes the Riss Glacial to be synchronous with the raised beach of Sangatte, the *presle*, and the terrace of Menchecourt; so that the Chellean, in part at least, can be referred to the Mindel-Riss Interglacial Epoch with the ancient fauna of *Elephas meridionalis, Rhinoceros etruscus, Equus stenonis*, etc.

Wiegers is inclined to class the loess formations as purely glacial, basing his contention on the character of both fauna and flora. Penck found that the higher of his two terraces had been subjected to a long period of weathering and erosion before the loess mantle had covered it. According to Wiegers, this long period was an interglacial epoch. Thus the gravels of the Günz, Mindel, and Riss Glacial Epochs would be covered by the loess mantles formed during the Mindel, Riss, and Würm Glacial Epochs, respectively. For the Alps he would have the following:

WÜRM GLACIATION: moraines and lower terrace gravels with reindeer fauna; formation of the recent loess with arctic-alpine fauna on the moraines and terrace gravels (*Hochterrasse*) of the Riss glaciation.

RISS-WÜRM INTERGLACIAL: denudation of the moraines, erosion of the valleys, weathering of the gravel beds; formation of the interglacial deposits with warm fauna and flora (Flurlingen tufa, etc.).

RISS GLACIATION: moraines and gravel terraces (*Hochterrassenschotter*) with reindeer fauna.

The chief difference between the views of Commont and Wiegers with respect to the deposits in the valley of the Somme hinges on the age of the old loess of the middle (second from the river) terrace. Commont links it with the Riss-Würm Epoch; for Wiegers it is of Rissian age, the Riss-Würm being represented by the old land surface separating it in time from the superposed recent loess, and by the Mousterian with warm fauna discovered by Commont in the gravel pit of Boutmy-Muchembled at Montières. The Warm Mousterian cannot well be subsequent to the Riss-Würm Interglacial Epoch; unless the Warm Mousterian and the Chellean both belong to the same interglacial epoch, the Chellean would have to be referred to the Mindel-Riss Interglacial, which is precisely what Penck, Wiegers, and Bayer have done.

According to Wiegers, the fauna of the loess includes the following:

SAINT-ACHEUL NEAR AMIENS (recent loess with Mousterian): *Elephas primigenius, Rhinoceros tichorhinus, Cervus tarandus, Bison, Equus* (small variety), *Spermophilus, Lepus timidus, Canis lagopus.*

HUNDSTEIG NEAR KREMS (recent loess with Aurignacian): *Elephas primigenius, Rhinoceros tichorhinus, Cervus tarandus, C. elaphus, C. canadensis, Bos primigenius, Bison priscus, Ovibos moschatus, Capella rupicapra, Capra ibex, Equus caballus, Canis lupus, C. vulpes, Felis spelaea, Gulo borealis, Spermophilus rufescens, Lepus variabilis, Lagopus albus.*

PŘEDMOST, MORAVIA (Solutrean): *Elephas primigenius, Rhinoceros tichorhinus, Canis lagopus, C. vulpes, Cervus tarandus, C. corax, C. canadensis, Bison priscus, Bos primigenius, Ovibos moschatus, Capella rupicapra, Felis spelaea, F. pardus, Gulo borealis, Lepus variabilis, Ursus spelaeus, Hyaena spelaea, Myodes torquatus, Vultur fulvus, Lagopus albus.*

ANDERNACH BELOW COBLENZ (Magdalenian): *Cervus tarandus, C. elaphus, Canis lagopus, C. lupus, Bos primigenius, Equus caballus, Lepus variabilis, Lagopus albus, Felis lynx, Arvicola amphibius, Mustela vulgaris, Mus musculus, Talpa europea, Sciurus vulgaris.*

DURATION OF THE ICE AGE

The studies of Penck and Brückner led them to conclude that the alternating glacial and interglacial epochs were not equal in length. Assuming that a ratio exists between the amount of geologic work done during the several interglacial epochs, they estimated that the Third Interglacial was three times as long as the post-Würmian Epoch; while not so long as the Second, the First Interglacial Epoch was at all events much longer than the Third.

Estimates for the lengths of the glacial epochs are not so trustworthy as those for the interglacial epochs. Judging from the amount of deposits they left, it would seem that the Mindel and the Riss were each longer than the Würm Epoch. There was apparently no glacial epoch comparable in length with the Second or Mindel-Riss Interglacial, but it is not possible to say whether there is any definite relation between the duration of glacial and interglacial epochs.

Penck estimates the time that has elapsed since the maximum Würm glaciation in the Alps by means of cultural data: some 3,500 years ago they were either inhabited or visited throughout their extent by Bronze Age races; bronze weapons, left after the retreat of the ice, were found in the Flüela pass which was invaded by ice during the Daun phase. Prehistoric copper mines, dating probably from near the close of the Neolithic Period, have been discovered in certain localities of the Austrian Alps. The Mitterberg mine, southeast of Hochkönig, is at an elevation of 1,500 meters (4,925 feet), near the present timber line; it could not have been worked when the snow line and timber line were 300 to 400 meters (985 to 1,313 feet) lower than at present. On the Kelch alp, southeast of Kitzbühel, Dr. Much found an ancient copper mine at a height of 1,900 meters (6,238 feet), that is, very near the snow line of the Daun stage.

According to Penck, the lake dwellings of Switzerland are all subsequent to the Daun stage; hence, the later, or typical Neolithic, must be of post-Daun age. The time that has elapsed since the Daun stage can thus be estimated at 7,000 or 8,000 years. Adding to this the duration of the Gschnitz, Bühl, and Achen stages a period of 25,000 to 30,000 years would probably be required for the whole of post-Würmian time.

Magdalenian industry is found in Switzerland well within the area covered by the maximum Würm glaciation; but this industry has not been found within the area covered by the Bühl stage. The Magdalenian Epoch would, therefore, seem to lie somewhere between the Bühl stage and the maximum Würm advance. The rock shelter of Schweizersbild was occupied by Paleolithic man after the Würm ice had retreated across the Rhine from Canton Schaffhausen (Fig. 13). The Paleolithic layers were covered in turn by successive deposits belonging to the Neolithic, the Bronze Age, and the Roman Epoch. Taking the thickness of the deposit left

FIG. 13. THE ROCK SHELTER OF SCHWEIZERSBILD NEAR SCHAFFHAUSEN, SWITZERLAND.

This rock shelter was occupied by Paleolithic man after the ice of the Würm glaciation had retreated across the Rhine. The lowest culture is Magdalenian; above the Paleolithic deposits are layers containing remains of Neolithic, Bronze Age, and Iron Age races. From the thickness of the deposit accumulated above the remains of the Roman Epoch, the period required for the accumulation of the whole series of deposits is estimated at 24,000 years—an example of the methods of arriving at the chronology of prehistory.

since Roman times as representing 2,000 years, the time required for the whole series of deposits is estimated at 24,000 years. By this method also one arrives at about 30,000 years as the length of time that has elapsed since the maximum Würm glaciation.

Penck's figures for the Riss-Würm and the Mindel-Riss Interglacial Epochs are 60,000 and 240,000 years respectively; and for the whole of the Ice Age from 500,000 to 1,000,000 years. Geikie does not believe Penck's estimate to be excessive. Chamberlin and Salisbury are inclined to hold a similar view regarding the Ice Age in America. Their estimate is as follows:

Climax of the Late Wisconsin	20,000 to	60,000	years	ago
Early Wisconsin	40,000 to	150,000	"	"
Iowan	60,000 to	300,000	"	"
Illinoian	140,000 to	540,000	"	"
Kansan	300,000 to	1,020,000	"	"
Jerseyan	y	z	"	"

According to Pilgrim, the duration of the Ice Age was approximately 1,290,000 years, the lengths of the various epochs being estimated as follows:

Fourth Glacial Epoch	190,000 years
Third Interglacial Epoch	130,000 "
Third Glacial Epoch	230,000 "
Second Interglacial Epoch	190,000 "
Second Glacial Epoch	170,000 "
First Interglacial Epoch	80,000 "
First Glacial Epoch	300,000 "

It will be seen that Pilgrim assigns the greatest duration to the First Glacial Epoch; like Penck, he considers the Second to have been the longest of the interglacial epochs, although he does not give it the dominant place in the system that Penck would.

Hildebrandt's figure for the total length of the Ice Age is 530,000 years. Obermaier believes that 300,000 years would suffice to cover the period. It is the opinion of nearly all glaciologists that the Ice Age was wholly post-Pliocene. Some writers however, including two well-known paleontologists, Boule and Schlosser, are still inclined to place the Günz Epoch in the Pliocene.

PLEISTOCENE HISTORY

During the Pliocene Period, the one immediately preceding the Pleistocene, the climatic conditions of Europe differed somewhat from those of to-day. Judging from the flora, the climate must have been both genial and equable. There were more genera and species than in the existing flora of Europe; moreover, the same species of trees lived contemporaneously in Italy and central France. Some of the species have become extinct, others belong to genera that are now exotic; still other species survive in the more southern and eastern regions of Europe and some are represented by allied species. The mammalian fauna included the mastodon, the dinotherium, several species of elephant and rhinoceros, hippopotamus, horse, saber-toothed tiger, and other forms. The mollusks then found in the seas bathing the shores of northwestern Europe are now confined to the seas of more southern latitudes.

The various phases of the Ice Age furnished the necessary setting for a remarkable series of faunal and floral migrations. As the climatic scenes in western Europe shifted, so did the life; the result in cross section, therefore, forms a complex representing the fauna of both hemispheres, the continents of Australia and South America excepted. In their turn came the hippopotamus, rhinoceros, and southern elephant from Africa and southern Asia; the bison, deer, and horse from the forests and meadows of Eurasia; the ibex, chamois, and ptarmigan from central Asia; and the arctic fox, musk ox, and reindeer from the Arctic tundras. One striking faunal feature was the adaptation of two huge pachyderms to arctic conditions—the mammoth (*Elephas primigenius*) and the woolly rhinoceros (*Rhinoceros tichorhinus*).

The First Glacial Epoch.—The Pleistocene Period began with the approach of a glacial epoch—the Günz Epoch of Penck, the *Scanian* of Geikie. It is represented in England by such deposits as the Chillesford clay and Weybourn Crag of Norfolk. The morainic accumulations of this First Glacial Epoch are met with in Scandinavia, northern Germany, and the Alps; they have not yet been located in England. During this epoch the ice sheet covered Scandinavia, occupied the basin of the Baltic, and extended as far south as Hamburg and Berlin. The arctic fauna migrated southward, remains of the musk ox, for example, being found in southern England. In the Alps the glaciers filled all the great mountain valleys, extending well out into the foothills, and the snow line dropped to some 1,200 meters (3,940 feet) below its present level. In America this glacial epoch is called the *Jerseyan;*

to it belong the Jersey drift, the old drift of the Allegheny drainage basin, and the Pre-Kansan or Nebraskan drift.

The First Interglacial Epoch.—This epoch is the Günz-Mindel of Penck, the *Norfolkian* of Geikie, and the *Aftonian* of America (from the type localities between Afton and Thayer, Iowa). Its deposits consist of fresh-water estuarine beds which overlie marine strata containing arctic shells. The position of the Forest Bed of Norfolk indicates that during the initial stages of this epoch the sea had retired from the English coast, causing the southern part of the North Sea to become dry land, Great Britain being joined to the continent. Under these conditions the Thames became a tributary of the Rhine, whose mouth was then much farther north than at present. The climate was probably somewhat warmer than to-day since the fauna resembled that of the Pliocene, including the hippopotamus, *Elephas meridionalis, E. antiquus,* saber-toothed tiger, *Rhinoceros etruscus,* beaver, *Equus stenonis,* deer, and bison. Deposits left by this epoch are met with somewhat rarely on the continent. They are recognized chiefly through their mammalian remains, which include the Norfolkian (Cromer elephant bed) and the *Paludinenbank* of Berlin and vicinity. The sand bed of Mauer near Heidelberg, from which Schötensack obtained a human lower jaw (Chapter VIII), would be placed in this epoch by Geikie, but Penck is inclined to place it in the Second Interglacial Epoch.

The lignites found in association with hippopotamus, *Elephas meridionalis,* etc., at Leffe in the Alps of Bergamo, Italy, are attributed by Obermaier to the First Interglacial Epoch. Above them are clays with *Picea balsami, P. seriana, Phragmites communis, Juglans bergamensis, Æsculus hippocastanum, Vitis neuwirthiana, Trapa heeri,* and *Andromeda polyfolia* (A. Portis). In this class belong the clays of Durfort near Sauve (Gard), containing *Fagus, Parrotia, Planera, Quercus,* and *Zelkova,* among which are some extinct species. According to Gaudry, the fauna of Durfort includes *Elephas meridionalis* and *Hippopotamus major.* To the same epoch also probably belongs Tegelen, province of Limburg, Holland, where have been found *Abies pectinata, Stratiotes websteri, Nuphar luteum, Vitis vinifera, Cornus mas,* etc., in association with remains of *Hippopotamus amphibius* and *Rhinoceros etruscus* (E. Dubois).

The Second Glacial Epoch.—This epoch is the Mindel of Penck, the *Saxonian* of Geikie, and the *Kansan* of the American geologists. It is represented by the younger outwash of the Alps, the old drift of England, the lower diluvium of northwestern Europe, and the drift of the Keewatin field of America. The marine deposits overlying the Norfolk Forest Bed indicate a submergence of the land amounting to some 15.3 meters (50 feet). The marine molluscan fauna, similar to that now found off the English coast, was gradually replaced by arctic mollusks; later the sea retired, making possible a fresh-water deposit overlying the arctic shell beds. The lowering of the temperature amounted to at least 20 degrees Fahrenheit.

The arctic fresh-water beds were followed by morainic accumulations during this epoch of maximum glaciation. The center of distribution was over the Gulf of Bothnia, where the ice must have reached a great thickness. It spread in every direction, not only to the east, south, and west, but also northward into the Arctic Ocean. The ice spread over the North Sea, the British Isles, and seaward perhaps to the margin of the Atlantic shelf. It descended southward to the Riesengebirge, Erzgebirge, and Carpathians, leaving erratic blocks at a height of 400 meters (1,313 feet) on the northern face of the Harz Mountains. It spread eastward to the foot of the Timan Mountains; these, with the Ural Mountains, formed a secondary center of dispersal. The southernmost extension of the ice in Russia was down the Dnieper valley beyond the 50th degree of latitude. The Pleistocene marine beds of Italy and Sicily, characterized by a northern fauna, belong to this epoch. As the great glaciers began to retire, tundra vegetation sprang up and the reindeer, musk ox, and mammoth roamed over the lowlands of central Europe.

The Second Interglacial Epoch.—This epoch is the Mindel-Riss of Penck, the *Tyrolian* of Geikie, and the *Yarmouth* of America. The deposits include the lower terrace beds of the Alps, the interglacial beds of central Russia, the Rixdorf horizon of northern Germany, and the Pre-Illinoian loess and silt deposits of America. Great Britain was still connected with the continent. The fauna and flora are characterized by southern types; the hippopotamus, *Elephas antiquus, Rhinoceros merckii,* deer, and ox roamed over

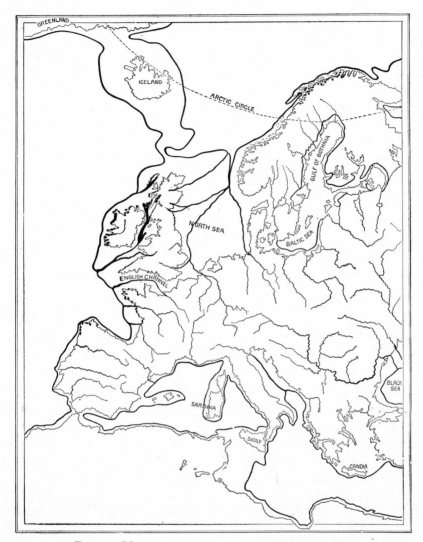

FIG. 14. MAP OF EUROPE DURING INTERGLACIAL TIMES.

The heavy lines represent the greatest extension of land which is inferred to have taken place in interglacial times, indicating the position of the land bridges which united Europe and Africa and the connections of Iceland and the British Isles with the Continent. This greatest extension probably took place during the Second and Third Interglacial Epochs. Modified from Geikie.

the greater part of Europe, as did also man. Penck would place the man of Heidelberg (Chapter VIII) in this epoch. Implements of the Chellean type date from this period. It is probable that some of the caves (Victoria Cave and others) were occupied, since they contain fossil remains representing a warm fauna—hippopotamus, rhinoceros, *Elephas antiquus,* hyena, etc.

Before the close of the Mindel-Riss Interglacial Epoch, the sea invaded the lowlands about the Baltic and central and northern Britain; there was also a land subsidence in the region of the Mediterranean, submerging the land bridges between Europe and Africa. This was apparently the longest as well as the warmest interglacial epoch, as indicated by the nature of the fauna and flora as well as by the geographic changes involved. The great length of time is especially attested by the large amount of denudation which took place. The whole land surface was lowered in some places from 15.25 to 30.5 meters (50 to 100 feet); while the excavation of many of the river valleys was practically completed.

The Third Glacial Epoch.—This is the Riss Epoch of Penck, the *Polonian* of Geikie, and the *Illinoian* of America. To it may be referred the high terrace deposits of the Alpine region, the middle drift of the north German lowland, and the Illinoian drift of America, including the so-called Illinoian of the Keewatin field west of Hudson Bay. According to both Geikie and Penck, the first Mousterian races lived during the third glaciation, being contemporaneous in England, Belgium, France, and Germany with the mammoth, woolly rhinoceros, reindeer, glutton, arctic fox, lemming, and tundra fauna in general. The continental ice sheet of this epoch was almost as extensive as during the Second Epoch, although the direction of the ice flow differed in some respects from that of the former. The ice sheet in England did not extend beyond the midland, while during the preceding epoch it had reached the valley of the Thames. In the Alps the Rissian glaciers of certain valleys were even larger than those of the Mindelian Epoch; while in other valleys there is evidence that the Mindelian glaciers were larger than the Rissian. The snow line was not so low as it was during the Mindelian Epoch. The Riss Epoch has left extensive terminal moraines in the mountain valleys of central and southern Europe. The rock of Gibraltar was elevated and a new formation of breccia took place; in regions outside the limits of the ice fields, outwash deposits were formed.

The Third Interglacial Epoch.—This is the Riss-Würm of Penck, the *Dürntenian* of Geikie, and the *Sangamon* of America. The loess of western Europe, that of central United States, and the Sangamon soils of the Labrador field belong to this epoch. According to Geikie, remains of *Elephas antiquus* and *Rhinoceros merckii,* both southern forms, occur in several interglacial deposits of this age. The bones of rhinoceros have been found at Pianico in northern Italy and at Flurlingen near Schaffhausen, while the lignite beds of Dürnten near Zurich contain the bones of both rhinoceros and elephant. The mildness of the climate made it possible for man to occupy the cavern of Wildkirchli at a height of about 1,500 meters (over 4,800 feet) on the Ebenalp near Säntis, and even the cavern of Drachenloch at an elevation of 2,445 meters (8,028 feet); artifacts of stone and of bone, representing what is supposed to be a Mousterian industry, were found in both of these caverns in association with fossil remains of the cave bear, cave lion, wolf, ibex, stag, etc. The human lower jaws, the flint artifacts, and the associated fossil animal remains in the lower travertine at Ehringsdorf near Weimar date from this epoch.

The Fourth Glacial Epoch.—This epoch is the Würmian of Penck and the *Mecklenburgian* of Geikie. Both Penck and Leverett have failed to correlate it with the Fourth, or *Iowan,* in America; they believe that it corresponds to the Fifth or *Wisconsin* stage. To this epoch belong the low terraces of the Alpine region, the upper diluvium or young drift of northern Germany, and the younger morainic drift of England. The greater part of the glaciated regions was submerged, and the British Isles were smaller than at present. The raised beaches of the period are conspicuous in the valleys of the Tay, Forth, and Clyde. The lowlands of Britain were treeless tundras which supported an arctic flora including the mountain avens,[2] polar willow, dwarf birch, and other northern forms. In the fresh-water lakes flourished a phyllopod crustacean (*Apus glacialis*) which now occurs only in Greenland and Spitzbergen.

This epoch is marked in the Alps by the presence of great moraines and fluvio-glacial gravels. The ice flows were not so

---

[2] A **prostrate** rosaceous flowering shrub (*Dryas octopetela*).

great as those of Mindelian and Rissian times. While large terminal moraines were piled up in Denmark, Schleswig-Holstein, and
the northern part of Germany and of Russia, the Scandinavian and
British ice sheets were not sufficiently developed to become con-
fluent as they were in earlier glacial epochs. The Aurignacian,
Solutrean, and Magdalenian races lived during this epoch.

Post-Würmian Climatic Oscillations in Scotland

It is probable that the Mecklenburgian of Geikie corresponds to
only the first maximum advance of Penck's Würm glaciation.
After reviewing the evidence bearing on the subsequent climatic
oscillations in Scotland as set forth by Geikie, an attempt will be
made to correlate these with the Alpine climatic oscillations connected with the Würmian Epoch in the Alps.

Lower Forestian (warm phase).—According to Geikie there
was a land uplift after the Mecklenburgian glaciation, causing the
sea to retreat until the British Isles were not only united into a
single land mass, but were also connected with the continent by
a land bridge. The climate grew mild and a thick growth of forest
not only covered the lowlands of Scotland, but also reached to
elevations which at present, do not admit of such growth. This
epoch has left its trace in the lower forest zone of the Scottish
peat mosses, hence the term Lower Forestian Interglacial chosen
by Geikie for the fourth warm or interglacial phase.

Lower Turbarian (cold phase).—Then followed a fifth stage
of cold, the Lower Turbarian, coincident with a land subsidence
that caused a complete separation of the British land mass from
the continent. Under the cold, wet conditions the forests disappeared, bog mosses flourishing in their stead. To this stage belong
the lower peat of Scottish inland bogs and certain moraines and
fluvio-glacial gravels as well as the formation of the 13.7 to 15.2
meters (45 to 50 feet) beach.

Upper Forestian (warm phase).—The fifth warm phase, the
Upper Forestian, was a period of land elevation, though less pronounced than that which occurred during the Lower Forestian;
it was apparently not sufficient to bring about land connection with
the continent. Forests again spread over the land, clothing even

the highlands to a height of 450 meters (1,500 feet) above the present limit for pine and birch. To this stage belong the upper forest zone of the inland peat and the trees found beneath the deposits of the 7.6 to 9 meters (25 to 30 feet) beach.

Upper Turbarian (cold phase).—Scotland was visited by a sixth and final period of cold, the Upper Turbarian of Geikie. Once more there was a partial subsidence. The forests again decayed and were buried under peat mosses. While the latest beach can be correlated with the upper peat, it is nowhere associated with moraines or glacial gravels. The mountains were glaciated at this time, but the actual presence of glaciers and snow-fields has not been determined. That cold, wet conditions did exist is fully substantiated by the fact that the Upper Turbarian covers the Upper Forestian in exactly the same manner that the Lower Turbarian covers the Lower Forestian.

After the Upper Turbarian cold phase, the sea gradually retreated to its present level coincident with an amelioration of the climate. That climatic oscillations of a marked degree took place in Scotland after the Fourth Glacial (Mecklenburgian) Epoch is seen in the alluvial terraces of the larger river valleys. The genial epochs coincident with land elevations were periods of rapid erosion; the rivers not only swept away the fluvio-glacial deposits that had been left in their paths, but also widened and deepened their valleys. The cold phases were periods of land subsidence and hence of abundant sedimentation. Proofs of climatic oscillations are also to be gathered from a study of the peat mosses. Dr. Francis J. Lewis has demonstrated that these confirm the geologic evidence previously gathered by Geikie.

CORRELATION OF CLIMATIC OSCILLATIONS

If the four glacial epochs of Scotland can be correlated with Penck's four glacial epochs of the Alps, it is probable that there is an agreement between the climatic oscillations connected with the Fourth Glacial Epoch in Scotland and in the Alpine region. It will be recalled that during the Alpine Würm glaciation there was a warm phase, the Laufen retreat; this was followed by a recrudescence of the Würm glaciation before the first step in its final

(Achen) retreat was taken. Successive forward and backward steps followed: the first forward step was the Bühl phase of cold; the following backward, the Bühl-Gschnitz warm phase. The next two steps marked the Gschnitz period of cold and the Gschnitz-Daun period of mild climate. The final effort to advance, the Daun stage, was weak in comparison with the preceding.

If the Laufen retreat had left as distinct a record in Scotland as it left in the Alps, the proper place for it would be the Lower Forestian of Geikie. This would cause the next advance of the Würm glaciation to coincide with the Lower Turbarian in Scotland, the Achen retreat with the Upper Forestian, and the Bühl stage with the Upper Turbarian. Thus the Gschnitz and Daun stages and their preceding warm stages would have no distinct counterparts in Scotland (Alternative I). Another alternative (II) would be the nonexistence in Scotland of the Laufen stage as well as the Gschnitz-Daun warm phase and the Daun cold interval. But Geikie is inclined to correlate the Gschnitz with the Lower Turbarian and the Daun with the Upper Turbarian. In

CORRELATION OF THE PENCK AND GEIKIE SYSTEMS OF
CLIMATIC OSCILLATIONS

	PENCK–BRÜCKNER	GEIKIE	ALTERNATIVE I	ALTERNATIVE II
	Postglacial	Recent and present	Recent and present	Recent and present
	Daun advance	Upper Turbarian		
	Gschnitz–Daun	Upper Forestian		
	Gschnitz advance	Lower Turbarian		Upper Turbarian
	Bühl–Gschnitz	Lower Forestian		Upper Forestian
	Bühl advance		Upper Turbarian	Lower Turbarian
PLEIS-	Achen retreat		Upper Forestian	Lower Forestian
TOCENE	IV Glacial (Würm)	Mecklenburgian	Lower Turbarian	Mecklenburgian
	Laufen retreat		Lower Forestian	
	IV Glacial (Würm)		Mecklenburgian	
	Riss–Würm	Dürntenian		
	III Glacial (Riss)	Polonian		
	Mindel–Riss	Tyrolian		
	II Glacial (Mindel)	Saxonian		
	Günz–Mindel	Norfolkian		
	I Glacial (Günz)	Scanian		
PLIOCENE				

that case the Bühl-Gschnitz warm phase would correspond to the Lower Forestian, and everything from the Laufen stage to the Bühl stage would be merged in the Mecklenburgian glacial, completely suppressing two warm stages.

An attempt at correlating the Penck and Geikie systems in the present state of our knowledge must, therefore, allow for two alternatives not mentioned by Geikie. These are indicated in the table on page 60.

Difficulties likewise arise in an endeavor to correlate the various phases of the Ice Age in North America with those in the Alps. The number and character of the phases agree as far as, and including, the Bühl cold interval. This would leave the subsequent Alpine oscillations to be merged in the American glacio-lacustrine and Champlain sub-stages. On the other hand, as might be expected, the American and Scottish systems would seem to be in complete accord if one makes use of Alternative I as noted on page 60.

CORRELATION OF EUROPEAN AND AMERICAN SYSTEMS
OF CLIMATIC OSCILLATIONS

	PENCK–BRÜCKNER	GEIKIE	CHAMBERLIN–SALISBURY
	Post–Bühl	Recent and present	Postglacial
	Bühl advance	Upper Turbarian	Late glacial
	Achen retreat	Upper Forestian	
	IV Glacial (Würm)	Lower Turbarian	Later Wisconsin glacial [3]
	Laufen retreat	Lower Forestian	Unnamed interglacial
PLEIS-	IV Glacial (Würm)	Mecklenburgian	Earlier Wisconsin glacial
TOCENE	Riss–Würm	Dürntenian	Peorian interglacial
	III Glacial (Riss)	Polonian	{ Iowan glacial { Sangamon interglacial { Illinoian glacial
	Mindel–Riss	Tyrolian	Yarmouth interglacial
	II Glacial (Mindel)	Saxonian	Kansan glacial
	Günz–Mindel	Norfolkian	Aftonian interglacial
	I Glacial (Günz)	Scanian	Jerseyan glacial
PLIOCENE			

[3] It should be recalled that both Penck and Leverett are inclined to correlate the Earlier and Later Wisconsin with the Würm glaciation.

That the uncertainties of correlation are great must be acknowledged. Discordances occur even in adjacent fields of glaciation. In the Labrador and Keewatin fields of North America, as pointed out by Leverett, one ice sheet experienced its maximum extension during the Second Glacial Epoch while the other was greatest during the Third Glacial Epoch. As yet the reason is not apparent. It is not strange, therefore, that attempts to correlate the phases of the Ice Age in three such widely separated fields as America, Scotland, and the Alps should meet with obstacles.

The Penck-Brückner system for the Alpine region has been confirmed in the main by Depéret, who has gone a step further by connecting the river-terrace levels of the Alpine foothills with the marine-terrace levels of the Mediterranean. Depéret believes the ancient marine shore lines to be the logical basis of classification for the Pleistocene in western Europe. His conclusions are based principally on data furnished by the Mediterranean region. The four lowest shore lines are referred to the Pleistocene. The highest of these, at an elevation of 90 to 100 meters (295.5 to 328.3 feet), represents the *Sicilian* stage; next below, at a height of 55 to 60 meters (180.6 to 197 feet), is the shore line referred to as the *Milazzian* stage. The third shore-line level, at a height of 28 to 30 meters (91.9 to 98.5 feet), is called the *Tyrrhenian* stage; the final stage is the *Monastirian,* with a shore line 18 to 20 meters (51.9 to 65.7 feet) above the present sea level. Oscillations of sea level are in accord with corresponding displacements of stream level. With land elevation (sea-level recession) the streams erode their beds; with land depression (rise of sea level) streams fill their beds. Thus the four river terraces correspond to the four ancient marine shore lines and both are linked up with the four glacial epochs of Pleistocene time.

Pleistocene History of the Somme Valley

Since the first appearance of man in northern France, the valley of the Somme has been worn down to a depth of 50 meters (164 feet). It is impossible to state how many thousands of years were necessary to accomplish this, but it was a sufficiently long period to allow for a change of fauna three times. The first fauna,

characterized by *Elephas meridionalis, E. trogontherii, Rhinoceros
leptorhinus, Hippopotamus major, Equus stenonis, Machaerodus,
Cervus solilhacus, C. somonensis,* etc., was contemporary with the
earliest population (Pre-Chellean); the second fauna, comprising
Rhinoceros merckii, Hippopotamus, Elephas antiquus, and the large
horse of Saint-Acheul, with the Chelleans, Acheulians and the first
Mousterians; and the third, or mammoth and reindeer fauna, with
the later Mousterians, the Aurignacians, Solutreans, and Magdalenians.

During this time, the configuration of the country changed
greatly. Land oscillations produced alternately Channel separation
from, and land connection with, England. Land movements at
the close of the Eocene had folded the earth's crust, especially the
Chalk, giving birth to parallel ridges which have ever since determined the direction of all the small streams of the region. During the Pliocene, France was separated from England by the
Diestian Sea, whose shores can be traced in the form of sands and
ferruginous grit at Noires Mottes near Sangatte at an altitude of
143 meters (469.5 feet), at different points in Boulogne, in the
high forest of Eu, and on the North Downs. At the close of the
Pliocene and the beginning of the Pleistocene, land connection between France and England is attested by the presence in the Cromer
Forest Bed of the same fauna that occurs in the second, or next
to the oldest, terrace at Abbeville, namely, *Elephas meridionalis,
Rhinoceros leptorhinus, Rh. etruscus, Hippopotamus major, Tro-
gontherium cuvieri, Equus stenonis, Cervus solilhacus, Machaerodus,* etc.

The ridge from Calais to Dover constituted a watershed dividing the waters of the northeast from those of the southwest. On
the north the Thames, Lys, Schelde, Meuse, and Rhine fed one
great watercourse, the *North Sea* river; on the south the Somme,
Bresle, Seine, etc., and the streams descending from the Weald,
including the Ouse, in whose valley the man of Piltdown was
found (Chapter VIII), were tributaries of the great *Channel* river.
The course of this fossil river is traceable on the Admiralty charts
by what are now known as the Hurd Deeps, between the Anglo-
Norman Isles and the Isle of Wight. According to Dollfus, the
old pre-Pleistocene Somme was the Avre, which, higher up, was

prolonged by the Aisne descending from the Ardennes highlands. Later, the Oise captured the Aisne from the Somme. This hypothesis was recently verified by the finding of fragments of quartzite from the Ardennes in the valley of the Somme.

A study of the Somme terraces leads to the conclusion that at the close of the Chellean Epoch a new invasion by the sea from both the northeast and the southwest took place. At the summit of the alluvions in which Chellean cleavers occur, Baillon and Prestwich found marine shells, which could have been carried there only by the sea. Moreover, traces of this invasion by the sea are seen in the raised beaches of Sangatte, Wissant, and Étaples on the French coast, and of Brighton, Selsey Bill, etc., on the English coast. This Pleistocene invasion by the sea seems to coincide with one of the glacial epochs, and it is evident that the valley terraces and the raised beaches are to be correlated with advances and retreats of the glaciers. The establishment of an exact correlation will have an important bearing on Pleistocene chronology.

An attempt has been made by Commont to correlate the valley terraces of the Somme with the raised beaches and sand dunes of the Channel region. He finds that the recent loess of the Somme valley is of the same age as that of the Rhine; at its base occurs a Mousterian industry associated with a cold fauna—*Rhinoceros tichorhinus* and reindeer. In both valleys the recent loess covers the alluvions of the lowest terrace, which, toward Basle, are in relation with a fluvio-glacial terrace; the recent loess is supposed to be postglacial, and the pebbly layer at its base to be synchronous with the Würm glaciation. The ancient loess of both valleys rests on alluvial deposits of the third, or next to the youngest, terrace, which (further up the valley) is connected with the fluvio-glacial terrace contemporary with the Riss glaciation; it is also connected with the invasion by the sea marked by the level with marine shells at Menchecourt and Abbeville, and with the raised beaches of the French coast containing crystalline rocks transported by floating ice. The ancient loess, which is interglacial, is characterized by a warm fauna (*Elephas antiquus*) and Acheulian industry in the Somme valley.

The ancient strand at Sangatte is covered by two deposits (calcareous) of different ages, corresponding to the ancient and recent loess. The maritime plain of Picardy is separated from the Channel by the same cordon of dunes that exists to the north of Dunkirk. The line of dunes nearest the coast was formed after the close of the Pleistocene, since it contains *in situ* cultural remains of Neolithic age; at some distance back of this line are the ancient dunes of Pleistocene age, characterized by the presence of *Cyrena fluminalis*. This shell was found at Menchecourt at the summit of the fluvial alluvions (Chellean) of the third terrace, which were crowned by ancient and recent loess; it was also found in the alluvial deposits of the Thames associated with *Elephas antiquus* and *Hippopotamus major,* characteristic of a warm fauna. It may thus be concluded that the Pleistocene marine invasion noted at Menchecourt is synchronous with the raised beaches of Sangatte and the ancient dunes extending as far as Ghyvelde (Belgium); it can be placed immediately after the Riss Glacial Epoch.

Soundings made in the Channel reveal the position of the Channel river at the close of the Pleistocene. Remains of mammoth are brought up by fishermen from the sea bottom between Calais and Dover; farther north, about 16 kilometers (10 miles) from Dunkirk, there is a veritable graveyard of mammoths. Remains are also found on the English coast from Crossier to the Isle of Wight, in a submerged forest near Torquay, and at many other points. Hence, the mammoth (and woolly rhinoceros) must have found pasture in the valley of the Channel river; and since remains of mammoth, woolly rhinoceros, and reindeer are usually associated with Mousterian, Aurignacian, Solutrean, and Magdalenian industry, it follows that these peoples may also have inhabited the territory now covered by the Channel.

The question of the date of the final separation of England from the Continent admits of an approximate answer. Soundings at the point of Hourdel disclose a Neolithic bog at a depth of 24 meters (78.8 feet) below sea level, while the fluvial gravels are found at an altitude of 28 meters (92 feet). A study of the profile of the river indicates that at that time the Somme emptied much further out and to the west of its present mouth. The depth of the Channel between Calais and Dover is 50 meters (164 feet); hence the Strait of Dover was barely open at the beginning of the Neolithic Period. Moreover, the similarity between the bronze industry of the Somme valley and England indicates that exchange between the two countries was possible even with feeble barks. The conclusion is, therefore, that the final opening of the Channel took place at the very close of the Paleolithic Period.

The final land depression producing the Channel diminished the pitch of the Somme and caused a filling process to take place in its valley, large deposits of tufa and peat being formed under humid climatic conditions. This fill amounted to 20 meters (65.7 feet) at Abbeville and 10 meters (33 feet) at Amiens; thus Magdalenian stations situated near the Somme were buried to considerable depths and consequently escape notice.

A study of the alluvial deposits of the various tributaries of the Somme indicates that these tributaries are not of the same age; some are as old as the Somme itself, others are much younger. The terraces of the Scardon at Abbeville correspond exactly to those of the Somme, indicating that the Scardon was already a tributary at the beginning of the Pleistocene; on the contrary, the alluvial deposits of the Ingon have furnished no industry older than the Mousterian. In addition to many streams which are still active, there are a great number of wind valleys which belonged to active affluents in prehistoric times. The valley terraces are largely due to stream action, although eolian action contributed in a large measure to the formation of the loess when France, connected with England, had a more continental climate than at present. At the close of the Paleolithic and the beginning of the Neolithic, the watercourses underwent a regressive evolution which has continued during historic times.

Late Pleistocene History of Fenno-Scandia

Detailed studies of the last glacial and the Postglacial Epoch have been made in Fenno-Scandia, where the records of the lapse of time, of changes of level, climatic evolution, and changes and movements of flora and fauna are probably more complete than anywhere else.

Late Glacial Time.—The ice recession has been studied in detail by Baron de Geer and his pupils by means of the varved, or seasonally banded, clays. These clays were deposited in fresh water off the receding edge of the ice sheet, one silty and one greasy layer each year, the two together forming a well distinguished annual layer or *varve;* the varves cover each other as shingles on a roof. Since their thickness varied from year to year, depending upon different summer temperatures, the varves can be recognized from widely separated localities both in the direction of the ice recession and at angles to it. By measuring series of varves at a number of localities, plotting graphs, and matching them, the

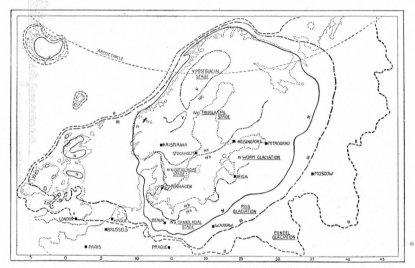

FIG. 15. MAP OF NORTHERN EUROPE IN GLACIAL AND POSTGLACIAL TIMES.

II, the Mindel Glacial Epoch; *III*, the Riss Glacial Epoch; *IV*, the Würm Glacial Epoch, including *IVa*, the Daniglacial stage, *IVb*, the Gotiglacial stage, and *IVc*, the Finiglacial stage; *V*, the Postglacial stage. Adapted from de Geer, Penck, Geikie, *et al.*

rate of ice recession can be determined and the ice edge mapped in detail.

The map (Fig. 15) shows some important positions of the ice border. The retreat over Denmark is called the *Daniglacial* stage; the recession over southern Sweden, or Gotaland, the *Gotiglacial* stage; the last stage is called the *Finiglacial* stage. These three stages comprise late glacial time. The Postglacial Epoch began when the ice margin retreated past Ragunda in northern Sweden and parted in two (Position V).

The rate of the ice recession outside central Scania is not yet

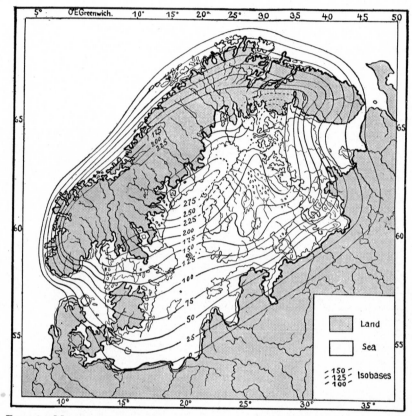

FIG. 16. MAP OF THE SUBMERGENCE OF FENNO-SCANDIA BY THE LATE GLACIAL SEA.
After Sederholm.

definitely known, but it was very slow. The retreat from central
Scania up to the Fenno-Scandian moraines, which marks the beginning of Finiglacial time, took some 3,000 years; the Finiglacial
sub-epoch represents some 2,000 years.

Glacial varve clay ceased to be formed with the disappearance
of the last ice, but another type of seasonally banded clay has been
deposited in the fiords of northen Sweden. By means of this clay,
Ragnar Lidén has been able, in the valley of the Ångermanälven
River, to extend the chronology over postglacial time practically
up to the present. The length of the Postglacial Epoch is found
to be about 8,500 years; the retreat of the ice over Scania, consequently, began about 13,500 years ago.

After the retreat of the ice, Fenno-Scandia, northern Denmark, and the region east of the Baltic were submerged, and increasingly toward the center of the previous ice sheet (Fig. 16). Generally, the highest shore line was developed coincident with the disappearance of the ice. Outside the zero isobase, which marks the outer limit for the later elevation, the land was higher than it is at present. Northern Germany, whose highest elevation was more than 35 meters (115 feet) above the present elevation, probably reached this maximum shortly after release from the ice. Since then the region seems to have undergone an almost uninterrupted slow submergence down to its present level, which was reached at the climax of the Littorina subsidence in Scandinavia. These conditions have recently been discussed by Antevs.

Inside the zero isobase the uplift, except at the very center, has several times been interrupted by subsidences. Vertical movements, particularly during the disappearance of the ice, were not contemporaneous all over Scandinavia, but undulating movements of the earth's crust are assumed to have advanced from all sides toward the center of the district of upheaval. The map (Fig. 16), therefore, does not show distribution of land and water at a certain time, but the amount of submergence of every single point at the time of its uncovering. When the ice border receded over southern Sweden, a land bridge connected Sweden with Denmark and Germany; and when the ice retreated from northern Sweden, the sounds across central Sweden no longer existed.

Since the connections of the Baltic with the ocean were of short duration, and since the water in the Baltic basin originally was entirely fresh (derived from the melting glacier), the main mass of the late glacial Baltic remained fresh. Some brackish water entered the basin, however, particularly through the central Swedish sounds which were both wide and deep. With the brackish water came a scanty arctic brackish-water fauna and flora which spread over the regions inside the sounds. Among the forms may be noted the bivalve *Portlandia (Yoldia) arctica* and the crustaceans *Idothea entomon* and *Gammaracanthus loricatus*. *Portlandia* was represented by a dwarf form, showing that the water was too warm even at the very edge of the ice and too diluted. It was limited to the Mälar (lake) depression and existed for only a short

time (less than 100 years at Stockholm). Hence, the late glacial
Baltic should not be called the Yoldia Sea.

Climatic conditions are revealed by the rate of the ice retreat
and by movements and changes of flora and fauna. The climate
in the southwestern Baltic region during Daniglacial time, or the
uncovering of the Danish Islands and southwestern Scania, was
very remarkable. It is, as yet, little known, but for some thousands
of years periods with continental temperate climate seem to have
alternated with periods of arctic climate. Then marked amelioration set in, as shown by the rapid ice recession from central Scania
northward. In southern Sweden, the uncovered land was first
inhabited by an arctic to sub-arctic *Dryas* flora, soon followed by
birch and pine, which were abundant before the extinction of the
arctic flora. The temperature was comparable with that now prevailing in northern Scandinavia. Somewhat later came alder, elm,
linden, and hazel, which occurred all over southern Sweden about
2,000 years after the ice release, indicating temperature conditions
like those of the coast regions of central Norrland at present. In
Norrland, the arctic belt, which had followed the receding ice,
was reduced to a minimum, so that the land from the beginning
was taken possession of by a temperate pine flora, largely similar
to the present vegetation of the region.

The earliest, or Gotiglacial, molluscan fauna of the Swedish
west coast was sub-arctic, characterized by large thick-shelled
Saxicavae and Balanidae. In Finiglacial time, from which the
oldest shell beds in southeastern Norway date, the arctic forms
gradually decreased in size and number, and several temperate
species immigrated.

Ancylus Time.—Ancylus time comprises the latter part of Finiglacial time and the first part of the Postglacial Epoch. The name
is derived from a characteristic fresh-water snail, *Ancylus fluviatilis*. The view held concerning the late Pleistocene changes of
level in Fenno-Scandia has undergone fundamental alterations in
recent years, as will be seen by consulting the table on page 74.
Antevs has shown that the Ancylus Lake was a fresh-water inland
sea. This Ancylus Sea was a direct continuation of the late-
glacial Baltic Sea which, as mentioned, was partly brackish. The
ingress of the water to the southern and central Baltic during

Ancylus time was due to sinking of the land. The southwestern Baltic region, though standing somewhat higher than at present, was not so elevated as has been previously supposed by many students, but there was connection with the ocean through the Fehmarn Belt and the Öresund.

The climate during Ancylus time, according to the latest researches of Lennart von Post, was not continental as has been supposed, but maritime or Mediterranean in type. As a consequence of the anticyclone over the remaining ice, the summers were dry and warm, but the winters were rather mild with much precipitation.

Littorina Time.—In late Ancylus time the *Darsser Schwelle,* the threshold between the Danish Islands and Germany, had sunk to about 15 meters (49.2 feet) below sea level, so that normally a salty undercurrent could force its way through the sound into the Baltic. In Littorina time it reached its present depth, which is 18 meters (59.1 feet). Also, in late Ancylus time Fenno-Scandia underwent an upheaval followed by a new submergence. During this subsidence, the crest of the bottom in the southern Öresund finally reached a depth of 13 meters (42.7 feet), that is, it stood 5 meters (16.4 feet) lower than at present (see Fig. 16). Consequently, a permanent salt undercurrent entered the Baltic through this sound, through which at present sea water comes into the basin only occasionally. Because of the submergence of the northern part of the Danish peninsula, the salinity of the waters of the Cattegat was considerably greater than at present. Gradually, the fresh water of the Ancylus Sea became salt, finally reaching a salinity twice that at present. This stage is called the Littorina time after the periwinkle, *Littorina littorea* and *L. rudis,* which existed in Baltic waters including the southern part of the Bothnian Sea.

The correlative of the Littorina stage in the Baltic was the Tapes stage on the Swedish west coast and in Norway; it was characterized by a molluscan fauna requiring salter and warmer waters than those which to-day wash the Scandinavian shores. In late Littorina time the country rose to its present level, which, on the Swedish west coast, was reached at the beginning of the Iron Age, or about 100 B.C.

As the Danish peninsula and the Öresund rose, and, as a consequence, the salinity of the Cattegat diminished and a permanent undercurrent could no longer enter through the Öresund, the salinity of the Baltic diminished so that *Littorina* and other marine forms had to retire towards the mouths of the inland sea.

The amelioration of the climate continued, and during this time (called the Atlantic period) the temperature became higher than at present. Because of the extension of the sea the climate was typically maritime.

Post-Littorina Time.—In post-Littorina time, Fenno-Scandia slowly rose to its present level, and the climate gradually became decidedly continental with slight summer precipitation. This was the Sub-Boreal Period. The maximum postglacial temperature, which allowed both plants and animals to spread farther northward than at present, was reached during this time, as shown by Sernander, von Post, Sandegren, and others. The temperature during midsummer, according to Samuelsson, was then about 1.5° C. higher and the vegetation period about fifteen days longer than now. About 100 B.C., or at the beginning of the Iron Age, a very rapid drop in temperature occurred, initiating the Sub-Atlantic Period. This drop in temperature was connected with an increase in moisture; later, the temperature rose slowly to that now prevailing. During historic time, the climate has undergone fluctuations, the most notable of which was the severely cold, wet period during the fourteenth century, but there has been no appreciable permanent change.

CORRELATION OF SCANDINAVIAN POSTGLACIAL AND CULTURAL CHRONOLOGY

The oldest cultural remains of Scandinavia belong to the Maglemose Epoch, which is discussed fully in Chapter X. The Maglemose culture dates from Ancylus time. The next cultural epoch is that to which the kitchen middens (*kjoekenmoeddinger*) belong. The relation of the kitchen middens to the general postglacial submergence south of central Denmark has been studied with interesting results by Holst. These refuse heaps occur at intervals along the coast from Brittany to Jutland; they belong to the Campignian Epoch. One of the most typical implements of this culture is the hatchet-shaped flint paring knife (French, *tranchet;* Norwegian, *skivespalter*), which has been traced from northern Italy through France and Belgium, thence along the coast to Denmark and Scandinavia. Another type is the flint pick representing the Nöstvet culture.

In northern Jutland all the kitchen middens occur on the highest strand lines of the Tapes submergence and at some distance from the present shore line; both they and the Tapes strand lines drop gradually as one passes southward along the coast, until they disappear beneath the sea in southern Jutland and Schleswig-Holstein to reappear once more on the Belgian coast. The older Danish kitchen middens containing Campignian artifacts are synchronous with the maximum postglacial Littorina-Tapes submergence. This type of implement had its beginning even in late Ancylus times, although then imperfect and rare.

Brögger has made a valuable study of the relation between the various phases of the Neolithic Period and the stages of land elevation at Christiania from the maximum Littorina-Tapes submergence. His subdivisions of the period are as follows:

4. Recent stage, corresponding to a land uplift of 8 meters (26.2 feet) to the present level. Appearance of *Mya arenaria.* Climate similar to the present.

3. Younger Tapes stage (Ostrea level), corresponding to a land uplift of from 19 to 8 or 10 meters (62.4 to 26.2 or 32.8 feet) below present level. Climate warmer than at present. Shell banks containing warm fauna no longer found living in Christiania Fjord.

2. Middle Tapes stage (Trivia level), during which the land rose from 45 meters (147.7 feet) to some 19 meters (62.4 feet) below present level. Climate cool. Isocardia clay.

1. Older Tapes stage, during which the land rose from 70 meters (229.8 feet) to about 45 meters (147.7 feet) below its present level. Climate mild with a summer temperature somewhat higher than at present. The older part of the Isocardia clay; shell banks with *Tapes decussatus* (a bivalve) and other southern forms no longer living in the Fjord.

The roughly chipped flint axes of the Nöstvet culture correspond in position, for the most part, with the shore line of the maximum Littorina-Tapes depression. Very few Nöstvet dwelling

Late Pleistocene History of Fenno-Scandia

	Chronology (Gerard de Geer and Ragnar Lidén)	Changes of Level in Western Sweden, in Bohuslän. (Ernst Antevs)	Stages in the Baltic. (Modified from H. Munthe and others)	Climatic Periods. (R. Sernander, L. von Post, and R. Sandegren)	Human History. (Adapted from O. Montelius)
0	A.D. 1900				Historic time
1,000				Sub-Atlantic cold, moist	
			Limnaea: salinity decreases		Iron Age
2,000	Christ			Drop in temperature	
		Present level reached / Elevation			Bronze
3,000				Sub-boreal: continental; warm; temperature maximum	Age
4,000	Post-				Stone cists
					Passage graves
5,000	Glacial				Dolmens
		Submergence (Tapes)	Littorina: salinity twice that at present; submergence	Atlantic: warm, maritime	Pick
6,000					Kitchen middens
		Elevation			
7,000			Ancylus: fresh-water inland sea; submergence		
8,000		Submergence		Boreal: warm, maritime	
		Elevation			Maglemose Epoch
9,000	Fini-	Bisection of ice remnant at Ragunda / Ice retreat from the Fenno-Scandian moraines (inclusive) to Ragunda	Sea connection through central Swedish straits / Drainage of last Baltic ice lake at Billingen		
10,000	Glacial	Submergence		Arctic	
		Elevation	South Baltic ice lakes and fresh-water inland seas		
11,000		Ice retreat from south Scania, the Danish Islands and north Germany to the Fenno-Scandian moraines / Submergence			
12,000		Elevation			
	Got	Central Scania released from the ice about 13,500 years ago	Depressed to marine limit, 141 m. (463 ft.) as released from the ice, but rising		
13,000					
13,500	Glacial	Time represented in the outer part of the zone unknown			

sites occur above this level, and none below it. The Nöstvet culture of Norway is synchronous with the shell heaps of Denmark and was, no doubt, an outgrowth from the latter; it is followed by a second and a third Stone Age epoch.

The early part of the second Stone Age epoch was characterized by a variety of ax with pointed pole, a transition form; it never occurs at a lower level than the beach line marking the close of the early Tapes period. The latter part of the epoch is characterized by a flat-poled ax. Before the close of this epoch the habits of the people changed. Their attention was turned more and more toward agricultural pursuits and the domestication of animals. At the close of this epoch the beaches at Christiania were from 23 to 26 meters (75.5 to 85.4 feet) above the present level.

The third epoch of the Stone Age is marked by the appearance of the thick-poled and the shaft-hole axes, types which are found in the Scrobicularia clay deposits as well as in graves. The position of some of these graves is such as to lead to the conclusion that when they were built, the beach line at Christiania was not more than 13 to 15 meters (42.7 to 49.2 feet) higher than it now is. According to Rygh, the oldest names of farms in Norway are associated with the word *Vin* (pasture land). Brögger's studies have brought out the interesting fact that the Vin colonization took place in southeastern Norway about the beginning of the third Stone Age epoch.

The presence of a cairn at Slagen near Tönsberg, on the Christiania Fjord, proves that toward the end of the older Bronze Age the sea was not more than 3 to 5 meters (9.8 to 16.4 feet) above its present level at this point. In his attempt to measure the lapse of time since the maximum postglacial submergence, Brögger assumes that: (1) the rate of elevation was about the same at the beginning as at the close; (2) the rate during the middle period of elevation was greater than at the beginning or the close; (3) the determining of the position of the beach lines at the beginning and end of the closing epoch of the Stone Age, compared with the estimates of archeologists as to the absolute length of the Bronze Age and the last epoch of Stone, gives a standard of measurement for the rate of elevation during the latter epoch. His results follow:

Stone Age:

First epoch	4900 B.C. to 3900 B.C. or	1,000	years
Second epoch	3900 B.C. to 2400 B.C. or	1,500	"
Third epoch	2400 B.C. to 1900 B.C. or	500	"
Bronze Age	1900 B.C. to 500 B.C. or	1,400	"
Iron Age	500 B.C. to 1900 A.D. or	2,400	"

Total................................... 6,800 "

In the accompanying tentative correlation of cultural and post-glacial chronology, the cultural data, adapted from Montelius, have not been so recently revised as the other data.

Correlation of Cultural and Ice Age Chronology

Europe was inhabited by man throughout a long series of epochs, during the Pleistocene and at least a part of the Tertiary. The successive steps in his cultural evolution can be traced, and the problem is to correlate these with the geologic and paleontologic data. During the Pleistocene there were four glacial epochs alternating with interglacial epochs, leaving two kinds of deposits and two types of fauna. Climatic oscillations are registered by the character of the accompanying fauna, which is of great assistance if properly interpreted. The data bearing on cultural evolution come largely from stations in regions that were at all times free from the great ice sheets, hence affected only by varying degrees of glacial contiguity; there are other stations which, because of their location and the nature of the relics they contained, could not have been inhabited during epochs of maximum glaciation.

Switzerland has furnished evidence of interglacial occupation of three caverns by Paleolithic man—Cotencher (Neuchâtel), Wildkirchli (Appenzell), and Drachenloch (St. Gallen). All three yielded remains of ibex, chamois, and cave bear, the latter averaging from 95 to 99.5 per cent of the entire faunal remains; also a primitive atypic Mousterian industry of stone and bone.

The cavern of Wildkirchli (Fig. 17) is situated near the top of the Ebenalp at an altitude of 1,477 to 1,500 meters (4,849 to 4,925 feet), some 200 meters (656.6 feet) higher than the maximum level attained by the local Würm glaciers descending from the Säntis range. The Ebenalp may be regarded, therefore, as a *nunatak* [3] and not habitable during a glacial epoch. Even more remarkable, in respect to altitude, is the cavern of Drachenloch (Figs. 18 and 19) at a height of 2,445 meters (8,028 feet), near the top of Drachenberg, southerly from Ragatz. Both of these stations are in the very heart of the Alpine field of glaciation. The nature of the animal and industrial remains excludes all possibility of their being post-

Fig. 17. Summit of Ebenalp, near Appenzell, Switzerland, showing the entrances to the cavern of Wildkirchli.

In the center of the bare escarpment is the lower entrance to the cavern, beside the little church from which it took its name; to the right of the center is the large upper opening, at an altitude of some 1,500 meters (4,925 feet). Stone and bone implements of the cave-bear hunter found in the cavern are evidence that the wide variation in climatic conditions of the Pleistocene Period permitted Paleolithic man to inhabit this cavern during the Warm Mousterian Epoch (Riss-Würm Interglacial). After Bächler.

Würmian; they must, therefore, be referred to the Riss-Würm Interglacial Epoch.

The cave of Cotencher is situated in the Gorges de l'Areuse at an altitude of 650 meters (2,134 feet), 400 meters (1,313 feet) below the maximum elevation attained by the Rhône glacier of the Würm Epoch, and more than a kilometer (nearly a mile) within the extreme limit of the Würm moraines. It would not have been

[3] An isolated hill rising like an island out of a sea of ice.

FIG. 18. DRACHENBERG, NEAR RAGATZ, ST. GALLEN, SWITZERLAND, NEAR THE
TOP OF WHICH IS THE CAVERN OF DRACHENLOCH.

The black spot near the summit, just above the snow field, is the entrance to the cavern, at an elevation of 2,445 meters (8,028 feet). The remains found in this remarkable abode of primitive man (see Fig. 19) indicate that it was occupied when the climate there was much more temperate than at present. This period has been identified as the Riss-Würm Interglacial Epoch. After Bächler.

chosen as a place of habitation during the period of maximum Würm glaciation, but it was habitable in postglacial times as well as during the last interglacial epoch. Since both the cultural and faunal remains are the equivalent of those from Wildkirchli and Drachenloch, the cave of Cotencher is referable to the Riss-Würm Interglacial Epoch.

The small cave of Bouicheta, half way up the steep slope of the Soudour, near Tarascon (Ariège), also throws some light on the

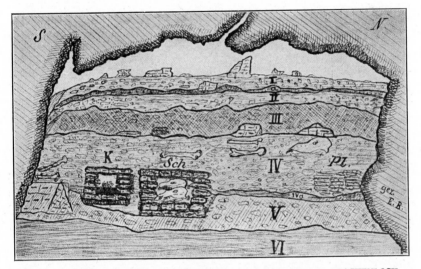

FIG. 19. SECTION THROUGH THE DEPOSITS IN THE CAVERN OF DRACHENLOCH.

Of the six numbered layers, only *II*, *III*, *IV*, and *V* are relic-bearing. At *K* were hearths with charcoal; at *Pl.*, an assemblage of flat stones. *Sch.* is a unique altar on which cave-bear skulls were piled, bearing witness to the existence of a cult of the cave bear during the Riss-Würm Interglacial Epoch. Stone implements and many bone implements made from the proximal (upper) end of the cave-bear fibula prove that the cavern was occupied by cave-bear hunters, 60,000 to 100,000 years ago. After Bächler.

relation of man to the Ice Age. The maximum altitude attained by the Rissian ice in the valley which the cave overlooks must have been greater than the altitude of the cave itself, since it left a deposit on the floor of the cave. Above this glacial deposit is a relic-bearing Mousterian horizon which is post-Rissian. Had the Würm ice, at its maximum, reached the altitude of the cave, it would have left a deposit overlying the Mousterian horizon.

Important evidence bearing on the problem of correlation is

afforded by river terraces in nonglaciated regions. The terraces of the Somme are the best known. At the Debary sand pit, Montières, in the first or oldest terrace with an altitude of 75 meters (246 feet), Commont found Neolithic and Mousterian industry on the surface and Acheulian industry at the base of a deposit of vegetal earth. Immediately below the Acheulian horizon was a deposit of calcareous loess, at the base of which implements of the Chellean type occurred; below this was a still older loess with Pre-Chellean or Eolithic industry. Then came gravels resting on the chalk. A section of the fourth (lowest) terrace at Montières as revealed in the Boutmy-Muchembled pit is reproduced in Figure 20.

FIG. 20.　THE BOUTMY-MUCHEMBLED SAND AND GRAVEL PIT AT MONTIÈRES, NEAR AMIENS, FRANCE.

This excavation is in the fourth or lowest of the alluvial terraces of the Somme valley; elevation above sea level, 20 to 28 meters (65.7 to 91.9 feet), the 8 meters representing the thickness of the deposit. The sequence is as follows: 1, coarse gravels with Chellean industry; 2, layer of whitish sands and gravels containing an ancient Mousterian industry and remains of a warm fauna (*Elephas antiquus, Rhinoceros merckii,* and *Hippopotamus*); 3, sterile layer of fine gravels; 4, thick deposit of recent loess with two horizons of Mousterian industry associated with remains of a cold fauna. See also Fig. 34. Photograph by the author.

The Bultel-Tellier sand and gravel pits at a height of 42 to 45 meters (138 to 147.7 feet) in the third terrace (second from the river) at Saint-Acheul present one of the finest Quaternary sections in Europe; they form the basis for Commont's type section described as follows:

14. Holocene or recent clay with Neolithic industry at its base; Gallo-Roman coins, traces of intrenchments, horse shoes, stirrups, coins of Louis XIV, Louis XV, and Louis XVI.

13. Brick earth formed by the decalcification and oxidation of the recent loess which lies beneath it; Aurignacian industry.

12. Upper recent loess.

11. Pebbly layer of flints with white patina; Upper Mousterian industry with La Quina facies; no cleavers.

10. Middle recent loess with a ruby-colored zone at the top; traces of an old land surface.

9. Pebbly layer analogous to 11, but more important; Mousterian industry, including one cleaver.

8. Red, sandy, lowest layer of recent loess.

The three divisions (12, 10, and 8) of the recent loess with small concretions are quite distinct at the pit of Bultel; but at the Tellier pit 10 and 8 are not separable, and 8 is replaced by a thin sandy bed with the same ancient Mousterian industry as in 7. The fauna of the recent loess includes the mammoth, woolly rhineroceros, and reindeer.

7. Pebbly layer, more important than 11 and 9; Mousterian industry with many cleavers.

6. Sandy brick earth; Upper Acheulian industry with white, lustrous patina in the upper part; fauna does not include the reindeer.

5. Calcareous loess resembling recent loess; ancient loess with large concretions; fauna includes *Cervus elaphus, Lepus cuniculus,* large horse, large lion, large Bovidae.

4. Sandy loess with black particles of manganese and pebbly layer at base; Acheulian industry at two levels.

3. Compact red clay sand, base of the ancient loess; Acheulian industry; fauna includes *Elephas antiquus,* large Bovidae, large horse, *Cervus elaphus,* and *Belgrandia marginata.*

2. Fluviatile sands; typical Chellean industry; *Elephas antiquus* of an archaic type.

1. Gravels composed of flints, Tertiary pebbles, coarse sands, and chalk worn by transport; rude Pre-Chellean industry; no fauna (corresponding beds of the same terrace at Abbeville have yielded a fauna with Pliocene affinities: *Elephus trogontherii, Hippopotamus major, Rhinoceros merckii, Rh. etruscus, Equus stenonis, Machaerodus*). The beds 2 and 1 constitute the fluviatile deposits of the third terrace.

A photograph of the Tellier pit, with the eight industrial levels indicated, is reproduced in Figure 21.

Another point of contact between cultural and glacial phenomena is afforded by the stations of Taubach and Ehringsdorf near Weimar, where a lacustrine deposit, known as lower traver-

tine, rests on Riss gravels. This travertine has yielded an interglacial fauna as well as human skeletal and cultural remains which
probably represent a very early phase of the Mousterian Epoch.

FIG. 21.   THE BULTEL-TELLIER SAND AND GRAVEL PIT AT SAINT-ACHEUL, NEAR
AMIENS, FRANCE.

This excavation is in the third terrace of the Somme valley; elevation above sea
level, about 42 meters (138 feet).   The section represents eight industrial levels, of
which the sequence is as follows: L, gravels with Chellean implements, resting on chalk;
F, ancient sandy loess with Acheulian workshops at two levels; E, ancient loess with
large concretions; D, brick earth of the ancient loess with Upper Acheulian lanceolate
types near the top; B', lower part of recent loess, with Lower Mousterian industry,
including many cleavers, at level c'; B, upper part of recent loess with Upper Mousterian
industry at level c; A, brick earth with Solutrean industry about midway and Neolithic
near the surface.   The weathered portion (D) of the ancient loess, known as *limon rouge*,
has a considerable thickness, indicating that a long period elapsed between the deposition
of the ancient loess and that of the recent loess. See also Fig. 46.   Photograph by
Commont.

In the interpretation of the evidence, differences of opinion have
arisen; but these differences are more apparent than real and are
destined to disappear with a more thorough study of the old and

the accumulation of new data. Some of the difficulties arise from the fact that the geographic area covered is large, including both glaciated and nonglaciated regions. Difficulties in the way of a satisfactory correlation multiply in direct ratio with the increase of area to be considered. The causes of cultural and climatic sequence even in a restricted area are not identical, although they may overlap to a certain degree.

The contrast between the geologic and the archeologic point of view is illustrated by a comparison of Schmidt's interpretation of certain paleolithic stations with that of Wiegers.

The gravel pit at Markkleeberg (Saxony), south of Leipzig, has yielded an Acheulian industry. According to Schmidt, the relic-bearing horizon belongs to the Riss-Würm Interglacial Epoch because it is capped by a Würm glacial deposit; but Wiegers finds convincing proof that this Acheulian deposit belongs to the Mindel-Riss Interglacial Epoch.[4]

These two authors differ quite as much in regard to the age of the cleaver from the loess at Sablon (Montigny), near Metz. Schmidt considers it Lower Acheulian because he believes the deposit to be of Riss-Würm Interglacial age; Wiegers declares the deposit must be referred either to the Riss Glacial (Upper Acheulian) or to the Würm (Upper Mousterian).

Schmidt considers the *Pariser* formation at Ehringsdorf to be Riss-Würm Interglacial loess (ancient), hence the relic-bearing lower travertine covered by it would be of Acheulian age. According to Wiegers, the Pariser formation is not loess, and the lower travertine, while of Interglacial age, contains industry of the Warm or Lower Mousterian Epoch.

It is evident that the last word has not yet been said on the subject of comparative chronology. But there is no longer any reason to doubt that man lived in Europe at least as long ago as the Mindel-Riss Interglacial Epoch, where practically all authorities have agreed in placing *Homo heidelbergensis.* The character of the dentition of the Heidelberg lower jaw points unmistakably to a tool user, even to one whose ancestors had for ages been tool users. It can safely be assumed, therefore, that man antedates the Mindel-Riss Epoch.

---

[4] Wiegers says "first interglacial," but he probably means the Mindel-Riss.

CORRELATION OF ICE AGE AND CULTURAL CHRONOLOGY

ICE AGE CHRONOLOGY	CULTURAL CHRONOLOGY	
	Post Daun	*Neolithic*	Campignian
	Daun Advance		Maglemosean
	Gschnitz Advance		Azilian-Tardenoisian
Holocene		*Mesolithic*	

	Bühl Advance		Magdalenian
	Achen Retreat		Solutrean
IV Glacial	⎧ Würm Advance		Upper Aurignacian
(Würm)	⎨ Laufen Retreat		Lower Aurignacian
III Interglacial	⎩ Würm Advance		Upper Mousterian
(Riss–Würm)			Lower Mousterian
III Glacial			Upper Acheulian
(Riss)			Lower Acheulian
II Interglacial			Upper Chellean
(Mindel–Riss)			Lower Chellean
II Glacial			
(Mindel)			
I Interglacial			Pre-Chellean
(Günz–Mindel)			
I Glacial			
(Günz)			
Pleistocene		*Paleolithic*	

Pliocene	*Eolithic*	Foxhallian
		Ipswichian
		Cantalian (?)

BIBLIOGRAPHY

ANTEVS, Ernst, "Senkvartära nivåförändringar i Norden," *Geöl. Fören. Förhandl.*, xliii, 642–652 (Stockholm, 1921).

—— "The Recession of the Last Ice Sheet in New England," *Amer. Geogr. Soc. Research Ser.*, No. II (1922).

—— "On the Late Glacial and Postglacial History of the Baltic," *Geogr. Review*, xii, 602–612 (1922).

BAYER, Josef, "Identität der Achenschwankung Pencks mit dem Riss-Würm Interglazial," *Mitt. Geol. Ges. Wien.*, i, 195 (1914).

BRÖGGER, W. C., *Strandliniens beliggenhed under stenalderen I det Sydöstlige Norge* (Christiania, 1905).

BROOKS, C. E. P., *The Evolution of Climate* (London, 1922).

CHAMBERLIN, T. C., and SALISBURY, R. D., *Geology*, III (1906).

THE ICE AGE 85

DEPÉRET, Charles, "Essai de coordination chronologique genérale des temps quaternaires," *C. R. Acad. Sci.*, Vols. 166, 167, 168, 170, 171 (Paris, 1918–1920).

DE GEER, Gerard, "A Geochronology of the Last 12,000 Years," *C. R. Congr. Géol. Intern. à Stockholm*, 1910, 241–253 (Stockholm, 1912).

—— "Om naturhistoriska kartor över den baltiska dalen," *Pop. Naturvet. Revy*, 189–200 (Stockholm, 1914).

GEIKIE, James, *The Antiquity of Man in Europe* (New York, 1914).

HUNTINGTON, Ellsworth, and VISHER, S. S., *Climatic Changes* (New Haven, 1922).

LEVERETT, Frank, "Comparison of North American and European Glacial Deposits," *Zeitschrift für Gletscherkunde*, iv, 241–316, 5 maps (Berlin, 1910).

MUNTHE, Henrik, "Studies in the Late Quarternary History of Southern Sweden," *Geol. Fören. Förhandl.*, xxxii, 1197–1292 (Stockholm, 1910).

OSBORN, H. F., and REEDS, C. A., "Old and New Standards of Pleistocene Division in Relation to the Prehistory of Man in Europe," *Bull. Geol. Soc. Amer.*, xxxiii, 411–490 (1922).

PENCK, Albrecht, and BRÜCKNER, Eduard, *Die Alpen im Eiszeitalter*, three vols. (Leipzig, 1909).

SEDERHOLM J. J., "Sur la géologie quaternaire et la géomorphologie de la Fennoscandia," *Bull.* (No. 30) *Com. géol. de Finlande*, 66 pp. and 6 maps (Helsingfors, 1911).

WRIGHT, W. B., *The Quaternary Ice Age* (London, 1914).

CHAPTER III

CHIPPED FLINTS FROM THE PLIOCENE AND MIOCENE

The discovery by Desnoyers of incised fossil bones in the Pliocene sand and gravel beds at Saint-Prest, near Chartres, first served to awaken interest in the probable coexistence of man with *Elephas meridionalis,* the remains of which occur in these beds. Since no flint implements had been found, the incisions were regarded as having been produced by the teeth of "some extinct rodent of the beaver family"; but in 1867 flints which he believed to have been intentionally chipped were found in these beds by the Abbé Bourgeois. They were obtained at various depths in the high-level gravels and did not include the cleaver of Chellean type which at that time was supposed to represent the earliest industry in stone.

The Abbé Bourgeois' researches, which were soon extended to the Miocene at Thenay (Loir-et-Cher), formed the subject of important communications to the International Anthropological Congresses of 1867 and 1872. At the latter Congress, held in Brussels, a committee of fifteen was appointed to report on the chipped flints submitted by Bourgeois. Nine of the members of the committee (de Quatrefages, d'Omalius, Cartailhac, Capellini, Worsaae, Valdemar Schmidt, de Vibraye, Franks, and Engelhardt) pronounced in favor of certain specimens; five (Steenstrup, Virchow, Neyrinckx, Fraas, and Desor) found no evidence of intentional shaping; and one (Van Beneden) was unable to decide. DeMortillet remained to the last a champion of the Thenay specimens, some of which are preserved in the National Museum at Saint-Germain. At the same Congress, a paper was presented by Ribeiro on chipped flints from the Upper Miocene and Pliocene deposits in the valley of the Tagus, near Lisbon. Similar specimens were discovered later by Delgado in the Upper Miocene at Otta; but the consensus of opinion is that these specimens, as well as those from the valley of the Tagus, are not artifacts.

A somewhat different fate was reserved for the discoveries by Rames, from 1869 to 1878, of chipped flints in the Tertiary deposits at Puy Courny near Aurillac (Cantal). All the flints possessed a brilliant black or dark-yellow patina; the retouches and marks of utilization were "most convincing." After a careful study of the pieces, de Mortillet, Cartailhac, Chantre, Capellini, and de Quatrefages declared that if these flints had been found in Quaternary deposits, no one would hesitate to regard them as having been chipped intentionally. Later, more than 7,000 chipped flints were collected at Puy Courny by Westlake, the chipping on many of which it would be very difficult to explain without invoking the element of intention. The Upper Miocene alluvial beds at Puy Courny, resting on Miocene basalt, contained the following fossils as determined by Gaudry: *Dinotherium giganteum, Mastodon (angustidens* or *longirostris), Rhinoceros schleiermacheri, Hipparion gracile, Tragoceros,* and *Gazella deperdita.*

More recently, Puy Courny and other stations near Aurillac, especially Puy de Boudieu, have been explored by a number of archeologists. A. de Mortillet was the first to discover chipped flints at Puy de Boudieu, and was followed by Capitan, Klaatsch, Verworn, Westlake, and Mayet. Puy de Boudieu has an altitude of 839 meters (2,755 feet). The gravels which contained the chipped flints were covered by a thick deposit of Miocene basalt. The flints were more plentiful, less worn, and larger than those from Puy Courny, dark in color, and deeply patinated.

In the opinion of Mayet, the Miocene chipped flints from the region of Aurillac may be accounted for through natural causes without invoking the interposition of man or his precursor. At the same time, he does not feel justified in denying the existence of intentionally chipped flints elsewhere and in other horizons. It has been suggested that the chipping on many of the specimens may have been due to pressure of the overlying beds. Such a result might be possible where unstable beds contained a sufficient quantity of flint nodules and chips pressing against each other, but at Puy de Boudieu the chipped flints were separated by masses of tufa, loam, and sand. For this reason, Verworn believed that many of the Aurillac specimens were genuine artifacts (Fig. 22). He held that no single character was a sufficient basis for declaring that a given flint was, or was not, an artifact; that each specimen should be subjected to a systematic diagnosis, as is a case of fever by a physician.

In observing a number of Paleolithic (or Neolithic) scrapers which were made from flakes retouched on a single side only, one finds that the direction from which the retouching took place is almost always oriented in the same manner with respect to the sides of the flake. If one calls the under, or bulb, side of the flake the front, and the outer side the back, it is seen that the blows or pressure which produced the marginal working were executed, almost without exception, from the front toward the back; that the tiny scars left by the chipping begin at the margin and extend over the back. The chipping, therefore, is visible only from the back. The one-sided marginal working of a flake is thus found to be the rule.

Verworn chose two series universally recognized as artifacts, compared them with each other, and then with a series of chipped flints from Puy de Boudieu. His results follow:

Locality	Total Number of Pieces Examined	Number that Follow the Rule	Number Opposed to the Rule
Vézère valley (Dordogne)....	686	654 or 95.3%	32 or 4.7%
Tasmania................	92	88 or 95.7%	4 or 4.3%
Puy de Boudieu..........	121	115 or 95 %	6 or 5 %

Klaatsch, who was familiar with the chipped flints from Aurillac, saw a similar stage in the culture of the Tasmanians who have recently become extinct, and also in a primitive culture of which traces occur throughout Australia. This is especially true of southwestern Australia, where the primitive stone industry consists, for the most part, of chips produced intentionally, utilized, then retouched, and finally cast aside or lost after a brief period of service. The workshops are characterized by the presence of great quantities of nonutilized chips, retouched flakes being rare. The same industry has been discovered at Ljelimane (Haute-Senegal) by de Zelt-

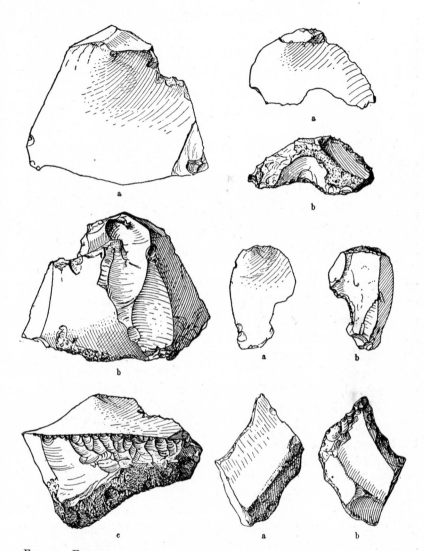

FIG. 22. EOLITHS FROM PUY DE BOUDIEU, CANTAL, FRANCE, TERTIARY PERIOD.

The large figures on the left are three views of a scraper; the two in the lower right-hand corner are views of a point; the others are views of two spokeshaves. These all show how the marginal chipping was done on one face only, in order to obtain a sharp, straight edge which would be useful for cutting, cleaving, and scraping. Scale, ¾. After Verworn.

ner. The people who left the industry have disappeared, apparently in recent times. In view of the primitive character of their stone implements, or *tronattas,* Noetling thinks it is not strange that believers in eoliths should cite the Tasmanians as a nineteenth-century example of an Eolithic race (Fig. 23).

CHALK-MILL EOLITHS

That natural processes such as torrential action, pressure of overlying beds, change of temperature, and wave action might suffice to produce eoliths, was thought to have been demonstrated by the appearance of flints that had passed through a chalk mill. The first observation of this kind was made by Laville, preparator at the *Ecole des Mines* (Paris), at a cement factory southeast of Mantes, near Paris.

In extracting chalk from the quarry, most of the flint nodules are cast aside; some, however, pass unnoticed by the workmen and are carried with the chalk to the factory. These, together with a certain amount of clay, are emptied into circular basins or diluters. In their passage through the basins, the flint nodules receive thousands of knocks, some mutual, some from the iron teeth. At the end of twenty-nine hours the machinery is stopped, and the nodules are removed, washed, and piled up to await their ultimate use as a by-product. It was in one of these piles that Laville's discovery was made. Later, he revisited the place in company with Boule, Cartailhac, and Obermaier.

According to Boule, the flints that had passed through the machine had all the characters of the ancient river gravels. Most of them had become rounded pebbles, but some were chipped in a manner to resemble true artifacts.[1] He and his companions were able in a few minutes to make a "superb collection including the most characteristic forms of eoliths, hammerstones, scrapers, spokeshaves," etc.

Boule does not pretend that all eoliths have a natural origin more or less analogous to those made by machinery. He does claim "that it is often impossible to distinguish between intentional rudimentary chipping and that due to natural causes." In his opinion, the artificial dynamics of the cement factory are comparable in every respect with the dynamic action of a natural torrent.

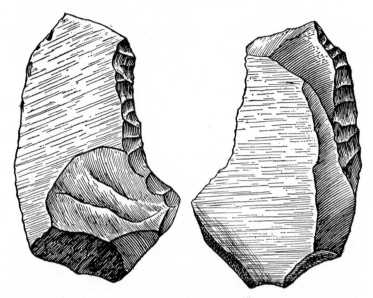

FIG. 23. CHIPPED FLINT IMPLEMENT ("TRONATTA") FROM SHENE, TASMANIA.

Implements of this Eolithic type were used for cutting, cleaving, and scraping by native Tasmanians some fifty years ago. Natural size. After Noetling.

After a careful comparison of machine-made eoliths from both Mantes [2] and Sassnitz with eoliths from Belgium, Hahne's conclusions are as follows: (1) the chalk-mill flints are all scratched and otherwise marked by the iron teeth of the mill; (2) the sides of all the larger pieces are bedecked with scars from blows that were not properly placed to remove a flake; (3) almost every piece shows more or less of the original chalky crust of the nodule; (4) anything like a systematic chipping of an edge or margin is never found, except for a very short stretch, where one would expect it to be carried along the entire margin; this is quite different from the long retouched margins of most eoliths; (5) the same edge is often rechipped first on one side and then on the other, absolutely without

---

[1] The author had occasion to examine a pile of flints that had passed through a chalk mill at Chipstead (Kent), but failed to find anything that could be mistaken for a utilized flint chip.

[2] According to Verworn, a fundamental difference exists between the machine-made eoliths from Mantes and the eoliths he found at Puy de Boudieu. The corners and edges of the former are worn, while those of the latter are not.

meaning or purpose (the "reverse working" of true eoliths is quite another thing) ; (6) in the mill product, coarse chipping alternates with fine retouches along the same margin, while on the eolith there is a regularity and orderly sequence of chipping; (7) the repeated rechipping of the same edge, while others are left untouched, does not occur in machine-made eoliths; (8) the chief difference is between the haphazard and meaningless on the one hand, and the purposeful on the other. The most prominent and easily breakable parts suffer most in passing through the mill. They are often retained intact, or only slightly altered to serve as a handhold on the eolith, and there is a logical relationship between the worked and unworked portions.

The Chalk Belt of England

Benjamin Harrison of Ightham, Kent, an enthusiastic naturalist who had been collecting paleoliths from the river drift of the neighborhood for years, extended his field of research in 1885 to include the summit of that portion of the Chalk Plateau which lies between the valley of the Darent on the west and that of the Medway on the east. There, at heights of from 122 to 183 meters (400 to 600 feet), he discovered flints supposed to have been fashioned by the hand of man. In 1888 his researches attracted the attention of Prestwich, whose country-seat was near-by.

The specimens found by Harrison on the North Downs were uniformly and deeply stained to a warm ocherous-brown color, precisely as the natural flint fragments associated with them were. The coloring matter comes from the red-clay matrix which is found in patches capping the summits of the Chalk Plateau. Associated with this red clay is a southern drift which was carried there from higher elevations to the south at a time when the chalk bridged the present fertile valleys of the Weald (woodland), connecting the North Downs of Kent with the South Downs of Sussex.

According to Rupert Jones, the implements are always accompanied by chert and ragstone from the outcrop of Lower Greensand on the side of the old Wealden range that once rose 610 to 915 meters (2,000 to 3,000 feet) over what are now the Crowborough and other Sussex hills. The red clay that stained the implements is, on the contrary, of local origin and occurs over other areas as well as those reached by the southern drift containing paleoliths.

The southern drift on the summit of the plateau is older than the great chalk escarpment or the valleys of the Darent and Medway which drain the Wealden district and, on their way northward to the Thames, cut the Chalk Plateau into three sections. The escarpment and the broad valleys of the present drainage system are older than the gravel terraces occurring at various levels in the valleys. But according to Prestwich, all of these terraces up to the height of 104 meters (340 feet) above sea level are of Pleistocene age and contain flint implements of the Paleolithic type.

The paleoliths associated with bones of the mammoth and woolly rhinoceros found in the gravel pits at Aylesford, only a few feet above the present bed of the Medway, are later than those found in the high-level valley terraces; these, in turn, are subsequent to the great denudation that swept away the chalk bridge spanning the Weald and uniting the North and the South Downs; and finally, from the very nature of things, this enormous denudation must have taken place subsequent to the time when the southern drift was carried northward and deposited with the red clay on the summit of the North Downs where patches of it still remain.

Prestwich thought the southern drift to be of later date than the locally derived red clay with which it is so intimately associated; both are older than the northern drift or boulder clay, and newer than the outcrop of Tertiary strata that caps the chalk at Swanscombe Hill. Prestwich called them simply "preglacial"; Rutot places them in the Middle Pliocene. The geological age of the plateau drift could be determined more definitely were it not for two missing links in the chain of evidence: first, the Tertiary series of deposits are not all present; second, there are no faunal remains, the property of the infiltrating waters being such as to dissolve all calcareous elements. However, as soon as the high-level river terraces are reached, the older type of paleoliths are found in association with a fauna in part now extinct. A good example is afforded by the Shelly gravel pit at Swanscombe.

There is no doubt as to the great age of the plateau deposit of red clay with flints and southern drift, even though the fauna has not been preserved. Two other questions, however, remain to be disposed of, namely: Do the specimens bear marks of use by man or of design in form? Are they as old as the patches of clay and drift on the summit of the plateau? Prestwich answered both of these questions in the affirmative.

Harrison's first plateau discoveries were made on the surface, in shallow plow furrows or in trenches and roadside cuttings, but their deep stain led Prestwich to believe that the specimens had been imbedded in a deposit beneath the surface. The finding of similar stain on an implement from a post hole at Kingsdown, on one from a hole 61 centimeters (2 feet) deep for tree planting on Parsonage farm, on a third from 61 centimeters (2 feet) or more beneath the surface in a bank of red clay at the side of a pond, and on a fourth from an equal depth in red clay at the Vigo gap, served to strengthen Prestwich's view.

In 1894 the British Association for the Advancement of Science appointed a committee "to investigate the nature and probable age of the high-level flint drift in the face of the Chalk escarpment near Ightham, which appears to be productive of flakes and other forms of flint probably wrought by the hand of man." A grant was placed at the disposal of this committee, which consisted of Sir John Evans (chairman), Prestwich, Seeley, and Harrison (secretary). Pink, the owner of Parsonage farm, Stanstead, had previously sunk a pit in the drift in which he had found plateau implements at a depth of 1.8 to 2.1 meters (6 to 7 feet). Adjoining this site, the committee sank its first pit through 76 centimeters (2.5 feet) of "humus and drifted material, white flints, pebbles, and many ocherous flints worn and worked"; through 1.07 meters (3.5 feet) of "gray loam, with scattered small pebbles, and a few small worked ocherous flints throughout"; and 30.5 centimeters (1 foot) of compact gravel with many worked flints (Fig. 24). A second pit sunk near-by to a depth of 7.9 meters (26 feet), most of the way through Lower Tertiary pebbles, without reaching the Chalk, revealed "precisely similar conditions." No implements were found below a depth of 2.4 meters (8 feet).

Soon after Harrison's first discoveries on the North Downs, chipped flints of a similar type were found by Shrubsole in Berkshire; by Martin on the South Downs at Beachy Head near Eastbourne, Sussex; by Blackmore, Bullen, and others near Salisbury,

Wilts; in Dorset, Surrey, Hampshire, the southern part of Essex, and Norfolk.

The deposits investigated by Shrubsole consist of preglacial gravel beds from 1.5 to 3 meters (5 to 10 feet) in thickness which cover "the summit of an elongated plateau stretching from East-hampstead, Berks, to Ash Common, near Aldershot." They had their origin in the heights that once rose over what is now the Wealden district to the south and southwest; they are composed of

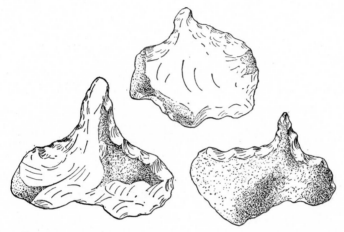

FIG. 24. POINTED TYPES OF EOLITHS FROM THE CHALK PLATEAU, KENT, ENGLAND.

These were found in pits sunk by a committee of the British Association for the Advancement of Science. They were chipped and utilized by man or his precursor who lived 1,000,000 to 500,000 years ago. Such flints were probably used for punching holes and boring. Scale, ½. After Harrison.

the same southern drift that has furnished the implements found on the North Downs.

The gravel-capped plateau rises to an average level of about 122 meters (400 feet) above the sea and "forms the highest ground between the rivers Wey and Blackwater." The specimens described came chiefly from Finchampstead, Easthampstead, and from near Bagshot. They present precisely the same general aspect as those from the North Downs. Shrubsole believes them to be as old or older than the gravel beds. He bases his opinion on their mineral condition, and on the fact that he himself took them "from the gravel freshly fallen from the face of the pits, or from the heaps of screened gravel in the pits." No artificial flakes and no implements of the Chellean type have been found in these gravels.

Prior to 1900 many chipped flints had been discovered in the Alderbury gravels near Salisbury, at all levels to a depth of 4.3 meters (14 feet) by Blackmore. These gravels were classed as southern drift by Prestwich; they rest upon the Bagshot sands and "are at a much higher level than the river drift, which furnishes both flint Paleolithic implements and a very good list of Pleistocene mammals and shells." But like the deposits on the Kent plateau, these gravels are not fossiliferous.

Blackmore wrote: "Being very anxious to fix the Pliocene age of these eoliths, rather more than a year ago I went down to Dewlish, in Dorset, with the express purpose of carefully examining the gravel which had furnished the remains of *Elephas meridionalis,* as this was the one spot in the South of England which was regarded as a patch of Pliocene gravel." . . . (As early as 1813, Blackmore's grandfather had found a molar of *E. meridionalis* in this patch of gravel, and he himself had been present at the discovery, in 1887, of the remains of *E. meridionalis* now in the Dorchester Museum.) "A trench was opened through the deposit of gravel," he continues, "and there was no difficulty in finding eoliths stained like the gravel at the same level and associated with the elephant bones. This was to me most satisfactory and conclusive."

The Chalk cliffs at Beachy Head are familiar to every Channel voyager. The Chalk suddenly disappears at Eastbourne and does not reappear until one is opposite Dover, a distance of more than 80 kilometers (50 miles). The cliffs at Beachy Head and Dover are the bases of a great anticlinal fold, whose axis passes from Dungeness in a westerly direction through Hampshire. The crest of the fold, including not only the Chalk beds but also the underlying strata of Upper Greensand, Gault, Lower Greensand, and Weald, has disappeared. If, before it disappeared, the old drift with eoliths was transported northward and left on the North Downs, the same old drift with eoliths must have been carried southward and deposited on the South Downs. This old drift with eoliths has been found on the South Downs at Beachy Head and Eastbourne by Percival Martin.

The North and South Downs are but slender tongues from the

great Chalk plain of Dorset, Wiltshire, and Hampshire, the tip of one being at Dover, that of the other at Beachy Head. The Chalk is continuous all the way from Dorset and Salisbury Plain in a northwesterly direction to Cromer on the Norfolk coast. At the southwestern extremity of the Chalk belt, Blackmore found eoliths associated with the remains of *Elephas meridionalis;* at its northeastern extremity, Abbott found a like association in the estuarine deposits of the Forest Bed, dating from the Günz-Mindel Interglacial Epoch. The results of the researches of Worthington G. Smith at Caddington near Dunstable, about midway between Dewlish and Cromer, and the remarkable discovery at Piltdown Common (Sussex) are indications of the possibilities of the entire Chalk belt.

The Pliocene of East Anglia

The discoveries of chipped flints in the Pliocene of East Anglia by Reid Moir have attracted considerable attention. These flints occur *in situ* in two horizons of the Red Crag in Suffolk, and in the detritus bed at its base. The author first visited East Anglia during the summer of 1912 and examined the Pliocene flints and the beds from which they came, in company with Moir. In 1922 the author again visited East Anglia for the purpose of studying all available specimens and of making a more thorough examination of the beds from which they came. He visited five sites in the vicinity of Ipswich—Bolton and Laughlin's brickfield, Coe's pit at Bramford, Thorington Hall, Foxhall Hall, and Foxhall Road (also known as Derby Road).

At Bolton and Laughlin's, Coe's pit, and Thorington Hall, worked flints are found in the detritus bed at the base of the Red Crag; some of the flints from Coe's pit have been subjected to the action of fire. Burnt, as well as worked, flints occur in the upper part of the Red Crag of the Foxhall Hall coprolite pit at a depth of 4.9 meters (16 feet) from the present surface. This horizon, which is obviously a habitation level, is referred to as the "16-foot level." Moir has found worked flints in another horizon 61 centimeters (2 feet) deeper (Figs. 25 and 26).

The Red Crag and the detritus bed at its base are regarded by geologists as unquestionably of Pliocene age. It is difficult to invoke natural agencies, such as pressure within the deposit, to account for worked flints at three different levels in the Pliocene of East Anglia, for the reason that flints, both worked and unworked, are scarce. These levels resemble the flinty levels at Saint-Acheul and are to be accounted for in the same way (Chapter IV). Both represent definite habitation levels. At Coe's pit, Bramford, and Foxhall Hall burnt flints and chips with cores to which the chips belong have been found in the habitation levels.

That the Red Crag has not been derived from older deposits is indicated by the presence of nonremanié marine shells of Pliocene age, especially abundant in the Thorington Hall pit. The lower

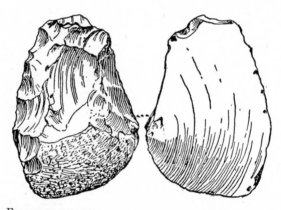

FIG. 25. EOLITH OF THE SCRAPER TYPE FROM THE PLIOCENE OF EAST ANGLIA.

This was found in the detritus bed at the base of the Red Crag, Pit No. 1, Bramford brickfield, Suffolk, England. After Reid Moir.

horizons have yielded remains of the three-toed *Hipparion,* hippopotamus, tapir, mastodon, and *Cervus caprea.* A more temperate fauna including *Elephas meridionalis* and *Equus stenonis* is found in the upper levels.

The Norwich Crag, although a later formation than the Red Crag, is classed as of Pliocene age. W. G. Clarke has found chipped flints at the base of the Norwich Crag, in the Eaton pit near Norwich. The thin flinty horizon is at a depth of 9.15 meters (30 feet) from the surface and contains fossil mammalian remains.

Some of the chipped flints from the Pliocene of East Anglia belong to a distinct type which has been described by Lankester as *rostro-carinate* or eagle beak. This type seems to have persisted during subsequent epochs; it has been reported by Moir from the midglacial [3] sands as well as from the Chalky Boulder Clay (Pleistocene) of East Anglia, especially in the vicinity of Ipswich. In order to demonstrate the artifact nature of rostro-carinate flints, Lankester selected as an example a specimen found by Clarke, in 1911, in a pit at Whitlingham near Norwich. The bed in which this specimen was found is below the Norwich Crag and is considered to be of Pliocene age; it also contained the teeth and bones of *Mastodon arvernensis*, a Pliocene type.

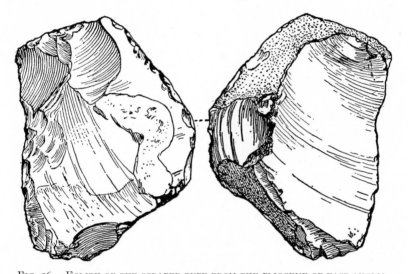

FIG. 26. EOLITH OF THE SCRAPER TYPE FROM THE PLIOCENE OF EAST ANGLIA.

This was discovered in the 16-foot level of the coprolite pit at Foxhall Hall, Suffolk, England. After Reid Moir.

The Norwich test specimen showed two roughly parallel cleavage planes—a ventral and a dorsal. The secondary flakings were large, but more concave than the primary flakings; even the direction of the blows which produced the secondary chipping could be determined. The beak, or keel, was produced by smaller fractures. Lankester did not believe that a single one of all these fracture surfaces could be regarded as accidental or due to a "fortuitous concourse and reciprocal battering of flint nodules." The specimen is now in the Sturge collection at the British Museum, Bloomsbury.

---

[3] Term used in the sense of interglacial.

The case of the Norwich test specimen is strengthened by the fact that Evans, always a conservative in prehistoric archeology, gives an illustration in the second edition of *Ancient Stone Implements* of a typical rostro-carinate presented to him by Canon Greenwell, who had procured it from a Lakenheath workman (Fig. 27).

FIG. 27. FLINT IMPLEMENT OF THE ROSTRO-CARINATE (EAGLE-BEAK) TYPE, FROM LAKENHEATH, SUFFOLK, ENGLAND.

This implement is believed by Sir John Evans to be a true artifact. It closely resembles the rostro-carinates found later by Reid Moir in Tertiary deposits. Scale, ⅓. After Evans.

The exact spot where it was found seems to be in doubt, though it certainly came from Suffolk and not more than 56 kilometers (35 miles) from Ipswich. This, however, does not minimize the fact that Evans never doubted its artifact nature. After describing the specimen, he added: "Until other specimens of the same form are discovered, it is hardly safe to regard this as furnishing an example of a new type of implement; yet its symmetry and character seem to prove that it was designedly chipped into this form, to fulfill some special purpose." If a special purpose can be invoked in this case, why not also in the case of the Norwich test specimen and others, is the contention of Lankester.

CHIPPED FLINTS FROM THE OLIGOCENE AND EOCENE OF BELGIUM AND FRANCE

In the light of our present knowledge of organic and cultural evolution, it is not surprising to find chipped flints which bear marks of utilization and intentional shaping in the Pliocene and Miocene, but when they are found in the Lower Tertiary, the problem takes on a different aspect.

Chipped flints have been found by de Munck at Boncelles near Liège and on the Hautes-Fagnes plateau (Belgium). The sand pit at Boncelles which yielded the flints is at an elevation of 265 meters (870 feet) above sea level and 185 meters (607 feet) above the bed of the Ourthe. Judging from the fossil shells in these sands, the formation belongs to the Upper Oligocene. The layer which contained the flints is somewhat older. Rutot figured a long list of what he believes to be artifacts from Boncelles: hammerstones, chopping blocks or anvil stones, retouching tools, scrapers, blades, scratchers, hollow scrapers or spokeshaves, points, drills, and sling or throwing stones. Several of the scrapers have well marked bulbs of percussion. The flints from Hautes-Fagnes plateau are of the same age as those from Boncelles.

The Abbé Breuil has found chipped flints resembling eoliths in deposits at the base of the Parisian Eocene (Thanetian) on the estate of Belle-Assise in the suburbs of Clermont (Oise). Having satisfied himself of the presence of chipped flints *in situ* in the pebbly bed at the base of the Thanetian sands, Breuil called several of his colleagues to the spot at various times for the purpose of confirming and controlling his own observations.

Some of the flints are worn, others patinated, indicating that the shaping had taken place prior to the deposition of the superposed Eocene sands. On the other hand, a large majority of the fractures are remarkably fresh in appearance. The sharpest edges, in such cases, are intact, indicating that these flints had never been subjected to the action of transport. Breuil illustrates a small scraper and a bladelike flake. The bulb of the scraper is quite distinct; the outer face is marked by fine and regular retouches, especially at the end and on the right margin. The bladelike flake resembles a micro-graver such as one might find at late Paleolithic stations.

Breuil also found flint chips and their parent cores in proximity; some of the chips were retouched, apparently by contact with their parent cores, the latter having served as a sort of anvil or chopping block, the pressure of the soil furnishing the force necessary for the operation. Whether or not this is the correct explanation of the mechanism that produced the chipped flints of Boncelles and of Belle-Assise, it is evident that some of the earmarks hitherto looked upon as evidence of intentional chipping may be counterfeited by Nature.

The discoveries at Boncelles and Belle-Assise have emphasized the difficulty of distinguishing between intentionally chipped and utilized, and natural and nonutilized, forms. Much of the difficulty met with in distinguishing between natural and intentional chipping is due, no doubt, to the peculiar properties of flint, its tendency to take on forms that would tempt man to test its utility; but the difficulty of drawing a hard and fast line of demarcation between the artificial and natural cannot, however, be regarded as either proof or disproof of the existence of man-used eoliths. The probabilities are in favor of them; the evidence against them is largely negative. The question of how far back in geologic time one must go before the probabilities reach a vanishing point and man-used flints no longer occur, is still an open one.

Artifacts with what has come to be known as *eolithic facies* are common to both Paleolithic and Neolithic horizons; they were in use even during the greater part of the nineteenth century among the native Tasmanians and Australians. Perhaps a majority of real eoliths are improvisations, and improvisations of one epoch are very much like those of another, which would account for the persistence of the eolithic type even to the present time. The difference between true eoliths and flints with eolithic facies is one of chronology. The true eolith belongs to an epoch antedating the appearance of the oldest type specimens of the Paleolithic Period; the flint with eolithic facies may belong to any period.

Since man, or his precursor, chipped and burned flint in Pliocene times, we have a pre-Paleolithic industry. Since a Tertiary industry occurs in East Anglia, we may reasonably expect to find it elsewhere. Its geologic position and its importance from the viewpoint of human physical and cultural evolution call for ample recognition in the system of prehistoric classification. The term *Pre-Chellean* can justly be applied only to the earliest epoch of the Paleolithic Period. Tertiary flint industry represents a period in itself, not an epoch, hence nothing could be more suitable for it than the term *Eolithic*.

CHAPTER IV

THE LOWER PALEOLITHIC PERIOD

There are certain broad distinctions to be drawn in a comparative study of industrial remains of the various periods. Eolithic industry consisted largely of improvisations, of primary tools or implements such as the hammerstone and the flint chip with utilizable sharp edge or point; secondary tools were few and simple, consisting largely of artificial chips. During the Lower Paleolithic Period the number of secondary tools was increased by the addition of the cleaver.

The Mousterians of the Middle Paleolithic Period made no great advances over their predecessors. They possessed an improved technique which is seen in the character of their nuclei and the well formed scrapers and points with carefully retouched margins; but so far as can be ascertained, they did not go beyond the making of secondary tools—that is to say, their secondary tools served directly an ultimate purpose; they were not used for the manufacture of tertiary [1] tools. The technical processes from Pliocene times to the close of the Middle Paleolithic Period (well along toward the close of the Pleistocene) remained relatively simple.

It was reserved for the Upper Paleolithic races to inaugurate a new era. This was made possible through improvement in the preparation of nuclei from which long, slender blades could be struck. The next step was important additions to their stock of secondary tools (various forms of the graver, microliths, small knives and awls), which enabled them to make extended use of bone, ivory, and reindeer horn, and led to two capital results, first,

---

[1] By *tertiary tool* is meant an implement which, in its shaping, requires the use of primary and secondary tools, and whose ultimate use is not in the shaping of implements; as an Upper Paleolithic example, the dart thrower may be cited.

the invention of a tertiary set of tools, and second, the dawn of the fine arts.

THE PRE-CHELLEAN EPOCH

The culture known as Paleolithic is coextensive with the Quaternary just as the culture known as Eolithic is referable to the Tertiary. The Neolithic and later cultures belong to the geologic

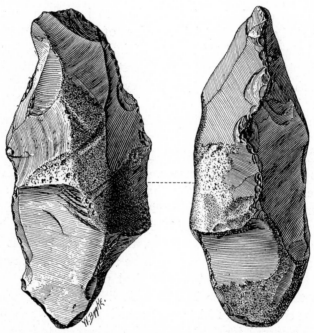

FIG. 28.   PRE-CHELLEAN FLINT IMPLEMENT FROM THE ROUTE DE BOVES, SAINT-ACHEUL, SOMME, FRANCE.

Commont found this crude prototype of the Chellean cleaver in the lower gravels of the second terrace of the Somme valley at an elevation of 55 meters (180.5 feet); probably 500,000 years old (Lower Pleistocene).   Scale, ½.   Yale University Collection.

recent. When G. de Mortillet announced (in 1869) his division of the Paleolithic into four epochs, he called the oldest *Acheulian*. But in 1878, after LeRoy and Chouquet had called his attention to the vast pits exploited at Chelles, de Mortillet changed the name of the first epoch to *Chellean*, reserving a place for the Acheulian between the Chellean and the Mousterian.

Later it was found that de Mortillet's conception of that which distinguishes the beginning of the Chellean Epoch, namely, the fully formed cleaver, does not date so far back as the beginning of the Quaternary. In order, therefore, to bridge the gap between the Chellean of the Quaternary and the Eolithic of the Tertiary, an initial epoch of the Paleolithic Period has been created; for this epoch, the name *Pre-Chellean,* already employed by a number of prehistorians, is eminently appropriate.

The industry of the Pre-Chellean Epoch consists, for the most part, of simple, utilized flint chips and rude beginnings of a form that later developed into the Chellean cleaver. It has been reported from various stations in France, Belgium, England, and Spain.

France.—The stratigraphic position of the Pre-Chellean industry in the valley of the Somme is the fluviatile gravels of the second and third terraces. It includes crude prototypes of the cleaver and numerous small implements derived from artificial flint chips. There is no accompanying fauna at Saint-Acheul; but at Abbeville the same deposits have furnished a fauna with Pliocene affinities—*Elephas t r o g o n-therii, Hippopotamus major, Rhinoceros merckii, Rh. etruscus, Machaerodus, Equus stenonis,* and *Cervus.*

In the gravels of the second terrace (third from the river) and

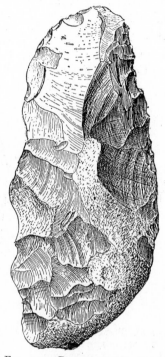

FIG. 29.   PRE-CHELLEAN FLINT KNIFE FROM THE RUE DE CAGNY, SAINT-ACHEUL, FRANCE.

Found at the base of the gravels in the third terrace of the Somme valley. Pleistocene Epoch. Scale, ⅓. After Commont.

at the base of the lower gravels of the third terrace at Saint-Acheul, Commont found a flint industry which he rightly calls Pre-Chellean. The implements are yellowish to yellowish-brown in color, their characteristic patina easily distinguishing them from the Chellean types. Commont made a special study of the Leclercq pit, Route de Boves, in the second terrace. He divided the implements from the lower gravels into three classes: (1) large flints rudely chipped, such as thick cleavers, disks, scrapers, hammerstones, and chopping knives (Fig. 28); (2) small implements belonging to definite forms, relatively few in number, such as scrapers and thick scratchers, the dominant type being the hollow scraper with one or more notches; (3) small chips without retouches which show marks of utilization as knives, saws, etc.   In the base of the gravels of the third terrace, Rue de Cagny, Commont found primitive cleavers, knives, and utilized flint chips, which are obviously referable to the Pre-Chellean Epoch (Figs. 29 and 30).

Other stations in the valley of the Somme which have yielded

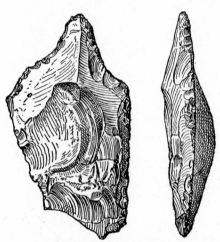

Pre-Chellean i n d u s t r y i n-clude the Bultel-Tellier pits in the third terrace at Saint-Acheul; probably the De-bary pit in the first terrace at Montières; and some of the pits in the vicinity of Abbeville. One of the hori-zons at Saint-Prest, near Chartres   (Eure-et-Loire), has also yielded a Pre-Chel-lean industry.

Belgium. — Pre-Chellean industry has been reported from the second terrace of the Haine valley at the Bois d'Epinois, east of Binche; and from the third terrace of the Trouille valley at two

FIG. 30.   PRE-CHELLEAN UTILIZED FLINT FLAKE FROM THE RUE DE CAGNY, SAINT-ACHEUL, FRANCE.

This flint could have been used as a knife, scraper, scratcher or borer.   Scale, ½.   After Commont.

sites, Mesvin (discovered in 1868) and Spiennes.

England.—Pre-Chellean flint implements were found *in situ* in the Elephant Bed of the Cromer Forest-Bed deposits (Norfolk) by Abbott prior to 1897.   In 1920 Reid Moir found what he believes to be vestiges of a workshop site on the foreshore at Cromer.   The site is some 78 meters (86 yards) removed from the base of a cliff under which lies the Cromer Forest Bed; a direct connection between the two is cloaked by intervening beach deposits.   If Moir's assumption is correct, this workshop is of Pre-Chellean age.

Rude flint implements, a human skull, and an apelike lower jaw, associated with a fauna for the most part extinct, were reported from a gravel bed at Piltdown Common (Sussex) by Dawson and Smith Woodward in 1912. Piltdown is not far from Beachy Head; it is only a short distance north of the South Downs and near the southern limits of the Weald. The Ouse takes its rise in the Weald, flows southward, and after cutting through the South Downs, empties into the Channel at New Haven. The Piltdown

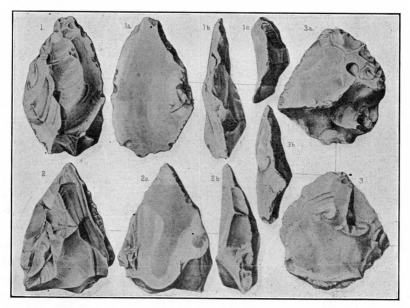

FIG. 31. PRE-CHELLEAN FLINT IMPLEMENTS FROM PILTDOWN, SUSSEX, ENGLAND.
These crude implements are chipped on only one face. Scale, $\frac{2}{5}$, except No. 3 which is $\frac{1}{5}$. After Dawson and Smith Woodward.

gravel bed is 24.4 meters (80 feet) higher than, and nearly 1.6 kilometers (a mile) distant from, the present stream bed of the Ouse. The physiographic features of this valley have not changed perceptibly since Roman times.

The relation of the present river bed to the one that existed when the Piltdown gravels were formed is such as to indicate great antiquity for the latter; it probably dates from the early Pleistocene, although it contains elements from an older (Pliocene) drift. The condition of the faunal and cultural remains supports this view.

The fossils representing a Pliocene fauna are rolled, worn, and highly mineralized, while those representing a Pleistocene fauna are not. The chipped flints of Pre-Chellean type have sharp edges and are, presumably, no older than the gravel deposit, while those resembling the eoliths from the Chalk Plateau are often much water-worn and may have been derived from some older geological formation.

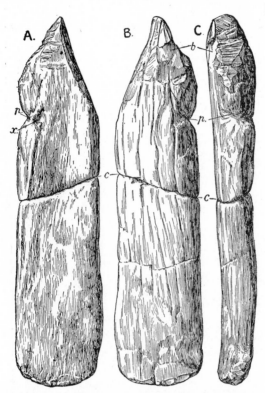

FIG. 32.   LARGE BONE IMPLEMENT OF THE LATE PLIOCENE OR EARLY PLEISTOCENE FROM PILT-DOWN, SUSSEX, ENGLAND.

The implement was made from the thigh bone of a very large elephant (*Elephas meridionalis* or *E. antiquus*), larger than the mammoth or any elephant of later date. One end of the implement was sharpened to form a point. Just beneath the point a perforation was made from which the outer wall has broken away, leaving the groove, *p*, which seems to have been worn smooth by the rubbing of a thong. *A* shows how part of the concave inner wall of the marrow cavity has been retained. The small letters indicate as follows: *b*, accidentally broken hollow; *c*, natural break by pressure in the gravel bed; *p*, perforation from which the outer wall has broken away; *x*, the beginning of another perforation. Scale, ¼. After Smith Woodward.

The rude triangular flint implement of Pre-Chellean type (Fig. 31) was found at only a slightly higher level than the human bones. A pointed bone implement made from the thigh bone of a gigantic elephant (either *Elephas meridionalis* or *E. antiquus*) was found in the yellow mud at the base of the gravel, near the human skull. The shaping process is quite distinct at the rounded base as well as at the point. This specimen is remarkable for its size, being 40 centimeters (16 inches) long (Fig. 32). Because of its discovery in *E. antiquus* beds, mention should also be made of a pointed wooden implement, almost as long (37.5 centimeters, 15 inches), found by Warren at Claxton-on-Sea.

At the Barnfield pit, Swanscombe (Kent), in the 30-meter (100-foot) terrace, there is a Pre-Chellean horizon in the lower gravel.

Spain.—Pre-Chellean flint implements of undoubted authenticity have been found recently by the Marquis of Cerralbo at Torralba (Soria). This station is an ancient camp site which has yielded an association of Eolithic and rude Chellean industry with the remains of an old southern type of fauna—*Elephas antiquus* (perhaps also *E. meridionalis*), *Rhinoceros etruscus, Equus stenonis,* and a large and a small deer.

THE CHELLEAN EPOCH

Following in the lead of geologists, prehistorians have named their epochs for localities where the culture of the epoch in question is well represented. The name *Chellean,* chosen by G. de Mortillet, is from Chelles (Seine-et-Marne), east of Paris on the Marne, where the ancient sands and river gravels form a terrace of considerable extent to the east of the city (Fig. 33). The deposits are about 8 meters (26 feet) thick; the typical Chellean horizon is only 4 meters (13 feet) above

FIG. 33. THE SAND AND GRAVEL PIT AT CHELLES, SEINE-ET-MARNE, FRANCE, THE TYPE STATION FOR THE CHELLEAN EPOCH.

The man is pointing to the Chellean horizon. The Lower Chellean Epoch was about 250,000 to 200,000 years ago; the Upper Chellean, from 200,000 to 150,000. Photograph by the author.

the Marne. The lower portion of the deposits is of Chellean age, the industry being associated with a warm fauna—*Elephas antiquus,*

Rhinoceros merckii, and *Hippopotamus amphibius.* Above the Chellean horizon occur Acheulian and Mousterian industries with the fauna of the mammoth and the reindeer. Since de Mortillet noted two horizons of the Chellean deposit in this terrace, he postulated a long duration for the Chellean Epoch.

The distinctive implement of the Chellean Epoch, the one on which de Mortillet based his classification, is amygdaloid in shape.

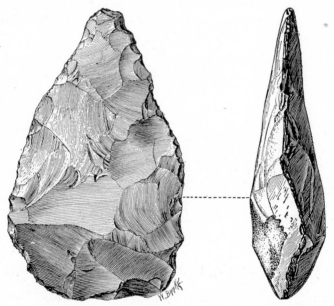

Fig. 34. Chellean flint cleaver from the Boutmy-Muchembled sand and gravel pit at Montières, Somme, France.

Cleavers were chipped on both faces to a shape resembling an almond, more or less pointed at one end; the opposite end was reserved for a handhold. The edges and the pointed end could have been used in a variety of ways—cleaving, skinning, splitting, scraping, puncturing, etc. The French call the cleaver *coup de poing,* the Germans, *Faustkeil.* Scale, ⅔. Yale University Collection.

It was first called *langue de chat* (cat tongue) by the workmen, and later, *hache de Saint-Acheul* (ax of Saint-Acheul). Since it was supposed to have been held in the hand while in use, de Mortillet renamed it *coup de poing,* a term which is still in use. Sollas has thought to improve on the name by changing it to *boucher,* in honor of Boucher de Perthes, who was the first to focus the attention of the scientific world upon flint implements of this type. However, two isolated discoveries of *coups de poing* had been made before the time of Boucher de Perthes (Chapter I). The German equivalent of *coup de poing* is *Faustkeil*. As an English equivalent, the author suggests *hand ax* or *cleaver*.

The Chellean cleaver is roughly chipped on the two faces; its base is enlarged, thick, and rounded, with often a bit of the nodular crust left intact. The upper two-thirds slope to a more or less well defined point and the margins are distinctly sinuous (Fig. 34). It varies in size and shape, and its uses must have been varied—cleaving, chopping, skinning, scraping, sawing, etc.; implements with sharp points might well have served as poniards or for boring purposes. Industrial forms associated with the cleaver include

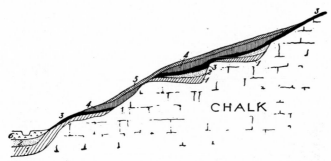

FIG. 35. COMPOSITE SECTION OF THE TERRACE DEPOSITS OF THE SOMME VALLEY AT SAINT-ACHEUL.

1, gravels with Pre-Chellean and Chellean industries; 2, sands (Chellean and Acheulian); 3, ancient loess (Acheulian); 4, recent loess or *ergeron* (Mousterian); 5, brick earth or weathered portion of the recent loess (Aurignacian, Solutrean, and Magdalenian); 6, recent alluvium. Adapted from Commont.

disk-shaped implements and numerous tools such as scrapers, scratchers, knives, spokeshaves, points, etc., evidently derived in part from the chips struck off during the manufacture of the cleaver.

In recent years the valley terraces of the Somme have attracted more attention than those of the Marne. Abbeville and Saint-Acheul first came into prominence with the discoveries of Boucher de Perthes, Rigollot, Prestwich, Gaudry, and others. Many of the implements which they found were of Chellean age. The industry is located in the lower gravels of the lowest or fourth terrace (the more highly developed forms), and in the fluviatile sands capping the lower gravels of the third terrace (the earlier, ruder forms), where it is associated with *Elephas antiquus,* large horse, large ox, and *Cervus elaphus.* Since Chellean artifacts occur in the third and fourth terraces, the cutting of the Somme valley from immediately below the second terrace (54 meters, 177 feet, above sea level at Amiens) to its bottom, now partly refilled, practically all took place during this epoch, which must have been a very long one. A composite section of the terrace deposits of the Somme valley is reproduced in Figure 35.

Chellean industry is especially abundant in the third, or so-called middle, terrace at Saint-Acheul. Several sand and gravel pits

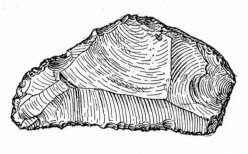

were being exploited when Rigollot was attracted by Boucher de Perthes' discoveries in similar deposits at Abbeville. In 1854, Rigollot found his first river-drift implements in the Tattegrain pit; later, he found many similar implements in two near-by pits, Fréville and Tellier père. This section, at the present time covered by buildings, was visited by Prestwich, Evans, and Gaudry.

An opportunity to test the richness of this section was offered to Commont in 1906 when a new structure was to be built at 54

FIG. 36. CHELLEAN FLINT KNIFE FOR THE RIGHT HAND.

Such an adjustment of the knife to suit the hand is rare in the Chellean Epoch. Scale, ⅔. After Commont.

Rue de Cagny, where the deposits were at a height of 42 to 45 meters (138 to 148 feet). These deposits were removed over an area of 1,752 square meters (1,962 square yards). Very few fossil remains were found—the molar of a large ox and several teeth of the horse; but only 150 meters (492.5 feet) removed from this spot, the same lower gravels furnished a molar of *Elephas antiquus.* During the excavations at the Rue de Cagny, Commont collected

540 cleavers, some of them crude, and
nuclei, utilized chips, hammerstones,
and pestles.

At the base of the gravels, Commont found thirty crude specimens
with dulled edges, but not worn by
transport; all were deeply patinated.
They were primitive cleavers, knives,
and utilized chips of various forms,
all obviously of the Pre-Chellean
Epoch (see Figs. 29 and 30).

From the upper part of the lower
deposits, consisting almost entirely of
gravels and white sands, Commont obtained 220 large implements; they were
not patinated, but still retained the
original yellow color of the flint. The
margins were not dulled or altered in any way. The cleavers were

450 specimens, including

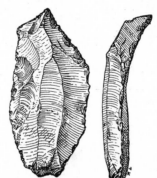

FIG. 37. CHELLEAN COMBINATION
GRAVER-SCRAPER.

The edge could have been used for
scraping and cutting and the point
for boring. Scale, ½. After Commont.

FIG. 38. CHELLEAN FLINT
CLEAVER FROM THE RUE
DE CAGNY, SAINT-
ACHEUL.

Scale, ¼. After Commont.

unusually thick at the base with some of the
nodular crust remaining; they were frequently elongate.

In addition, a good many utilized chips
were found; in fact all the chips present
bore marks of utilization as knives, scrapers,
etc., many of them having been used just
as they came from the parent core. Some had
well defined forms—rather thick scratchers
and scrapers, drills, spoke-shaves, and even
blades accommodated to the hand as knives.
One of them is illustrated in Figure 36.
Retouches of accommodation, as seen on
the knife, are, however, the exception rather
than the rule. One of the blades might
well have served as a combination graver-
scraper (Fig. 37). In the deposit of red sands
above the lower gravels, Commont found
300 implements of Acheulian workmanship.
Figure 38 is a good example of a Chellean cleaver from the Rue

de Cagny. The sinuous margins of the cleaver bear marks of utilization; they may have served for cutting purposes and would have answered also the purposes of either a saw or a scraper.

Torralba, on the eastern slope of the Sierra Ministra (Soria), is an old Chellean camp site in the open (Fig. 39). It was discovered in the building of a branch railway to Soria in 1888, at

FIG. 39. AN ANCIENT CHELLEAN CAMP SITE AT TORRALBA, SORIA, SPAIN.

The Chellean level is at the base of the light-colored deposit in the lower center (calcareous gravels). Here the Marquis of Cerralbo found many ancient Chellean implements. He also found in direct association with these implements the bones of animals which the Chellean hunters brought to this camp in the open, including the remains of a species of large elephant (*Elephas antiquus* or *E. meridionis*). Photograph by the author.

which time fossil bones of *Elephas* were gathered and removed to the School of Mines in Madrid. For its rediscovery (in 1907) we are indebted to the Marquis of Cerralbo, who has explored the site. Ancient Chellean (and perhaps Pre-Chellean) hunters left bones of the animals they hunted and fed upon—*Cervus elaphus, Equus stenonis, Bos, Rhinoceros etruscus,* and *Elephas,* remains of the latter being particularly abundant; they probably represent an early variety of *E. antiquus,* although certain authors would refer them to *E. meridionalis* (*trogontherii*).

By degrees a quantity of Triassic marl and pulverized limestone was washed from the near-by slopes, forming a deposit over the camp. Before this covering had been altered appreciably by weathering, a landslip occurred, burying the relic-bearing deposit under 2.5 meters (8.2 feet) of clay. The upper half of the clay bed has been decalcified and turned to a reddish color; the lower half, protected by the layer above, retains

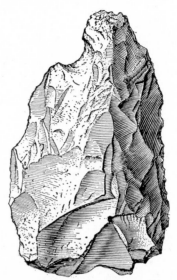

FIG. 40. CHELLEAN IMPLEMENT OF LIMESTONE FROM TORRALBA, SPAIN.

The notch near the point may have been used as a hollow scraper (spokeshave). Scale, ½. Yale University Collection.

its original bluish color. The relic-bearing deposit, composed largely of calcareous gravel, is grayish and averages 0.7 meter (2.3 feet) in thickness.

In addition to the faunal remains, the deposit contained stone implements of Chellean and Pre-Chellean types made of limestone (Fig. 40), chalcedony, and quartzite (Fig. 41). The limestone is local, but the chalcedony and quartzite were brought from a distance of 5 kilometers or more (over 3 miles). Similar implements of quartzite have been found in Pleistocene beds of India (Fig. 42).

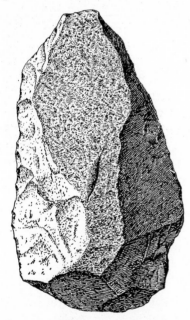

FIG. 41. CHELLEAN IMPLEMENT OF QUARTZITE FROM TORRALBA, SPAIN.

The bed of quartzite from which the material for the making of this implement was taken is 5 kilometers (3 mi.) from the camp site. Scale, ⅔. Yale University Collection.

A representative list of Chellean fauna would include: *Elephas antiquus, Rhinoceros merckii, Cervus elaphus, C. capreolus, C. megaceros, Bos priscus, Hyaena spelaea, Felis leo antiquus, Equus* (closely related to *E. stenonis*), *Corbicula fluminalis, Hippopotamus major*, and *Ursus spelaeus.*

The flora included the ash, oak, birch, linden, hazel, and mistletoe.

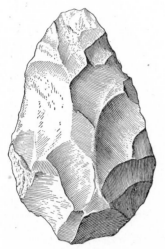

FIG. 42. CLEAVER OF QUARTZITE FROM PLEISTOCENE LATERITE BEDS, MADRAS PRESIDENCY, INDIA.

Note the remarkable similarity of this implement to the Chellean cleaver of quartzite found at Torralba, Spain (Fig. 41). Scale, ⅓. Yale University Collection.

THE ACHEULIAN EPOCH

The Acheulian Epoch, like the Chellean, is characterized by the cleaver. In both epochs chipping was by means of percussion, but the Acheulian cleaver is distinguished from the Chellean by a marked improvement in the art of chipping; the chips removed are smaller, thus making it possible to reduce the thickness of the implement as well as the sinuosity of its edges. The differences are well illustrated in Fig. 44, which gives the two views in profile.

The Lower Acheulian types are derived directly from the Chellean. The principal one of these is the *limande,* so-called from its resemblance in shape to a flounder; oval, flat, it is completely chipped on both faces, and even at the base or hilt (Fig. 45). Another, although comparatively rare type, is the twisted cleaver; one of the diagonally opposite margins rises appreciably above the common plane while the other drops below it. Pointed as well as elongate-oval forms also occur (Fig. 46).

Cleavers vary greatly in size; rare examples [2] attain a length of 30 centimeters (12 inches), while a minimum length of 53 millimeters (2.1 inches) has been reported.

---

[2] In 1921 L. Treacher presented to the Natural History Museaum, South Kensington, an Acheulian cleaver from the river gravel at Maidenhead (Berkshire), which was 31.5 centimeters (13 inches) long, 17.5 centimeters (7 inches) wide, and weighed 2.78 kilograms. (6 pounds 2 ounces).

The Upper Acheulian types represent the acme of workmanship reached in the production of cleavers; the specimens are thinner and lighter, and the retouching is carried further.   One of the characteristic forms is distinctly triangular; another is long and pointed (Fig. 47).

The Acheulian cleaver, like the Chellean, is accompanied by disk-shaped implements (probably derived from the cleaver) and a variety of artifacts consisting of flint chips or flakes, more or less

FIG. 43.   THE VALLEY TERRACE AT SAN ISIDRO, SPAIN, ACROSS THE MANZANARES RIVER FROM MADRID.

This is another important station for Chellean industry (see also Fig. 3).   The Chellean level, at a depth of about 14 meters (45 to 50 feet), has a thickness of about 3 meters (9.5 feet).   In this deposit were found many waterworn Chellean cleavers and fossils. Photograph by the author.

altered through retouching, use, and accommodation: knives, several kinds of scrapers, drills, etc. (Figs. 48 and 49).

Reboux was the first to call attention to implements made from large oval flakes which he found in Quaternary deposits only a few meters above the Seine at Levallois-Perret, near Paris.   On account of their relative frequency at this station, as well as their distinctive form, implements of this age and kind (late Acheulian and early Mousterian) have come to be classed as belonging to the Levallois type (Fig. 50). This type has been found at many Acheulian stations, including Saint-Acheul (Somme), Arceuil and Cergy (Seine-et-Oise), Villejuif (Seine), Montguillan and Grand-Bruneval (Oise).

The stratigraphic position of the Acheulian industry is immediately above that of the Chellean, occurring at two or three distinct levels in the ancient loess of the third terrace: the later, more highly perfected forms in the red loess, or weathered upper layer of the ancient loess; the earlier forms, including the *limande*, at two levels near the base of the ancient loess. Isolated discoveries of Acheulian implements have also been noted in the intermediate layers of this loess. Perhaps nowhere else is the industry so well typified, and its geologic horizon so distinctly indicated, as at Saint-Acheul; hence the appropriateness of the name Acheulian for the epoch.

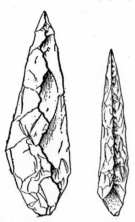

FIG. 44.   PROFILE VIEW OF A
CHELLEAN IMPLEMENT (LEFT)
AND AN ACHEULIAN IM-
PLEMENT (RIGHT).

The Chellean implement was from Chelles, the Acheulian, from Pontlevoy. The Acheulian implement maker learned how to reduce the thickness of the cleaver as well as the sinuosity of its edges by the removal of smaller chips and more carefully directed blows. Scale, ⅓. After de Mortillet.

The fauna associated with the Lower Acheulian culture at Saint-Acheul includes *Elephas antiquus*, a very large horse, a large ox, *Cervus elaphus*, and such shells as *Belgrandia marginata* and *Unio littoralis*. Toward the middle of the ancient loess occur the large horse (not unlike the small Mousterian horse), a species of large lion, *Cervus elaphus*, and *Lepus cuniculus;* the last two apparently point to a temperate climate. The fauna of the red loess at the top includes the mammoth and woolly rhinoceros, but not the reindeer.

Floors and Workshops.—Worthington G. Smith has been able to trace an old Paleolithic land surface or floor at Caddington (Bedfordshire), on which the Acheulians lived and left the products of their industry. The floor could be traced by a flinty seam

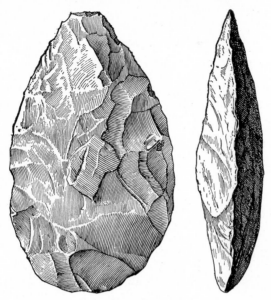

Fig. 45. Lower Acheulian flint implement of the "limande" (flounder)
TYPE.

This, the principal Lower Acheulian type of implement, is chipped on both faces and
even at the base or hilt. It is called the *limande* (flounder) by the French. Scale, $\frac{1}{2}$.
Yale University Collection.

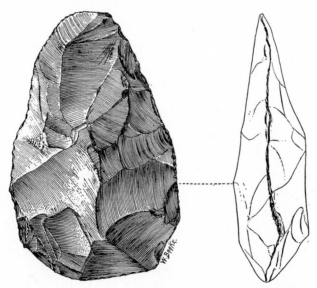

Fig. 46. Lower Acheulian cleaver from the Tellier sand and gravel pit,
SAINT-ACHEUL.

Found in the workshop level near the base of the ancient loess (see Fig. 21, *F*).
Scale, $\frac{2}{3}$. Yale University Collection.

in the ancient brick earth. The fact that Smith was able to replace flakes on flint implements and thus build up the parent cores proves that the spot was a workshop. The implements were lustrous, white, and sharp-edged, proof that they were exactly as left by the Acheulian workman (Figs. 51 and 52).

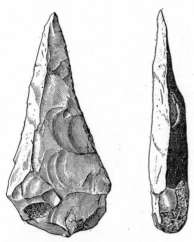

Smith reports a floor of similar age in the northeastern suburbs of London, a few feet beneath the surface, and extending for many miles on both sides of the Lea. This floor is usually just above the river gravels and at the base of the old brick earth. Sections of the floor were exposed by the cutting of new roads on the north side of Stoke Newington Common. At a depth of about 1.22 meters (4 feet), there was an immense accumulation of Acheulian implements, flakes, and

FIG. 47. FLINT CLEAVER OF THE POINTED TYPE FROM THE FALLEN ROCK SHELTER OF LA MICOQUE NEAR LES EYZIES, DORDOGNE, FRANCE.

This pointed Upper Acheulian cleaver, manufactured 125,000 to 100,000 years ago, represents the highest achievement in cleaver making. Scale, ½. Yale University Collection.

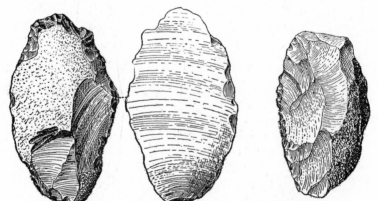

FIG. 48. LOWER ACHEULIAN FLINT SCRATCHER AND SCRAPER FROM RUE DE CAGNY, SAINT-ACHEUL.

These implements were made from chips struck off (probably) in the making of cleavers. Lower Acheulians (150,000 to 125,000 years ago) and Chelleans utilized in this way what otherwise would have been waste material. Scale, ½. After Commont.

cores, most of the implements being as "perfect as on the day they were made." About 2.44 meters (8 feet) deeper than the Acheulian floor level, and some 3.66 meters (12 feet) from the surface of this section in London, occurs the top of a stratum of Chellean age; occasionally this old gravel approaches the surface, and in rare instances are found Chellean implements which were picked up by the Acheulians and retrimmed.

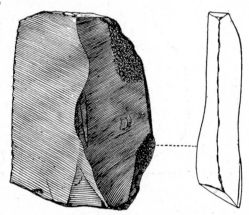

FIG. 49. LOWER ACHEULIAN FLINT KNIFE WITH ACCOMMODATED BACK, FROM SAINT-ACHEUL.

This shows how the Acheulians chipped the back to obtain a comfortable handhold. The implement was found in the workshop level near the base of the ancient loess in the Tellier pit at Saint-Acheul (see Fig. 21, F). Scale, ⅔. Yale University Collection.

An Acheulian workshop was found at Crayford (Kent) at a depth of from 11 to 12.8 meters (36 to 42 feet) by Spurrell; débris was spread over an area of some 143 square meters (150 square yards).

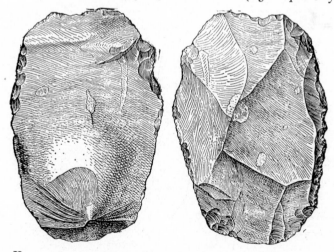

FIG. 50. UPPER ACHEULIAN FLINT IMPLEMENT OF THE LEVALLOIS TYPE FROM GRAND-BRUNEVAL, OISE, FRANCE.

Upper Acheulians (125,000 to 100,000 years ago) and early Mousterians utilized large oval flakes of this type. Scale, ½. After de Mortillet.

Spurrell succeeded in fitting back on one of the cleavers

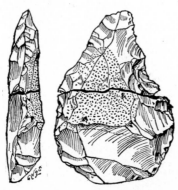

various flakes that composed the original block of flint. Many of these flakes had served as implements. When first taken from the sandy layer, the pieces were fresh and clean on the under side, but slightly discolored by dust or iron on the upper.

FIG. 51. ACHEULIAN FLINT IMPLE-
MENT FROM CADDINGTON, BEDFORD-
SHIRE, ENGLAND.

This implement was broken in Acheulian times. Both the pieces were recently found within three feet of each
other on the floor of a workshop. Scale,
½. After W. G. Smith.

A Paleolithic workshop floor was found at Creffield Road, Acton (Middlesex), in the high terrace at 30.5 meters (100 feet) by J. Allen Brown. In the third, or next to the lowest, terrace at Saint-Acheul, Commont found Acheulian workshop floors at two levels in the lower portion of the ancient loess (at level *F* in Fig. 21). Syria has furnished some important Lower Paleolithic workshops, including the station of Ain-el-Emir at Nazareth and Akbyeh on the seacoast where the workshop débris is spread over several acres.

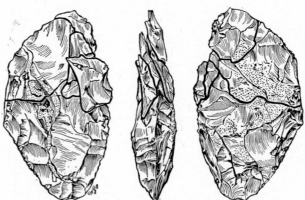

FIG. 52. ACHEULIAN FLINT IMPLEMENT WITH SEVERAL OF THE FLAKES PRODUCED
IN ITS MANUFACTURE REPLACED.

It is probable that an unfortunate blow broke this implement in two during the process of manufacture; the two pieces and the flakes struck from it by the workman were found on the floor of an Acheulian workshop at Caddington, Bedfordshire, England. Scale, ½. After W. G. Smith.

Geographic Distribution of Chellean and Acheulian Cultures

It is rather hazardous to draw any very definite conclusions from the geographic distribution of Chellean and Acheulian cultures, which means the distribution of the Lower Paleolithic culture complex, since their distribution covers that of the Pre-Chellean

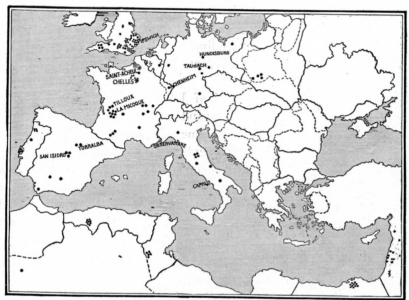

FIG. 53. MAP SHOWING THE GEOGRAPHIC DISTRIBUTION OF CHELLEAN AND ACHEULIAN CULTURES IN EUROPE AND THE MEDITERRANEAN BASIN.

The national boundary lines conform with the Treaty of Versailles of 1919. Boundaries which were then eliminated are represented by broken lines.

Epoch. In the first place, our knowledge is very incomplete in so far as large portions of Africa and especially of Asia are concerned. One can say from the data at hand that in Europe the cultures are best represented in France, England, Belgium, Spain, and Portugal; in Africa, Algeria and Tunis, Congo, and Egypt; in Asia, India and Syria. They occur to a lesser extent in Germany, Italy, Poland, and Monaco; Sahara and South Africa; Mesopotamia, Indo-China, and Japan. These countries are all situated in tropical to middle latitudes, but there are large areas in the same latitudes from which no trace of Chellean and Acheulian cultures has yet been reported—Austria, Hungary, Rumania, Czechoslovakia, and Jugoslavia. Switzerland has yielded only negative results, for which its high altitude may be responsible, just as high latitudes may be chiefly responsible for negative results in Ireland, Scotland, Fenno-Scandia, Holland, northern Russia, and Siberia. Broadly speaking, the Lower Paleolithic races avoided cold climate and ventured into middle latitudes only during interglacial epochs. This climatic preference, combined with the great length of time covered by the epochs in question, would account for a maximum longitudinal extension of the Chellean and Acheulian cultures in the Old World, namely, from Portugal to Indo-China and Japan.

The geographic distribution of Lower Paleolithic cultures is treated below under continents and nations, or political grand divisions. So far as it can be done with exactness, stations are grouped under cantons, departments, or provinces in any given nation. Both Chellean and Acheulian cultures occur in some of the stations; in others, only one of the cultures is represented. In a few cases the nature of the data at hand makes it difficult to determine which of the two is represented.

GEOGRAPHIC DISTRIBUTION OF CHELLEAN AND ACHEULIAN CULTURES

EUROPE

Belgium

Hainaut.—Helin (*A*); Mesvin (*C*); Spiennes (*C*); Saint-Symphorien.

Liège.—Sainte-Walburge.

Rutot reports both Chellean and Acheulian from Batignies-les-Binches, Bois d'Epinois, Cronfestu, Haine-Saint-Pierre, Ressaix, and Strépy.

England

Bedfordshire.—Biddenham; Biggleswade; Caddington; Kempston.

Cambridgeshire.—Barnwell; Kennett station; Traveler's Rest (*C-A*).

Derbyshire.—Robin Hood cave (?).

Devonshire.—Axminster; Brixham (?); Kent's Hole (*A*); Kentisbeare.

Essex.—Gray's Thurrock, north bank of the Thames below London (associated with *Elephas antiquus, Rhinoceros merckii, Hippopotamus,* horse, stag, ox, bear, *Corbicula fluminalis, Unio littoralis,* and *Paludina marginalis,* southern shells now living in Egypt and Syria); Quendon.

Hants.—Bournemouth; Fordingbridge; Hill Head; Maidenhead; Southampton.

Hertfordshire.—Folly and Ransom pits south of Hitchin.

Kent.—Aylesford; Barnfield pit; Canterbury; Crayford (*A*); Dartford Heath; Folkestone; Galley Hill; Hearne Bay; Ightham; Oldbury Hill; Reculver; Studhill; Swanscombe (*C-A*); Thanington.

Middlesex.—Creffield Road, Acton (*A*); Ealing Dean; Gray's Inn Lane; Highbury New Park; Lower Clapton; Stamford Hill; Stoke Newington Common.

Norfolk.—Bromehill; Redhill; Shrub Hill.

Somerset.—Wookey Hole.

Suffolk.—Bury St. Edmunds; Foxhall Road Brickfield, near Ipswich; Gravel Hill; High Lodge (*C-A*); Hoxne (*A*); Mildenhall; Nowton; Rampart Hill and Warren Hill, at Icklingham; Santon Downham; Westley.

Surrey.—Farnham; Limpsfield.

Warwickshire.—Saltley.

Wiltshire.—Milford Hill.

France

Ain.—Chelles de Bohan.

Alsace.—Achenheim (*A*).

Ardèche.—Saint-Just.

Charente.—Carmagnac sand pits at Les Planes (*A*); Marignac (*C-A*); sand pit at Tilloux.

Corrèze.—Chez-Pouré, near Brive (*A*).

Dordogne.—Bergerac; La Croix-du-Duc gravel pit, near Périgueux (*A*); Micoque (*A*); Pech de Bertrou (*A*); Rochette (*A*); Rodas sand pit, near Trélissac; La Vignole (*A*).

Haute-Garonne.—L'Infernet; Roqueville; Venerque.

Jura.—Conliège, near Lons-le-Saulnier.

Manche, La.—Bretteville, near Cherbourg.

Oise.—Saint-Just-des-Marais sand pit, near Beauvais (*A*).

Rhône.—Nety.

Saône-et-Loire.—Charbonnières; La Salle.

Seine.—Billancourt; Bois-Colombes; Colombes; Creteil; Villejuif (Gournay pit).

Seine-et-Oise.—Cergy (*C-A*); Moru (*C*); Rosny.

Somme.—Abbeville: Carpentier pit (*C*); Champ de Mars (*C*); Gamain (*C-A*); L'Heure (*C*); Leroy (*C-A*); Mareuil (*C*); Mautort (*C-A*); Menchecourt (*C*); Mercade gate; Moulin-Quignon (*C*).

Amiens: Buhant and Boutmy-Muchembled (*C*); Debary (*C-A*); Montières; Muchembled.

Amiens-Saint-Acheul: Bultel-Tellier (*C-A*); Fréville (*C-A*); Leclerq (*C*); Rue de Cagny (*C-A*); Tellier.

Aubercourt; Beaucourt; Boves; Buire-sur-Ancre; Contoire; Damery; Demuin; Epehy; Hangard; Longeau; Mézières; Mon; Montdidier; Moreuil; Morlancourt; Tertry; Thennes; Villers-aux-Erables; Villers-Faucon.

In discussing the distribution of Chellean and Acheulian culture, de Mortillet states that the cleaver had been reported, by the close of the nineteenth century, from 45 communes in the Somme basin, 292 in the Seine basin, 17 in Normandy and Brittany, 79 in the basin of the Loire, 25 in Vendée, Charente, and Charente-Inférieure, 14 in the Adour basin, 74 in the Garonne basin, 46 in the Rhône basin, and 2 in the Meuse basin. Since the cleaver persisted into the Mousterian Epoch, it is probable that this list of de Mortillet includes a number of Mousterian stations.

Germany

Bavaria.—Klause (*A*).
Leipzig.—Markkleeberg (*A*).
Pomerania.—Wustrow-Niehagen (*A*).
Reuss.—Lindental Hyena cave (*A*).
Rhine.—Kartstein (*A*).
Saxe-Weimar.—Taubach-Ehringsdorf (*A* or early Mousterian).
Saxony.—Hundisburg (*A*).

Italy

Campania.—Capri (*C*).
Chieti.—Vicinity of Maiella mountain (*A*).
Piacenza.—Gravel bed of the Tavo river, near Pianello.
Potenza.—Lacustrine deposit at Terranera near Venosa (*A*).
Umbria.—Terraces of the Tiber and its tributaries.

Monaco

L'Observatoire cave (*C*).

Poland

Kielce.—Korytanja hollow (station on the right bank of the Prad-nik); Stavna mountain and Stok miejski in the vicinity of Miechow (Acheulian cleavers from a considerable depth in the loess).

Smardzewitz Mountain.—Czarnowski found 100 cleavers, some of which date back at least to the early Acheulian, in a flinty layer beneath a deposit of loess.

Portugal

Coimbra.—Mealhada (alluvial station).

Estremadura.—Vicinity of Lisbon: stations in the open at Agonia; Alto do Duque; Amoreira; Bica; Boticaria; Casal das Osgas; Casal de Serra; Casal do Monte; Estrada de Aguda-Queluz; Moinho das Cruzes; Penas Alvas; Serra de Monsanto.

Northwest of Lisbon: Valley of the Alcantara (C); Furninha cave, peninsula of Peniche (60 kilometers, 35 miles, from Lisbon); Leiria (Chellean quartzite cleaver from surface soil).

Oporto.—Vicinity of Oporto: Castello do Queijo; Ervilha; Paços.

Spain

Cadiz.—Loma del Macharro, near Laguna de la Janda (*C-A*).

Cordova.—Pasadas (*C*).

Jaen.—Puento Mocho (*C-A*).

Madrid.—Arenero de Portazgo (*C-A*); Portazgo (*C-A*); San Isidro (*C-A*).

Soria.—Cerrada de Salana (*A*); Torralba (*C*).

AFRICA

Algeria

Oran.—Ain-el-Hadjar and Marhoum (cleavers found by Doumer-gue); Lake Karar, southeast of Montagnac (hundreds of stone imple-ments including cleavers made of quartzite and various flint chips retouched for points and scrapers. Associated remains of *Elephas atlanticus, Rhinoceros mauritanicus, Hippopotamus amphibius, Sus scrofa, Cervus elaphus,* and a species of ox that may be either *Bubalus antiquus* or *Bos primigenius* were found); Kolea, near Algiers; Oued Menöuel valley; Kabyles (quartzite cleaver); Ouzidan sand pit, north of Tlemcen (early discovery by Bleicher); Ternifine sand pit east of Mascara (quartzite cleavers associated with extinct fauna, dis-covered by Tommasini); Sporadic finds reported by Pallary.

Congo

Banja, between Loudima and Louvisy; Boma and Matidi, near mouth of the Congo River; district of the Cataracts; Kibanja; station on the Lubudi, toward the headwaters of the Congo and the Katanga (discovered by Cornet); Somaliland (A); Stanley Pool district.

Egypt

Environs of Cairo; stations near oasis of El Khargueh (reported by Legrain); Luxor (Sir John Lubbock in 1874 and Henry Haynes in 1878 were among the first to call attention to implements of Chellean and Acheulian types from Luxor, Abydos, and the environs of Cairo).

In an inventory published in 1897, de Morgan mentions as many as twenty stations between Cairo and El Kab where cleavers had been found, generally at the surface of gravelly alluvions bordering the Nile valley. The most important are: Abydos, Dahchour, Deir-el-Bahari, Esneh, Farschut, Gebelin, Hierakonpolis, Negadah, Thebes, and Toukh.

Sahara

Erg district; Timassinin, central Sahara (specimens in Saint-Germain Museum); Timbuktu (surface finds by Bonnel, often in groups as if marking camp sites).

South Africa

Valleys of the Caledon, Orange (gravel breccia), Vaal, Zambesi, Zwartkops at Algoa Bay (all reported by J. P. Johnson); valley deposits of the Umhltuzane at Marianhill, Natal (typical cleaver *in situ* at depth of 6 meters, 19.7 feet, reported by Otto and Obermaier).

Tunis

Gafsa; El Mekta (workshops); camp sites.

ASIA

Asia Minor

Surface finds of both Chellean and Acheulian have been reported.

India

Valley of the Ganges at Banda, Kaimur Range, Mirzapur, Jubbulpore, and Rajpur.

Valley of the Indus north of Hyderabad; valley terraces of the Narbada (associated with an extinct fauna).

Southern India: Ballari; Madras (Chellean and Acheulian implements *in situ*, see Fig. 42; the laterite beds overlook the sea at a height of 92 meters, 302 feet) ; Nellore; Nilgiri Hills, and elsewhere.

Mesopotamia

Abou-Shahrein in southern Babylonia; valley of the Euphrates; Dscharebis; several stations on the right bank in the alluvions of the Arabo-Syrian desert.

Syria

Adlun (station in the open); Ain-el-Emir spring, at Nazareth (workshops reported by Cazalis de Fondouce); Akbyeh on the sea-coast (a workshop or dwelling site, spread over several acres); Beit-Saour, near Bethlehem (discovered by Abbé Moretain prior to 1867); Dukha, between Lebanon and Anti-Lebanon; plain of Ephraim south of Jerusalem (industry at base of loess deposit near its contact with gravels); Lake Genezareth (Galilee) shores (several rich Acheulian stations); environs of Jerusalem; cave of Mogharat-el-Bzez, at Adlun; Ras-el-Kelb on the Syrian coast; region between Mount Tabor and Lake Tiberias (Galilee); Tyre.

CHAPTER V

THE MOUSTERIAN EPOCH

Relics of the Chellean and Acheulian Epochs occur chiefly in valley deposits; those of the Mousterian Epoch also occur in valley deposits, but they have been found chiefly in caves and rock shelters. The name of this epoch is derived from Le Moustier, a small village on the Vézère river in the commune of Peyzac (Dordogne).

FIG. 54. LE MOUSTIER, DORDOGNE, FRANCE, THE TYPE STATION OF THE MOUSTERIAN EPOCH.

A rock shelter nearly halfway up the escarpment, first explored in 1863 by Lartet and Christy, was found to be rich in the variety of implements which distinguish the Mousterian Epoch. This shelter, because of the number of its artifacts and because it was the first Mousterian station to be described, has been selected as the type station of this epoch.

130

In a cliff just back of the village is a rock shelter, one of many in the Dordogne partially explored by Lartet and Christy in 1862–63. The choice of this rock shelter as a type station is especially appropriate since it was the first of its kind to be described, and since it was very rich in artifacts. The escarpment in which the station occurs is seen in Figure 54; a cave, showing near the top,

FIG. 55.   THE LOWER ROCK SHELTER AT LE MOUSTIER, SHOWING A CROSS-SECTION OF THE DEPOSITS.

This station shows evidences of intermittent occupation by Mousterians and Aurignacians over a period of 60,000 to 70,000 years. It was here that Hauser found the skeleton of a Mousterian youth. The section in the center by which the woman is standing has been preserved to show the sequence of culture levels. The site has been classified as a national monument (see Appendix III).

has not yielded archeological remains, but the rock shelters, one about halfway up (the classic station), and the other near the bottom, were rich in relic-bearing deposits (Fig. 55). A section through the relic-bearing deposit of the lower shelter is shown in Figure 56; it was here that skeletal remains of Mousterian man were found by Hauser.

Prehistorians differ as to the position of the Mousterian Epoch in the Paleolithic series. Among those who would divide the Paleolithic into two primary subdivisions, some are inclined to place the Mousterian in the lower period, others in the upper. Since there is a Mousterian with warm fauna followed by a Mousterian with cold fauna, it is scarcely permissible to relegate the whole of the epoch to either the Lower or the Upper Paleolithic subdivisions. The difficulty can be obviated in two ways: (1) by dividing the epoch, giving the Warm Mousterian to the Lower, and the Cold Mousterian to the Upper Paleolithic; (2) by creating three subdivisions of the Paleolithic and considering the second as the equivalent of the whole Mousterian Epoch.

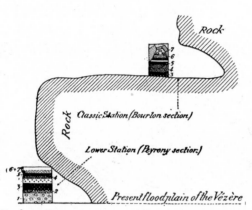

FIG. 56. SECTION THROUGH THE TWO ROCK SHELTERS AT LE MOUSTIER.

The upper is the type station for the Mousterian Epoch; the lower is the one in which was found the skeletal remains of a youth. The sequence of deposits is as follows: 1, Lower Mousterian; 2, Middle Mousterian with cleavers; 3, Upper Mousterian; 4, horizon with waterworn objects; 5, Final Mousterian; 6, Lower Aurignacian; 7, Middle Aurignacian. Horizon 4, with waterworn objects in the upper shelter is 11.6 meters above the same horizon of the lower shelter and 13.8 meters above the present flood plain of the Vézère. Redrawn from Peyrony.

The Mousterian method of chipping flint was quite different from the method which prevailed during all previous epochs. A distinguishing feature between the Mousterian and the Acheulian lies in the plane of percussion or striking platform. The Acheulian workman knocked off chips from flint nodules without any predetermined method; after the crust had been removed, an oblique blow was directed against an approximately plane surface resulting from the prior removal of a chip. The percussion point is marked by a semicircular surface, and the conchoid of percussion is very small (scarcely developed on the inner surface of the chip detached). Often a single blow did not suffice, as traces of two or three successive shocks indicate.

In Mousterian implements, the conchoid of percussion is very prominent, spreading over a considerable portion of the inner face. If one tries to detach chips from a block of flint that has not been prepared, he will obtain small chips with conchoids similar to the Acheulian; on the other hand, one can arrive at the results achieved by the Mousterians by preparing a nucleus and striking one of the facets of the polygonal contour. The chip produced will be larger, will have a more prominent bulb, and a faceted plane of percussion (the plane of percussion of the Acheulian chip is not faceted).

FIG. 57. MOUSTERIAN FLINT DRILL FROM THE CAVE OF LA COMBE, DORDOGNE, FRANCE.

Natural size. Yale University Collection.

The Acheulian workman selected from among the chips, which were short, thick, and of irregular shape, those which could be directly utilized and those which could be easily adapted to the manufacture of tools. The small implements, such as the scraper, at first glance resemble Mousterian implements, but careful observation brings out distinctions. Although the retouches along the margin are similar, the chip itself, with smooth, expansive plane of percussion and reduced conchoid, is quite different from the Mousterian scraper with small, faceted plane of percussion.

An inventory of Mousterian industry includes implements of both stone and bone—points, scrapers, scratchers, cleavers,

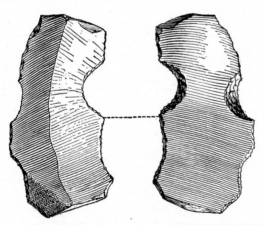

FIG. 58. MOUSTERIAN FLINT SPOKESHAVE FROM THE CAVE OF LA COMBE.

Scale, ¾. Yale University Collection.

Levallois flakes, disks, gravers, drills (Fig. 57), spokeshaves (Fig. 58), saws, hammerstones (Fig. 59), knives, and bolas (stone); compressors and skinning tools (bone). A few pedunculate points have been reported by Peyrony from the Dordogne; Reygasse has found many examples of this type in the Mousterian of southern Tunis. Paring knives have been noted at Le Moustier by Capitan and Peyrony.

The Mousterian Epoch witnessed the degeneration of the cleaver and its final disappearance. Chellean cleavers w e r e large, but an improved technique made it possible to reduce the size of the implement during the Acheulian Epoch; this reduction continued until they completely disappeared in Mousterian times. A good example of a Mousterian cleaver was found in the small cave of La Combe (Dordogne), excavated by the author in 1912 (Fig. 60). The flint, of excellent quality, was from the heart of a nodule, as indicated by a bit of crust near the base. Within a few centimeters another cleaver was found; it was larger, but of much c r u d e r workmanship, the flint from which it had been made being of poor quality, which accounts, in part at least, for the character of the workmanship.

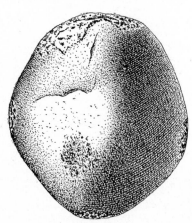

FIG. 59.   HAMMERSTONE OF QUARTZITE FROM THE CAVE OF LA COMBE, MOUS-TERIAN EPOCH.

Smooth stones similar to that shown were used by primitive man for many purposes, one of the most important being the chipping of flints.   This stone by its abrasions shows distinct signs of use. Scale, ⅔.   Yale University Collection.

The Levallois type, associated with the Upper Acheulian cleaver, persisted into the early Mousterian Epoch. Commont believed that the large disk-shaped nuclei found in the same deposits with Levallois flakes, were not sling stones, as some prehistorians have maintained, but were simply the parent cores of the flakes; the smaller disks could have been used as sling stones, but the number-less Tertiary pebbles would have furnished a ready-made supply.

Two principal types of flint implements especially characterize the Mousterian Epoch—the point and the scraper. The point was produced from a more or less triangular chip. The bulb end forms the somewhat irregular base, and the two margins, usually somewhat arched, slope away to a well defined point. The chipping and retouching show only on the outer face; the inner, or bulb, face presents a single surface of fracture (Fig. 61). The point might have been held between the thumb and flexed forefinger, though the intentionally reduced thickness of the base indicates that it might have been hafted. It served as a drill, awl, punch, and as the point of a knife or javelin; its margins were knife blades, saws,

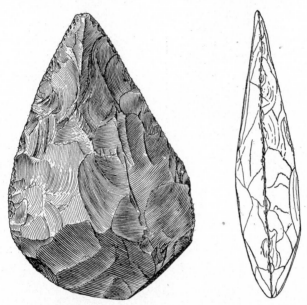

FIG. 60. MOUSTERIAN FLINT CLEAVER FROM THE CAVE OF LA COMBE.

Cleavers were largely displaced by other implements during the Mousterian Epoch. Consequently good examples like that above are rarely found.   Scale, ⅜.   Yale University Collection.

or scrapers. It was a prototype of the two-edged pointed knife blade, and possibly of the arrowhead and spearhead.

The scraper is closely related to the point. Given a large enough series of scrapers and points, the gradual merging of one type into the other is found to be complete. The scraper is somewhat larger, and was evidently produced from chips of sizes and shapes not easily converted into points; it occurs in relatively large numbers. The retouching is visible only on the outer surface; the inner consists of a single facture plane (Figs. 62 and 63). A single margin was usually selected for use as a scraping edge and retouched when necessary. Occasionally two opposite margins were converted into scraping edges, thus forming a double scraper.

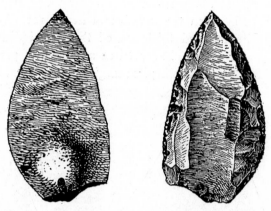

FIG. 61. MOUSTERIAN FLINT POINT FROM LE MOUSTIER.

The point is one of the two flint implements which especially characterize the Mousterian Epoch. For its manufacture the Mousterians chose a triangular flake which they chipped along two margins of one face only. The thickness of the bulb end which formed the base was often reduced, indicating that the implement might have been hafted. Not only was it useful as a point or awl, but its margins served as knife, saw, or scraper. Scale, ½. After de Mortillet.

Both single and double scrapers could easily have been evolved from the Levallois flake. The scraper was employed as a skinning knife and in cleaning skins; no doubt it served also as a carving

FIG. 62. MOUSTERIAN FLINT SCRAPER FROM THE PLATEAU DU ROCHER DE SOYON, ARDÈCHE.

Flint flakes which could not easily be converted into points were probably used in the manufacture of the scraper, a characteristic Mousterian implement. It is somewhat larger and less triangular than the point. One face only was chipped, and as a rule, along a single margin only. Scale, ½. After de Mortillet.

or chopping knife. It was sometimes converted into a kind of ax by chipping both faces along one margin; such implements might have been hafted (Fig. 64).

The presence of small spherical stones and their mode of occurrence in numerous Mousterian deposits point to the use of something resembling the bola. They were reported as early as 1866 by de Rochebrune; Henri-Martin has found seventy-six calcareous spheroids at La Quina (Charente); Pittard discovered a curious association of balls in groups of three at Les Rebières I (Dordogne); nearly all were of quartzite, although some calcareous spheroids, pecked into shape, occurred.

There are several theories regarding the use to which these

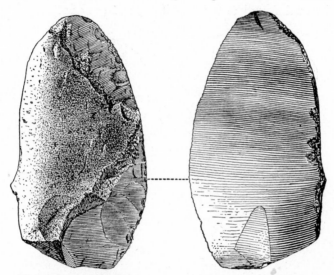

FIG. 63. MOUSTERIAN FLINT SCRAPER FROM THE CAVE OF LA COMBE, DORDOGNE.

The outer face has been chipped along one margin only; the remainder of this face is covered by the crust of the original nodule from which the implement was made. Scale, ¾. Yale University Collection.

spheroids were put. They may have been gaming stones, or they may have served a more practical purpose as sling stones or bolas similar to the weapon in use among the Point Barrow Eskimos and the natives of South America. Darwin describes two kinds of bolas: one consists of two round stones covered with leather and united by a thong about eight feet long; the other consists of "three balls united by thongs to a common center." The fact that balls have been found in groups of three favors the presumption that bolas of the triple-ball type were used by the Mousterians of western Europe. Further evidence in support of the existence of the bola has been obtained by Henri-Martin at La Quina, where he found two complete halves of a broken ball cemented together; a skin covering had presumably kept the parts conjoined until its use in that capacity was no longer needed.

A rude industry based on the use of bone first made its appearance during the Mousterian Epoch. In 1905 Henri-Martin reported the discovery at La Quina of bones (chiefly metacarpals and metatarsals) that bore marks of having served as chopping blocks or for the retouching of flint implements. The following year, Pittard found at Les Rebières, not only metacarpals, meta-

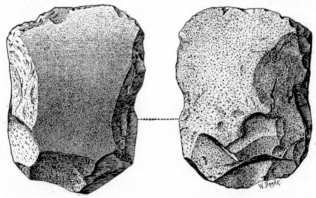

FIG. 64. MOUSTERIAN IMPLEMENT OF QUARTZITE FROM THE CAVERN OF CASTILLO
NEAR PUENTE VIESGO, SANTANDER, SPAIN.

The Mousterians occasionally chipped both faces of a flake along one margin, producing an implement which could have been used as a kind of ax; it may have been hafted. Scale, ⅜. Yale University Collection.

tarsals, phalanges, and fragments of diaphyses that had served as chopping blocks, but also some rude pointed bone tools (Fig. 65).

Crude bone tools have been found by Bächler in the caverns of Wildkirchli and Drachenloch (Switzerland). The Mousterian hunter improvized a bone tool, consisting of half of a cave-bear fibula, for skinning and preparing the hide of the cave bear. The bone was broken obliquely near the center and the broken surface polished for, and by, use; the proximal epiphysis was employed as a handle (Fig. 66). Bächler found thirty-one specimens arranged on a flat stone so as to bring the handles at the same end of the pile. Some of the specimens had seen service, as indicated by the wearing down of the broken end to a facet with an angle of from 32° to

36°. Since the proximal epiphysis of the fibula is more easily grasped in the hand, one would expect to find the distal halves also, but without marks of use, and such were found.

Bone splinters and canines of the bear, split longitudinally, were utilized as pointed implements. The innominate bone was made to serve a variety of uses through the removal of the distal ends

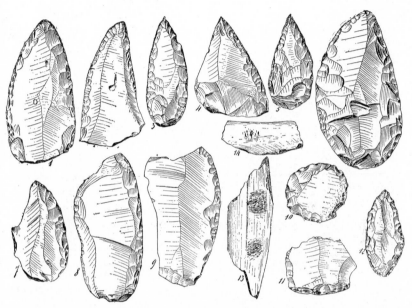

FIG. 65. FLINT IMPLEMENTS AND BONE COMPRESSORS OF THE MIDDLE MOUSTERIAN EPOCH FROM THE SECOND HORIZON IN THE ROCK SHELTER OF LA FERRASSIE, DORDOGNE.

Nos. 1–5, points; Nos. 6 and 12, oval scrapers; Nos. 7–9, scrapers; No. 10, disk. The two bone implements (Nos. 13 and 14) are scarred through use as chopping blocks or in the making of flint implements through pressure flaking. The Mousterians made frequent use of the toe bones of various animals as chopping blocks. Scale *ca.*, ⅓. After Capitan and Peyrony.

of the iliac, pubic, and ischiac portions; the margins often bear marks of use. Such a tool admits of service as a skin scraper, or as a vessel for holding water, blood, or oil (lamp). Specimens of this kind occur by the hundreds at Drachenloch, as many as twenty-five or thirty having been found in a single heap. They were found also at Wildkirchli.

A number of caves and rock shelters have a well defined Mousterian horizon with an archaic industry which may be characterized as Eolithic in type. This is not to be confused with the oldest stage of the Mousterian Epoch; it is rather to be explained as a reversion to, or the persistence of, a culture stage with a technique which prevailed during the Eolithic Period.

The type station of Le Moustier has yielded many archaic specimens which exhibit a worn condition; they have also been reported from La Micoque; from the base of the floor accumulations at the

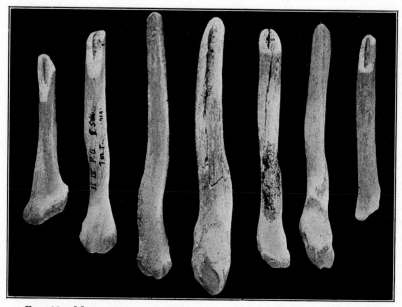

FIG. 66. MOUSTERIAN BONE TOOLS FROM THE CAVERN OF DRACHENLOCH, SWITZERLAND.

The Drachenloch (see Figs. 18 and 19) hunter learned by experience that the proximal or knee-joint end of the cave-bear fibula made a good handhold, so he broke the bone obliquely about midway and used the proximal end as a tool for skinning and preparing the hide of the cave bear; the broken surfaces of these tools are polished through use. With the exception of the large bone implement from Piltdown (see Fig. 32), these are the oldest known bone implements. Lower or Warm Mousterian dating from the Riss-Würm Interglacial Epoch (about 100,000 to 60,000 years ago). After Bächler.

entrance to Font-de-Gaume by Breuil; from Fond-de-Fôret by Rutot; from the lowest culture-bearing layer at Sirgenstein by R. R. Schmidt. In this category, no doubt, belong the specimens found at the open station of Venta de la Pasada de Gibraltar by Hernandez-Pacheco.

The lowest horizon at La Combe was characterized by archaic Mousterian flints. They are undoubtedly artifacts, since they were found *in situ* at a depth of 1 to 2 meters (3.3 to 6.6 feet) in a deposit to which they were foreign. The patination was pronounced and the angles often reduced by wear due to transport or use. If found in valley deposits, these flints would have easily passed for eoliths. Utilized flint chips were somewhat rare in comparison with nonutilized chips and bone fragments. The notched scraper or spokeshave was the most common type (see Fig. 58). This archaic industry resembles that in the lowest horizon at La Micoque, although the latter specimens are less worn; it resembles even more closely the specimens from the lowest layer in the classic station at Le Moustier. Neither the typical Mousterian scraper and point nor the cleaver occur in the archaic Mousterian at La Combe.

The race that left the archaic Mousterian industry was either careless or incapable of producing anything but indifferent results in the way of chipping flint. The nodular crust shows on many of the specimens and a lack of method is manifest in the retouching or working of the flint flakes. As a result, the chipping often shows on the bulb, or inner surface, of the flake, instead of the outer surface or back, where one would expect to find it. Certain nodules from which flakes have been detached resemble rude Chellean forms; they were found at the top of the archaic Mousterian layer. One rude implement made from a quartzite pebble, the only one of its kind, was found near the base of the archaic Mousterian deposit.

The Mousterian Epoch was of long duration, beginning during the Riss-Würm Interglacial and lasting well into the Würm period of glaciation. The early, or Lower Mousterian culture was associated with a warm fauna, the Upper (including Middle) with a cold fauna. The Lower or Warm Mousterian has been reported from (1) a number of stations in France, including Boutmy-Muchembled at Montières (Somme), Villefranche-sur-Saône (Rhône), La Ferrassie, La Micoque, and Laussel (Dordogne); (2) in Italy from the Grimaldi caves near Mentone; (3) in Jugoslavia from Krapina (Croatia); (4) in Germany from Taubach and Ehringsdorf near Weimar; and (5) in Switzerland from Cotencher, Drachenloch, and Wildkirchli. The stratigraphic position of the Lower Mousterian industry associated with a warm fauna was located by Commont at the Boutmy-Muchembled pit in the fourth terrace at Montières. The horizon is the thin bed of whitish sands and gravels below the recent loess and above the deposit of Chellean gravels (see Fig. 20).

The fauna of the Warm Mousterian includes *Alces palmatus, Bison priscus, Bos primigenius, Canis, Castor fiber, Cervus elaphus, C. capreolus, Elephas antiquus, Equus caballus, Felis spelaea, Hippopotamus, Hyaena spelaea, Rhinoceros merckii, Sus scrofa ferus,* and *Ursus spelaeus.*

The Upper or Cold Mousterian culture is better known and more widely distributed than the Warm. It differs from the Warm Mousterian in faunal association rather than in typology and technique. The chief cultural differences are: (1) a tendency in the Cold Mousterian toward emphasis on the scraper and the point; (2) the beginnings of an industry based on the flint blade; and (3) the gradual disappearance of the cleaver.

As typical stations with the Cold Mousterian there may be cited: La Quina (Charente) and Achenheim (Alsace) in France; Spy (Namur) in Belgium; Sirgenstein (Württemberg), Klause (Bavaria), Buchenloch and Kartstein (Eifel) in Germany; Gudenus (Lower Austria) in Austria; Šipka and Čertova-díra (Moravia) in Czechoslovakia; and Tata in Hungary.

The Cold Mousterian has been found at three horizons in the recent loess of the third and fourth terraces at Saint-Acheul and Montières, thus admitting of subdivisions (see Figs. 20 and 21). The lowest of the three horizons is the pebbly layer at the base of the recent loess,[1] where the Mousterian method of chipping is seen in the preparation of nuclei from which Levallois flakes could be struck off. Most of the flint chips from this horizon were retouched for the purpose of accommodation to the hand, or utilization as either simple scrapers or thick points. Their patina is, in general, bluish-white.

A small number of cleavers, typical of the ancient Mousterian, have the shape of an equilateral triangle; all three margins are retouched and the inner surface is almost a single plane. Some

---

[1] Commont noted in a section of the recent loess at Achenheim (the equivalent of the third terrace at Saint-Acheul) the same three subdivisions, the lowest containing hearths and industry.

of the cleavers, fashioned so as to give a lateral handhold (**Fig.** 67), could have been used in various ways—for skinning, cutting, sawing, and scraping. The cleaver may occur in any horizon from the Chellean to the Mousterian inclusive; it alone, therefore, is not a sure guide to the age of a deposit.

Commont found two cleavers that had been reutilized by the Mousterians; one side had the red Acheulian or Chellean patina, and the other, retouched by the Mousterians, had the characteristic white patina.

There are marked differences between the lower and upper horizons of the Cold Mousterian in the recent loess of northern France. Cleavers are almost completely wanting in the upper horizon, as exemplified at the station of Marlers, which as yet has not furnished a single cleaver. The industry is but little varied, consisting almost wholly of points and scrapers, to which could be added gravers, drills, and spokeshaves.

FIG. 67. **MOUSTERIAN FLINT CLEAVER** WITH LATERAL HANDHOLD AT THE BACK OR RIGHT.

From the sandy layer at the base of the recent loess at Saint-Acheul. Cleavers of this type might have served equally well for chopping, sawing, scraping, or skinning. Scale, ½. After Commont.

The fauna of the Cold Mousterian is the same in the Somme and Rhine valleys (recent loess), and in the caves of Belgium, the Pyrenees, and the valley of the Vézère (France). It includes: *Elephas primigenius, Rhinoceros tichorhinus, Rangifer tarandus, Equus caballus, Bos primigenius, Cervus elaphus, C. megaceros, Hyaena spelaea, Ursus spelaeus, Canis vulpes, C. lupus, Meles taxus, Arctomys marmotta, Arvicola amphibius,* and *Spermophilus rufescens.*

Mousterian mentality, in so far as it is reflected in the character of the cultural and skeletal remains, was not on a very different

plane from that of the Lower Paleolithic races. There must have
been some sort of social organization, although there is very little

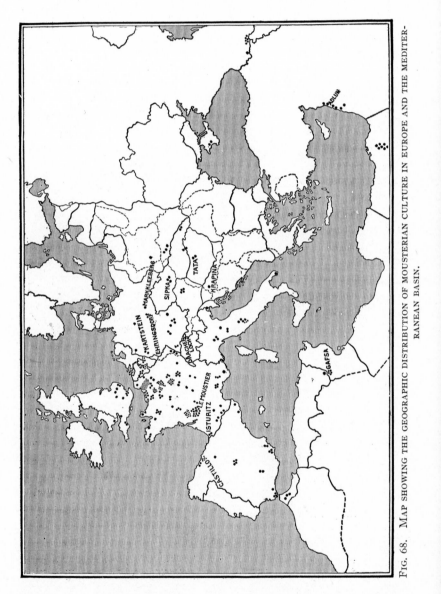

FIG. 68. MAP SHOWING THE GEOGRAPHIC DISTRIBUTION OF MOUSTERIAN CULTURE IN EUROPE AND THE MEDITER-
RANEAN BASIN.

evidence bearing on the subject aside from the extent and thickness
of relic-bearing deposits at some of the Mousterian sites.

Body covering for protection must have been confined to single pelts, since nothing that could have been used for the purpose of piecing skins together has been found. Body ornamentation was probably indulged in to some extent, since faceted chunks of oxide of manganese are sometimes met with, for example at La Quina; the coloring matter was apparently removed by means of flint scrapers and the powder applied either to the body or to the skin used as a covering, perhaps to both. It is hardly probable that the application of the paint took on pictographic forms, since nothing of the kind on imperishable material and no figures sculptured in the round have come to light.

The Mousterians may also be credited with some care for their dead. Examples of their burial have been found at several stations, including La Chapelle-aux-Saints, La Ferrassie, and Le Moustier.

GEOGRAPHIC DISTRIBUTION OF MOUSTERIAN CULTURE

In comparing the geographic distribution of the Middle Paleolithic culture with that of the Lower, there are points of resemblance as well as of contrast. The Mousterian culture is widespread, but not coextensive with the earlier cultures. It covers Europe more completely than does the Lower Paleolithic, having been found everywhere in middle latitudes except Bulgaria and Rumania. A few stations have been reported from Russia, but none from Finland, Scandinavia, Holland, Scotland, and Ireland. The Mousterians penetrated into Switzerland, but only during the last interglacial epoch. Their culture is most frequently met with in France, England, Belgium, Germany, Italy, Spain, and Portugal. A few stations have been reported from Austria, Hungary, Poland, Czechoslovakia, Jugoslavia, the Channel Islands, Malta, and Monaco. Mousterian culture also occurs in parts of Africa and Asia. It has recently been reported from China.

The Mousterians had a better inventory of tools than their predecessors had; this is especially true of their flint skinning knife and scraper, to which was added a skinning tool of bone. They were thus able to face with hardihood the rigors of the oncoming Würm glaciation, and finally gave ground only to the Aurignacians, their superiors in both material and mental equipment.

GEOGRAPHIC DISTRIBUTION OF MOUSTERIAN CULTURE

EUROPE

Austria

Lower Austria.—The loess stations of Autendorf, Drosendorf, and Trabersdorf; Gudenus cave.

Belgium

Hainaut.—Carrières du Hainaut; Helin; Leval-Trahegnies; Mesvin; Spiennes.
Liège.—Le Docteur; Fond-de-Fôret; Sainte-Walburge; Sandron.
Namur.—Furfooz; Goyet; Hastière; Magrite; La Naulette; Spy; Le Sureau.

Channel Islands

Jersey.—La Cotte de Saint-Brelade on the south shore; La Cotte de Saint-Ouen on the north shore.

Czechoslovakia

Moravia.—Čertova-dira; Kulna; Šipka.

England

Bedfordshire.—Caddington (floor levels).
Cambridgeshire.—Traveler's Rest.
Devonshire.—Brixham; Kent's Hole.
Glamorganshire.—Paviland.
Kent.—Baker's Hole, a workshop in the 15 meters (50 feet) terrace, at Northfleet; Barnfield pit in the 30.5 meters (100 feet) terrace, at Swanscombe; Globe pit in the same terrace at Greenhithe.
Middlesex.—Highbury New Park (London).
Somerset.—Wookey Hole.
Suffolk.—High Lodge; Ipswich (Bolton and Laughlin brickfield and Foxhall Road).

France

Aisne.—Coeuvres, a station in the open near Soissons; Cologne sand pits, explored by Gosselet and Pilloy.
Allier.—Sand pits at Pierrefitte-sur-Loire, explored by Bailleau.
Alsace.—Loess stations at Achenheim, Mommenheim, and Vögtlinshofen.

Ardèche.—Le Figuier; cave of Neron; Trou du Renard.

Ariège.—Cave of Bouichéta, near Tarascon; Le Portel.

Aube.—Loess stations at Troyes.

Aude.—Bize.

Basses-Pyrénées.—Caves of Isturitz and Olha.

Charente.—Borderies, near Cognac (in a reddish clay); Carmagnac sand pits at Les Planes; La Chaise; surface finds at Cognac; Gavechou; cave of La Gelie, at Edon; Hauteroche (Grotte-à-Melon); Montgaudier; Papeterie; Petit-Puymoyen; Le Placard; Les Planes; La Quina; Tilloux.

Corrèze.—La Chapelle-aux-Saints; Chez-Pouré, a protected site in the open, above Brive.

Côte d'Or.—Billiardes; Brèche de Genay, or Saint-Côme, on the southern slope of the Cars mountain (a tufaceous formation with associated remains of horse, ox, mammoth, and reindeer; Etrelles; Gray; Meilly-sur-Ronvres; Musigny (flint implements occur under 2 meters, 6.6 feet, of alluvial clays in contact with a bed exploited for phosphate of lime); rock shelter of Saint-Aubin.

Côtes-du-Nord.—Beach of Portrieux, at Saint-Quay.

Dordogne.—La Balutie; La Combe; Couze; La Ferrassie; Laugerie-Haute; Laussel; Le Moustier; La Mouthe; Pech de l'Aze; Les Rebières; Rey; Les Roches (Blanchard No. 2); La Rochette; Tabaterie.

Eure.—Amecourt gravel pit; Evreux; Radepont; Saint-Ouen-du-Tilleul brick works.

Eure-et-Loir.—Crécy-Couvé, near Dreaux (osseous breccia under a thick deposit of brick earth); La Hutte; Saint-Jean sand pit, at Châteaudun (reported by Ballet).

Finistère.—Guengat; Parcar-Plenen, at Guiclan.

Gard.—Aigueze; Froissac; Salazac; Serviers.

Gironde.—Bertonne; Marignac; Pair-non-Pair.

Haute-Garonne.—Gourdan; L'Infernet, valley station where, in 1851, Noulet found quartzite implements in gravel under a loam 15 to 30 meters (49.2 to 98.5 feet) above the Garonne.

Haute-Saône.—La Rochelle (surface workshop on top of a hill).

Hautes-Pyrénées.—Gargas.

Ille-et-Vilaine.—Mont Dol.

Indre-et-Loire.—Cave at La Roche-Cotard, east of Langeais.

Loir-et-Cher.—Artins (Pleistocene deposits, reported by Bourgeois); Pontlevoy (surface finds on the plateaus).

Lot.—Pis de la Vache; Roussignol.

Manche, La.—Bretteville.

Mayenne.—La Bigote, Cave-à-Margot, La Chêvre, Le Four, and Rochefort caves in the valley of the Eure, at Thorigné-en-Charnie.

Nièvre.—Breugnon and Chevroches, near the headwaters of the Yonne (explored by Darlet).

Nord.—A workshop covering several acres, explored by Pilloy.

Oise.—Montguillain sand pit, at Goincourt; Pont-l'Évêque; Saint-Just-des-Marais.

Pas-de-Calais.—Roellecourt (gravel pit).

Puy-de-Dôme.—Neschers.

Rhône.—Garret sand pit, at Villefranche-sur-Saône.

Saône-et-Loire.—At Les Bouleaux, 5 kilometers (3.1 miles) south of Mâcon, is a station where the Mousterians seem to have quarried flint.

Seine.—Asnières; Bois-Colombes; Chevaleret; Grenelle; Levallois-Perret; Villejuif.

Seine-et-Marne.—Bagneux, near Nemours; Breviandes; Chapelle-Saint-Luc; Château de Surville; Chelles; Isle-Aumont; sand pit near Montereau; Rochettes, between Lizines and Sognolles; Rosières; Saint-Julien; Saint-Leger-près-Troyes; Sainte-Savine.

Seine-et-Oise.—Bréval; Cergy; Cutesson; Eragny, near Pontoise; Isle-Adam; Mantes-la-Ville brickfield; Moru; Le Pecq; Rolleboise; Rosny.

Seine-Inférieure.—Bléville, Frileuse, and La Mare-aux-Cleres, near Le Havre; Saint-Pierre-d'Epinay (workshop discovered by Milet); Saint-Pierre-les-Elbeuf.

Somme (including the Somme valley and tributaries).—Abbeville (Carpentier, Leroy, Mautort, and Menchecourt pits); Amiens (Place Longueville, Rue de l'Union, Rue Saint-Dominique); Amiens-Montières (Boutmy-Muchembled, Debary, and Étouvy pits); Amiens-Saint-Acheul (Bultel-Tellier and Fréville pits); Broyes; Corbie; Courtemanche; Domiliers; Menez sand pit, at Sailly-Laurette, 8 kilometers (5 miles) from Corbie; Longeau; Longpré-les-Saint-Corps (gravel pits of Garçon and Dubourquet, at the mouth of the Airaines); Montdidier; Rivière des Trois Doms; Saint-Quentin.

Valley of the Ancre: Bazentin; Beaufils brickfield, at Miraumont.

Valley of the Avre: Boiteau, near La Boissière; Catigny; Damery-Goyencourt; Guerbigny; Pierrepont; Roey.

Valley of the Bellifontaine: Camp de Liercourt.

Valley of the Bresle: Incheville.

Valley of the Cologne: Busu, Mons-en-Chaussée, and Prusles on the heights separating the Cologne from the Omignon; Le Catelet sand pit on the left bank; Doublet brickfield near Roissel, at an altitude of 95 meters (312 feet) and 20 meters (65.7 feet) above the present bed of the river; workshop near Hargicourt, discovered in 1875.

Valley of the Germaine: Station between Ham and Saint-Quentin; Sancourt.

Valley of the Gézaincourt: Beauquesne; Beauval.

Valley of the Hallue: Molliens-au-Bois; Querrieu brickfield.

Dry valley of the Hem-Monacu: Monacu farm, above Cléry; Pécourt and Legrand brickfields, at Combles.

Valley of the Ingon: Ercheu; Étalon; Liaucourt-Fosse; Marais de Breuil; Marais de Rouy-le-Grand.

Valleys of the Liger and the Vimeuse: Délavier brickfield, at Maisnières; Heliecourt-Gamaches.

Valley of the Luce: Caix-en-Santerre; Cottinet; Guillancourt; Lihons-en-Santerre sand pit on the Rosières road; Rosières; Villiers-Bretonneux.

Valley of the Nièvre: Breugnon and Chevroches, at the headwaters of the Yonne; Naours; La Viccogne.

Valley of the Noye: La Faloise; Gannes; Remiencourt.

Valley of the Quiliennes: Rivière de Marieux.

Valley of the Rivière de Poix: Equennes; Fransure, at "Champ Flandrin"; Meigneux-Marlers.

Valley of the Scardon: Carpentier pit; Domquer; Gamain pit; l'Hermitage (corresponding to Menchecourt); l'Heure, near Caours.

Valley of the Selle: Crevecoeur-le-Grand; Gérard sand pit.

Valley of the Tortille: Bouchavesnes.

Var.—Château Double.

Vaucluse.—Cave of La Baoume dei Peyrards, rock shelter of Bau de l'Aubesier, and workshop of Deffend, all near Sault.

Vienne.—Caves of Les Cottés on the banks of the Gartempe; l'Ermitage, Lamartine (at Charroux), and Lussac-les-Châteaux.

Yonne.—Annay-sur-Serein brickworks; Arcy-sur-Cure (Grotte du Cheval, Grotte des Fées, Trou de l'Hyène, Grotte de l'Ours, Grotte du Trilobite); Auxerre Quaternary alluvions; Les Blaireaux; Chablis (at base of reddish clay); Le Mammouth; La Roche-au-Loup.

Germany

Bavaria.—Hasenloch, near Pottenstein; Hohlefels, at Hapsburg; Klause; Räuberhöhle; Schulerloch in the Altmühl valley.

Leipzig.—Markkleeberg.

Northern Germany.—Baumanshöhle, near Rübeland in the Harz mountains.

Rhine.—Buchenloch; Kartstein; Feldhofen cave (Neandertal skeleton).

Saxe-Weimar.—Ehringsdorf; Taubach; Weimar.

Württemberg.—Irpfelhöhle, near Giengen; Sirgenstein.

Hungary

Budapest.—Tata, right bank of the Danube; Kiskevély, left bank of the Danube.

Szinva Valley.—Miskolcz loess stations; Szeleta cave, near Miskolcz.

Italy

Bologna.—Middle terrace of the Santerno valley.

Emilia.—Cave of Santa Maria dei Bagni.

Liguria.—Grimaldi caves (Barma Grande, Cavillon, Grotte des Enfants, Grotte du Prince).

Otranto.—Romanelli cave.

Parma.—"La Grotta," near Cassino; cave of Scalea.

Trieste.—Cave of Pocola, at Aurisina (Marchesetti found two cave-bear skulls, each pierced by a stone implement).

Tuscany.—Cave of Cucigliana; Olmo.

Umbria.—San Egidio; various stations in the valley of the Tiber and its tributaries.

Jugoslavia

Croatia.—Krapina.

Malta

Cavern of Ghar Dalam.

Monaco

Cave of L'Observatoire.

Poland

Galicia.—Caves of Eulen and Zigeuner in the Smardzewitz mountains; Wierzchow cave.

Piotrokow.—Okienik cave in the district of Bedzin (explored in 1914 by Krokowski, who found a Mousterian industry of the Micoque type).

Portugal

J. Fontes reports Mousterian types from a number of surface stations.

Russia

Caucasus.—Ilskaja.

Crimea.—Labo and Wolf's cave near Simferopol.

Transcaucasia.—Alagheuz, region of Erivan (several stations explored by J. de Morgan).

Spain

Alicante.—Aspe, a station in the open; cave of El Cuerva.

Asturias.—El Conde; La Paloma.

Barcelona.—Romani cave, near Capellades-Igualada.

Burgos.—Rock shelter of Barranco del rio Lobo.

Cadiz.—Los Derrmaderos near Laguna de la Janda; Venta de la Pasada de Gibraltar.

Gerona.—Bañolas.

Gibraltar.—Forbes Quarry.

Jaen.—Aldeaquemada; La Puerta.

Madrid.—Arenero del Portazgo; Casa del Moreno; Portazgo tileworks; Posos de Feito.

Malaga.—Bobadilla.

Murcia.—La Bermeja; El Palomarico; Las Perneras.

Santander.—Castillo; Cobalejos; La Fuente del Frances; Hornos de la Peña; Morena; rock shelter of San Vitores.

Switzerland

Appenzell.—Wildkirchli.

Neuchâtel.—Cotencher.

Saint-Gallen.—Drachenloch.

AFRICA

Algeria and Tunis

Bir-el-Ater, a station in the open 90 kilometers (56 miles) south of Tebessa, Algeria; Fedj el Bottna, Algeria; Gabes; Gafsa; Lake Karar; El Mekta; Sidi Mansur.

Mauritania

Surface finds reported by Madame Crova.

Nile Valley

Many localities.

ASIA

Asia Minor and Syria

Adlun; Akbyeh, a workshop in the open, northeast of Ain-el-Kantara; Antelias; Nahr-el-Djoz, between Kefer-Hay and Keftun; Nahr-Ibrahim on the coast; Ras-el-Kelb on the Syrian coast.

China

Shensi.—P. Teilhard de Chardin has just discovered a number of stations dating probably from the Mousterian (or Lower Aurignacian) Epoch. They occur either as actual hearths or workshops in the loess (or formation of the same age), or as implements disseminated in the conglomerate at the base of the great loess formation of Shensi. In the cliffs of certain canyons he has found hearths at a depth of 15 meters (49 feet) and again at 58 meters (190 feet). The fauna is distinctly Quaternary—woolly rhinoceros (very abundant), mammoth (rare) hyena, ostrich, several varieties of deer, bison, gazelle, antelope (spiral horns), etc. With the exception of the cleavers, the implements (quartzite and silicified rock) are all made of flakes retouched on one face only.

GEOGRAPHIC DISTRIBUTION OF MOUSTERIAN BONE COMPRESSORS OR CHOPPING BLOCKS

Belgium

Goyet (Namur)	Magrite (Namur)
Hastière (Namur)	Spy (Namur)

France

Amiens (Somme)	Laugerie-Basse (Dordogne)
Boutmy-Muchembled	Olha (Basses-Pyrénées)
Bize (Aude)	Petit-Puymoyen (Charente)
Catigny (Oise)	Pont-Leveque (Oise)
Combe, La (Dordogne)	Pont-Neuf (Charente)
Ferrassie, La (Dordogne)	Quina, La (Charente)
Hauteroche (Charente)	Rebières, Les (Dordogne)
Isturitz (Basses-Pyrénées)	Roches, Les (Dordogne)

Germany

Räuberhöhle (Bavaria)   Sirgenstein (Württemberg)
Schussenquelle (Württemberg)   Wildscheuer (Nassau)

Italy

Grimaldi (Liguria)
Barma Grande

Spain

Castillo (Santander)

CHAPTER VI

Transformations in Cultural and Ethnic Elements

The Upper Paleolithic Period is divided into three epochs, the *Aurignacian* of Breuil, and the *Solutrean* and *Magdalenian* of de Mortillet. It is marked by a complete transformation of both the cultural and the ethnic elements. The cleaver and the scraper are replaced by a stone industry based on the evolution of the blade-like flint flake, and bone comes into more general use with improved methods of working it. The entire period was approximately coextensive with the last glacial epoch, so that in addition to flint and bone as raw materials on which to base a primitive culture, the hunter population of the time, being contemporary with the mammoth and the reindeer, soon learned to make use of two additional by-products of the chase—ivory and reindeer horn. For the working of these new materials (as well as of bone) various forms of the flint graver were perfected. This tool was well adapted to serve a double purpose, (1) the working of by-products of the chase, and (2) the execution of art objects, whether of bone, ivory, and reindeer horn, or of stone.

The perfecting and extensive use of the flint graver gave rise to a product which is common to all Upper Paleolithic horizons, the microliths or small slender chips struck off in the forming of the beveled point of the graver.[1] Many of these microliths were retouched and utilized in such delicate operations as the perforation of a tooth, shell, or stone pendant or the making of a needle's eye; for Upper Paleolithic culture was of such a nature as to require processes undreamed of by the primitive Mousterians.

Given a set of flint tools based on the bladelike flint flake, one

---

[1] Another source of slender microlithic chips might have arisen from the manufacture of the carinate (keeled) scraper, which made its appearance early in the Upper Paleolithic Period.

can postulate with a fair degree of accuracy the nature of the cave man's cultural output. During the middle of the Lower Paleolithic Period, the only primary shaping tools seem to have been natural flints with utilizable edges or points, and hammerstones; to these was added in Mousterian times the bone compressor. The list of secondary shaping tools produced by means of the primary ones was relatively restricted, as appears from the foregoing chapters, and these, with the possible exception of the hollow scraper, served an ultimate purpose. Upper Paleolithic culture was very early transformed through the addition of the secondary shaping tools produced from bladelike flint flakes. With this addition to the inventory of shaping tools, it

FIG. 69. PERFORATED CANINE TOOTH OF THE STAG FROM THE CAVE OF LA COMBE, DORDOGNE, FRANCE, AURIGNACIAN EPOCH.

It may have been used singly as a pendant or may have formed part of a necklace. Note the three tally marks. Natural size. Yale University Collection.

was possible for Upper Paleolithic man to produce an entirely new array of tools (tertiary), such as the bone needle, the javelin point of bone, ivory, or reindeer horn, the javelin shaft, the dart thrower, and the harpoon of reindeer horn; he was also able to embellish his dart throwers and satisfy a rapidly developing artistic sense by producing various objects of art and of personal adornment. Another product of improved secondary tools was the bone tally or hunter's record.

FIG. 70. INCISOR OF THE STAG FROM THE CAVE OF LA COMBE, AURIGNACIAN EPOCH.

The root end is marked by an artificial groove and pit. Natural size. Yale University Collection.

The inception, development, and decay of Quaternary art all took place during the Upper Paleolithic Period. The beginnings of sculpture, engraving, and fresco are traceable to the Aurignacian Epoch. Sculptured representations of the human form have been found in widely separated stations: Brassempouy, Grimaldi, Lespugue, Mainz, and Willendorf. These figures are remarkably uniform in type, a type to which the bas-reliefs from Laussel also conform. It is suggestive rather of a symbolic than a physical type. These figures belong to various Aurignacian levels.

During the Aurignacian Epoch, the love of ornament developed in conjunction with the decorative arts in general, as is witnessed

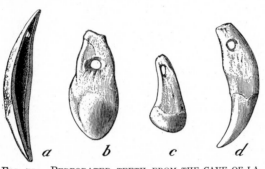

by the use of bone and ivory pendants, as well as perforated shells and animal teeth which served as necklaces, pendants, and trimmings (Figs. 69–73). The subject of Paleolithic art will be treated in detail in Chapter VII.

FIG. 71.   PERFORATED TEETH FROM THE CAVE OF LA MOUTHE, DORDOGNE, FRANCE, AURIGNACIAN EPOCH.

a, tooth of a wolf, perforated after having been split longitudinally, scale, ¾; *b* and *c*, elk, natural size; *d*, lynx, natural size.   Yale University Collection.

Polishing of Stone.—In comparison with the Middle and Lower Paleolithic races, the Upper Paleolithic were men of new ideas, practical as well as æsthetic. Among their innovations mention should be made of polished stone implements. In the early days of the science, it was customary to speak of the two grand divisions of the Stone Age as the period of chipped stone implements and the period of polished stone implements. Later the terms Paleolithic and Neolithic came into general use, and fortunately so, since the former terms had become misnomers for two reasons: in the first place perhaps more than half of the Neolithic stone implements were never polished; and secondly, stone objects of undoubted Upper Paleolithic age, bearing marks of a polishing process, have been reported from various stations. The polishing of bone by use is a phenomenon of common occurrence since Mousterian times and need not enter into the discussion here.

FIG. 72.   PERFORATED HUMAN LOWER LEFT FIRST OR SECOND MOLAR FROM THE CAVE OF LA COMBE, AURIGNACIAN EPOCH.

One of the rare examples of the human tooth artificially perforated for suspension. Natural size. Yale University Collection.

For some reason not easily explained, it never occurred to any of the Paleolithic races to polish flint implements; in fact, the polishing process in Paleolithic times, so far as stone is concerned, seems to have been wholly incidental and not a means to an end.

The abrasions on primitive hammerstones and anvil stones are the result of their use as such, not of their manufacture. This is also true of the polished facets on the upper and nether rubbing or grinding stone. The constant polishing of bone needles would finally leave grooves in the stone on which the work was done.

FIG. 73. PERFORATED SHELLS FROM THE CAVE OF LA COMBE, AURIGNACIAN EPOCH.

The shells are those of *Littorina obtusa*, a periwinkle. Perforated shells were extensively used as ornaments by the Aurignacians of some 30,000 years ago, especially on head-dresses and other apparel as well as in necklaces. Natural size. Yale University Collection.

FIG. 74. AURIGNACIAN COMBINATION RUBBING STONE, HAMMERSTONE, AND ANVIL OR CHOPPING BLOCK FROM THE CAVE OF LA COMBE.

A weathered, oval, slightly flattened, waterworn pebble from a rock of igneous origin. The pebble had been extensively employed as a rubbing stone, which process further flattened and polished its two faces; later its use as an anvil or chopping block left scars on the polished surfaces. In addition the battered ends of the pebble prove that it had also functioned as a hammerstone. Scale, ½. Yale University Collection.

A fine example of a rubbing stone (Fig. 74) was found *in situ* within the cave of La Combe at a depth of 50 centimeters (20 inches) from the surface of the cave deposit. It is a weathered, slightly flattened, oval, waterworn pebble from a rock of igneous origin. The opposite sides are much reduced by polishing, the pebble having been extensively employed as the upper, or active, rubbing stone. In the center of each polished surface are two groups of scars, indicating that the pebble had been used as an anvil stone or a chopping block. The battered ends and the bruised condition of one of the margins prove that the pebble had also functioned as a hammerstone. The relation of the contused to the polished surfaces establishes the fact that the implement had last served as a rubbing stone. A few weathered granite pebbles that had been employed as active, rubbing stones were found in the same station.

Ethnic Elements.—The ethnic elements revealed by the skeletal remains associated with Upper Paleolithic cultural remains are different in type from the apparently homogeneous Middle Paleolithic race. They will be described in detail in the chapter on "Fossil Man." There are two types: (1) the Grimaldi race with negroid affinities, and (2) the Cro-Magnon race; both are more closely related to Neolithic and later races than to the Mousterian race. There is no good reason to separate them either physically or mentally from *Homo sapiens.* The Upper Paleolithic ethnic and cultural elements were probably acquired in regions at a distance from the southernmost limits of the last glaciation, in the Mediterranean basin or farther south, under climatic conditions similar to those of central and western Europe to-day, and promptly spread northward with the first intimation of changing climatic conditions. Against their onslaught the primitive Mousterians could put forth but feeble resistance.

The passing from the Mousterian Epoch to the Upper Paleolithic Period is a mutation, or transition, rather than a gradual evolution; and there are not lacking both skeletal and cultural remains suggestive of intermediate stages, the results probably of intermixture. The Audi type of culture is an illustration. The intermediate physical types that may be cited include crania from Eguisheim, south of Strassbourg; Grenelle, near Paris; Marcilly-sur-Eure (Eure); Bréchamps (Eure-et-Loire); Tilbury, on the Thames below London; Brüx and Podbaba in Bohemia; and one of the skulls from Předmost in Moravia.

The Upper Paleolithic of Africa (Capsian).—During the Upper Paleolithic Period, Spain was the meeting place of two culture waves—one from the north spending its force before crossing the Cantabrian Mountains, the other from Africa dominant over the rest of the Iberian peninsula. For the African culture, contemporaneous with the Upper Paleolithic of Europe, Capitan has proposed the name Capsian (from Capsa, the Latin for Gafsa). The type station is El Mekta (Tunis), a rock shelter about 50 meters (164.2 feet) long by 4 meters (13.1 feet) deep; the overhanging rock had already crumbled away in prehistoric times. The floor deposits are rich in artifacts—knives, scrapers, scratchers, nuclei, hammerstones, bone objects, shells, burnt flints, the whole embedded in charcoal and ashes. When this rock shelter was inhabited by the Capsians, the country was not arid; during Roman times the aridity was already such that it became necessary to construct vast cisterns, the remains of which are still seen at the foot of the mountain. At present the region is completely dry.

The typical artifact of the Capsian is the flint chip finely retouched on one side only to form a knife. These knives are long or short, thick or thin, as the case may be, but all are of the same form. The scrapers of this epoch are in every way like those of the Aurignacian, Magdalenian, and even the Campignian of Europe. The hammerstones, very abundant, are also like those of Europe.

West of Gafsa a vast plateau extends to the Algerian frontier. Both Lower and Upper Paleolithic stations abound in this region and even beyond the Algerian frontier. All the Cretaceous chains of hills, both Algerian and Tunisian, contain workshops similar to that of El Mekta, but not so important. The lower Capsian is intermediate between the level of Châtelperron (Lower Aurignacian) and that of La Gravette (Upper Aurignacian); in France, Upper and Lower Aurignacian are separated by thick Middle Aurignacian beds. The upper Capsian is synchronous with the Solutrean and Magdalenian of Europe. It is subdivided into two regional groups, a western or *Ibero-Maurusian* and an eastern or *Getulian*.

THE AURIGNACIAN EPOCH

The Abbé Breuil, ably seconded by Cartailhac and Rutot, was the first to differentiate and firmly establish the Aurignacian Epoch, although the need of such a division, one which would include the stages represented in the caves of Montaigle and Hastière (Belgium), had been felt many years earlier by Dupont.

This culture was first reported by Lartet from Aurignac (Fig. 75), a cave in the flat-topped hill of Fajoles about 18 meters (60 feet) above the small stream of Rodes near the town of Aurignac (Haute-Garonne). The cave was used as a burial place in Neolithic times and artificially closed by means of a slab of limestone.

The entrance, completely buried beneath a talus slope, was discovered accidentally in 1852 by following up a rabbit burrow.

Before the arrival of Lartet, the Neolithic human skeletons, representing some seventeen individuals, had been removed and given a Christian burial by order of the mayor. The floor of the cave, on which the Neolithic burials rested, was found by Lartet to be nearly a meter (3 feet) thick. It was continuous with a

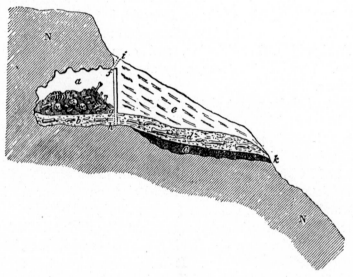

FIG. 75. SECTION OF THE HILL OF FAJOLES, PASSING THROUGH THE CAVE OF AURIGNAC, NEAR THE VILLAGE OF AURIGNAC, HAUTE-GARONNE, FRANCE.

This cave is the type station of the Aurignacian Epoch, first reported by Lartet in 1861. *a*, Neolithic burial containing seventeen human skeletons; *b*, *c*, and *d*, Aurignacian deposits; *e*, talus of rubbish washed down from the hill above; *f–g*, slab of stone set up to protect the Neolithic sepulture; *f–i*, rabbit burrow which led to the discovery of the cave (in 1852).

similar deposit covering a small terrace in front of the cave in which was abundant evidence of man's presence in the shape of artificially broken, charred, and incised animal bones, as well as implements and art objects now recognized as belonging to the Aurignacian Epoch. It also contained remains of the mammoth, woolly rhinoceros, cave bear, cave lion, cave hyena, bison, wild boar, etc.

Aurignacian stone industry is characterized by the evolution of various tools from the bladelike flint flake. The Acheulians took little trouble in preparing a nucleus from which they were able to strike off short, thick, irregular chips; the Mousterians showed more skill in the preparation of nuclei, the chips struck from them being larger, but still thick and irregular; but the Aurignacians produced nuclei from which comparatively long blades could be struck.

The Aurignacian Epoch admits of a triple division, Upper, Middle, and Lower. In passing from one to the next, marked differences are noted. In certain French stations, a transition from the Mousterian to the Lower Aurignacian occurs, as for example, at Le Moustier (Dordogne), La Verrière (Gironde), and especially at the rock shelter of Audi in the village of Les Eyzies. In comparison with Mousterian points, those of Audi are more slender and are slightly recurved. The convex margin is rendered blunt by re-

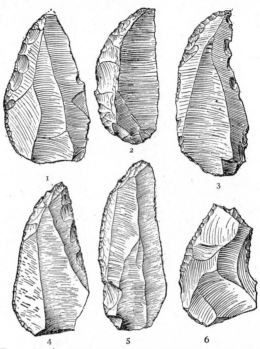

FIG. 76. AURUGNACIAN FLINT BLADES OF THE AUDI TYPE FROM THE ROCK SHELTER OF AUDI AT LES EYZIES, DORDOGNE, FRANCE.

This series illustrates the early phases in the evolution of the Aurignacian flint knife. By its ample breadth and slight incurvation, No. 1 reproduces a type frequent in the Upper Mousterian. In No. 2 the Mousterian aspect has disappeared: the breadth is reduced and the incurvation pronounced, while the retouching along the left margin and at the base is carried so far as completely to destroy the edge along the margin affected which was evidently intended to fit the curved forefinger. In Nos. 3–6 the left margin has also been intentionally dulled; the right margin or cutting edge was dulled by use in Nos. 3 and 4. The edge (right margin) in No. 5 was only slightly affected by use. The end of No. 6 was broken in the process of manufacture, or later, and retouched. Lower Aurignacian. Scale, ½. After Breuil.

touching so as not to injure the hand while using the opposite margin for cutting or other purposes (Fig. 76). Such a tool, as much a knife, or scraper, as a point, bridges the gap between the Mousterian point or double scraper and the Lower Aurignacian blades of the Châtelperron type. At Audi it is associated with small cleavers and disks, scrapers, spokeshaves, asymetric points, and scratchers. The Grotte des Fées at Châtelperron, though distinctly Aurignacian, is so closely related to the transition stage that the chronologic difference must be small (Fig. 77). An intermediate stage is recognizable at La Ferrassie (Dordogne).

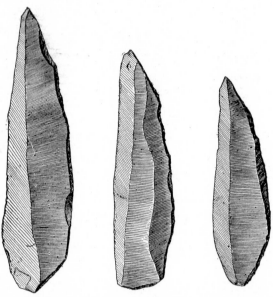

FIG. 77. AURIGNACIAN FLINT BLADES OF THE CHÂTELPERRON TYPE FROM THE CAVE OF LA COMBE.

Blades of this longer, more slender type gradually superseded the Audi blades. The right forefinger could press without discomfort against the retouched right margin of the blade while the left margin was made use of as a cutting edge. These three blades were either new when laid aside or else had been used with so much care as to escape injury. Châtelperron is the name of a cave in the department of Allier where many flint knives of this type have been found. Natural size. Yale University Collection.

The blades of the Lower Aurignacian are large and rather thick, with numerous simple marginal notches, either alternating or opposed. The latter are the so-called strangled blades (Fig. 78). Large, somewhat crude, keel-shaped or carinate scratchers (*grattoirs carenés*) and gravers with lateral bevel appear. New types in bone, ivory, and reindeer horn occur, including the flat Aurignac point, either long or short, and with or without a cleft base (Fig. 79).

Batons (*bâtons de commandement*) appeared as early as the Middle Aurignacian. Toward the close of this period, carinate [2]

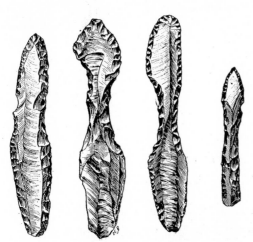

scratchers of divers forms with long, fine, parallel retouches multiply (Fig. 80). Gravers of many forms occur, the busked, or muzzle-shaped, graver predominating at first, then the graver with lateral bevel and the common or median graver (Fig. 81). These gravers were employed by the artists and by workers in bone, ivory, and reindeer horn. The bone industry grew to large proportions.

In the Upper Aurignacian, the Audi and Châtelperron types of blades gave place to the slender Gravette [3] point (Fig. 82). A similar slender blade, but with truncated end, also occurs in large numbers and in various sizes from microliths to pieces of considerable vol-

FIG. 79.   BONE POINT
WITH CLEFT BASE FROM
THE CAVE OF LA COMBE
(DORDOGNE).   LOWER
AURIGNACIAN EPOCH.

The Lower Aurignacians made two types of javelin point, one with and the other without a cleft base. The material from which they were made was usually bone, more rarely reindeer horn or ivory. The wedge-shaped end of a javelin shaft was fitted into the cleft of the point. Scale, ¾. Yale University Collection.

---

[2] Originally called *grattoirs Tartés,* from the station of Tarté (Haute-Garonne).

[3] So named from the station of La Gravette (Dordogne).

ume. Some of these points possess a lateral gibbosity, which foreshadows the Solutrean point with lateral notch at the base; this type is well represented at Willendorf and Grimaldi.

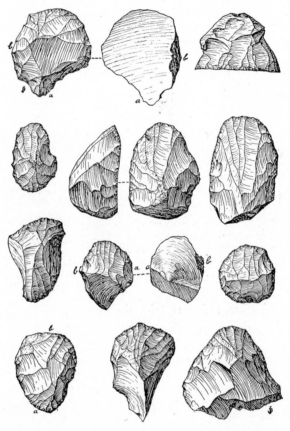

FIG. 80. CARINATE OR KEEL-SHAPED SCRATCHERS FROM THE LOWER HEARTHS OF LA COUMBA-DEL-BOUÏTOU, CORRÈZE. MIDDLE AURIGNACIAN EPOCH.

These implements were probably pushed as diminutive planes in skinning and in the cleaning of skins; they might also have been employed in retouching flint implements and as strike-a-lights. Scale, ½. After Bardon and Bouyssonie.

There is often a notch on both sides, giving rise to pedunculate forms known as *pointes de Font-Robert;* examples of this kind have been reported from Laussel (Dordogne), Font-Robert (Corrèze), and Spy in Belgium. The perforated baton of reindeer horn persists through this period and is passed on to later epochs.

In brief, Aurignacian typology may be expressed chronolog-
ically as follows:

3. UPPER AURIGNACIAN:
   Pedunculate points; Gravette blades; lateral, median, and
   busked gravers; scratchers.

2. MIDDLE AURIGNACIAN:
   Batons of reindeer horn; simple scratchers; busked or muzzle-
   shaped gravers; strangled blades; carinate scratchers.

1. LOWER AURIGNACIAN:
   Bone points with cleft base; gravers; strangled blades (rare);
   Châtelperron blades; Audi blades.

FIG. 81.   VARIOUS TYPES OF AURIGNACIAN FLINT GRAVERS FROM THE DEPARTMENTS
OF ALLIER, YONNE, AND CORRÈZE.

The graver was employed not only in the execution of works of art but also in the
making of implements of bone, ivory, and reindeer and staghorn.   The arrow in each
case shows the direction of the pressure which removed a sliver producing the beveled
point of the graver.   Scale, *ca.* ⅔.   After Breuil.

The Aurignacian fauna in Europe includes many extinct forms.
Among the nearly constant species are: *Ursus spelaeus, Hyaena
spelaea, Rhinoceros tichorhinus, Equus caballus, E. hemionus,* reindeer, mammoth, and bison. *Bos primigenius, Felis spelaea, F.
catus, F. lynx, Cervus elaphus, C. megaceros, C. capreolus,* and
Capra ibex are met with less frequently. Other species include:
Ovibos moschatus, Canis lupus, C. vulpes, C. lagopus, Gulo borealis,
and *Lepus variabilis.* This is distinctly a cold fauna. According
to Obermaier, *Rhinoceros merckii* is nearly always present in Lower
Aurignacian deposits of southern Europe.

Geographic Distribution of Aurignacian Culture

The distribution of Aurignacian culture is not quite so uniform over Europe as is that of the Mousterians and not quite so many stations have been reported. It agrees with the Mousterian in being best represented in France, Belgium, Germany, Spain, and Italy, with England, Russia, Poland, Czechoslovakia, Austria,

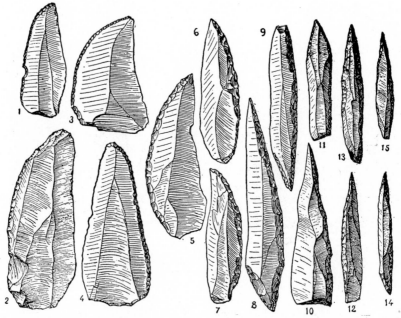

FIG. 82. AURIGNACIAN FLINT BLADES OF THE AUDI (1–5), CHÂTELPERRON (6–7), AND GRAVETTE (8–15) TYPES.

This succession covers the whole of the Aurignacian Epoch. The names were derived from the stations where characteristic examples of these types occur. Scale, ½. After Breuil.

and Hungary taking secondary rank. There are five countries (some of them very small, to be sure) in which Mousterian is represented and Aurignacian lacking, and two countries in which the reverse is true. Nor is the Aurignacian so widely distributed over Africa (where it is known as Capsian) and Asia. An industry similar to the Aurignacian occurs on the surface in southeastern India. The Aurignacian falls wholly within the limits of the last glacial epoch and as one might expect, is not represented in Switzerland, where the Cold Mousterian is also lacking.

The starting point of Aurignacian culture has not yet been determined. There is, according to Breuil, more evidence of an African than of an Oriental origin.

GEOGRAPHIC DISTRIBUTION OF AURIGNACIAN CULTURE

EUROPE

Austria

Lower Austria.—Aggsbach; Autendorf; Getzersdorf; Giesslingtal; Gobelsburg; Gruebgraben; Hundssteig; Lang-Mannersdorf; Trabersdorf; Willendorf; Zeiselberg.

Belgium

Liège.—Le Docteur; Engis; Fond-de-Fôret.
Hainaut.—Leval-Trahegnies.
Namur.—Font-Robert; Goyet; Hastière; Magrite; Princesse Pauline; Spy; Le Sureau.

Bulgaria

Teteven.—Morovitsa, cave near Gloźene.

Channel Islands

Jersey.—La Cotte de Saint-Brelade.

Czechoslovakia

Bohemia.—Generalka and Lufna, near Prague.
Little Carpathian Mountains.—Pallfy.
Moravia.—Brünn; Býčiskála; Joslowitz; Lautsch; Předmost.

England

Devonshire.—Brixham and Kent's Hole (according to Sollas and R. A. Smith).
Norfolk.—Grime's Graves (according to Smith).
Somerset.—Hyena Den, at Wookey Hole (Smith and Sollas).
South Wales.—Paviland.
Suffolk.—Ipswich (Bolton and Laughlin pit, Foxhall Hall coprolite pit).
Sussex.—Cissbury (Smith).

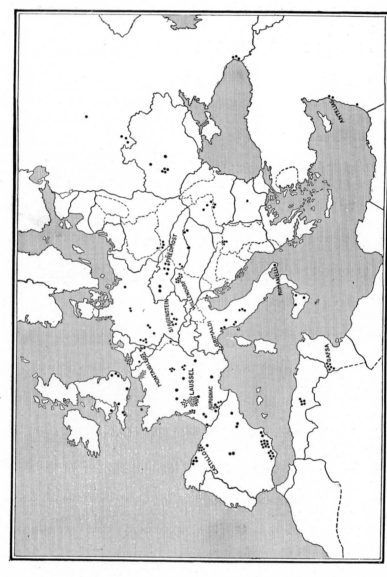

FIG. 83. MAP SHOWING THE GEOGRAPHIC DISTRIBUTION OF AURIGNACIAN CULTURE IN EUROPE AND THE MEDITERRANEAN BASIN.

France

Ain.—La Colombière, 48 kilometers (30 miles) southwest of Geneva.

Allier.—Grotte des Fées at Châtelperron.

Alsace.—Achenheim.

Ariège.—Trois-Frères; Tuc d'Audoubert.

Basses-Pyrénées.—Isturitz.

Charente.—Ammonite; La Chaise; Le Cluzeau; Hauteroche; Petit-Puymoyen; Les Planes; Pont-Neuf; La Quina (south station); Le Roc (commune of Sers); Les Vachons.

Corrèze.—Bos del Ser; Font-Robert; Font-Yves; Noailles; Planche-Torte (Bellet, Combe à Negre, Coumba-del-Bouitou, Lacoste, Les Morts, Pre-Aubert, Raysse).

Dordogne.—Audi; Combarelles; La Combe; Combe-Capelle; Cro-Magnon; La Croze de Tayac; Durand-Ruel, near Brantôme; La Ferrassie; Gorge d'Enfer (Galou, Lartet, Pasquet, and Poisson); La Gravette; La Grèze; Laugerie-Haute; Laussel; Masnaigre; La Micoque; Le Moustier; La Mouthe; Patary; Petit-Puyrousseau; Les Rebières; Rey; Le Roc de Combe-Capelle; Le Roc de Saint-Christophe; Les Roches de Sergeac (Assieur, Blanchard, Castanet, Delage, Labatut, and Landesque); La Rochette; Le Ruth; Tabaterie; Terme-Pialat.

Gard.—Grotte du Pont.

Gironde.—Bertonne; Haurets; Pair-non-Pair; La Verrière.

Haute-Garonne.—Aurignac; Marsoulas; Tarté.

Hautes-Pyrénées.—Gargas.

Indre.—Les Roches, commune of Pouligny-Saint-Pierre.

Landes.—Brassempouy; Rivière.

Lot.—Crozo de Gentillo.

Saône-et-Loire.—Four de la Baume; Germolles, at Mellecey; Solutré.

Somme.—Abbeville (Mautort, not the principal pit); Amiens (Boutmy-Muchembled, Bultel-Tellier, Étouvy); Boutillerie-les-Amiens; Renancourt-les-Amiens (Boutmy, Devallois, and Sauval pits); Ailly-sur-Somme (Sauval pit); Assainvilliers; Beaucourt-sur-Ancre; Belloy-sur-Somme (Le Plaisance); La Boissière; Bougainville; Boulogne-la-Grasse; Flixecourt; Guerbigny; Longpré-les-Corps-Saint; Querrieu; Rollot.

Vienne.—Les Cottés.

Yonne.—Arcy-sur-Cure (Grotte du Trilobite); Le Mammouth; La Marmotte; La Roche-au-Loup.

Germany

Bavaria.—Grosse Ofnet; Kleine Ofnet; Räuberhöhle.
Berlin.—Dahlem.
Brunswick.—Thiede; Westeregeln.
Hesse.—Mainz.
Nassau.—Wildhaus; Wildsheuer.
Rhine.—Buchenloch; Kartstein; Metternich; Rhens.
Württemberg.—Bockstein; Rosenstein; Sirgenstein.

Hungary

Budapest.—Kiskevély.
Bükk mountains.—Balla; Istalloskö.
Little Carpathian mountains.—Brasso.
Klausenburg.—Colnosky.
Transylvania.—Cioclovina.

Jugoslavia

Region of Belgrade.

Italy

Apuan Alps.—Goti.
Elbe Island.—Porto Longone cave.
Karst.—Gabrevizza and Pocolo, near Trieste.
Liguria.—Grimaldi caves (Grotte des Enfants, Cavillon, and Barma Grande).
Lucca Alpi.—All' Onda.
Otranto.—Romanelli.
Sicily.—Several finds according to Patiri and Schweinfurth.
Tuscany.—Cucigliana; Olmo.
Umbria.—Pila (Perugia), Branca and Ponte d'Assi (Gubbio), according to Bellucci.

Poland

Galicia.—Ciemna, near Ojcow; Koscinski, near Cracow; Wierzchow.
Piotrokow.—Nad-Galoska.

Rumania

Siebenburgen.—Valea Cremini (station in the open); Magyarbodza (loess station).
Transylvania Alps.—Malkata near Gložane-Teteven.

Russia

Caucasus.—Baratschwili; Uwarof; Virchow, near Kutais.

Central Russia.—Karatscharowo, between Moscow and Nijni Novgorod.

Chernigov.—Mezine (?).

Little Russia.—Gontzi, between Kief and Poltava; Koselsk, between Kief and Moscow; Kostionki, near Voronezh (late Aurignacian with Font-Robert types).

Ukraine.—Cyrill Street, Kief (Volkov mentions four other stations in about the same latitude, i.e., near the southern limits of the continental ice sheet).

Spain

Alicante.—Caves of Las Marivallas and El Parpillo (Capsian).

Almeria.—Ambrosio near Velez Blanco; Chiquita de los Treinta; La Fuente de los Molinos; Humosa; El Serron; Zajara (all Capsian).

Asturias.—El Conde; Cueto de la Mina; Peña de Candamo; Collubil (either Aurignacian or early Magdalenian).

Burgos.—La Aceña.

Lerida.—Cogul, near Lerida (Capsian).

Madrid.—Portazgo tile works; Las Carolinas tile works.

Murcia.—Ahumada; El Arabi; La Bermeja; El Palomarico; Las Palomas; Las Perneras; El Tesoro; Los Tollos; La Tozana (all Capsian).

Santander.—Camargo; Castillo; Hornos de la Peña; Salitré; probably the caves of El Miron and El Mar.

Teruel.—Rock shelter of Calapatá, near Cretas; Cocinilla del Obispo, a station in the open (both Capsian).

AFRICA

Algeria and Tunis

Various stations (Capsian).

ASIA

Syria

Antelias and Nahr-el-Kelb, Phoenicia; Nahr-el-Djoz, Palestine (transition between Mousterian and Aurignacian).

The Solutrean Epoch

The type station of the Solutrean epoch is Crot [4]-du-Charnier in the commune of Solutré, near Macon (Saône-et-Loire), explored by Arcelin (1866) and by Ducrost (1868). Before the discovery of Solutré, Lartet, Christy, and de Vibraye had explored a station of the same age, Laugerie-Haute, near Les Eyzies (Dordogne). As there was another Laugerie, though of later age, in

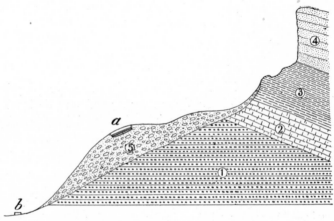

Fig. 84.   Geological section at Solutré, near Macon, Seine-et-Loire, France.

1, Upper Lias; 2, fucoid bearing limestone; 3, entrochitic limestone; 4, polyp-bearing limestone; 5, talus formation; a, excavation of Crot-du-Charnier, the type station of the Solutrean Epoch; b, spring.   After Ducrost and Lartet.

the same vicinity, it was thought advisable not to choose Laugerie-Haute for the type station.

The station of Solutré is on the south side, and near the base, of an eminence composed of Jurassic limestone (Figs. 84 and 85). Arcelin's first work was in a deposit of Aurignacian age; shortly thereafter, Ferry discovered the Solutrean deposits which were beneath a layer of humus. They were excavated in part by the Abbé Ducrost (1868) who found in them many laurel-leaf points, punches, polishers of reindeer horn and of bone, minerals for colors, perforated animal teeth, and figures carved in bone and stone. Bones of the reindeer were especially plentiful; those of

---

[4] Also written *Cro, Cros,* and *Clos;* it evidently means a place walled in, an enclosure.

the horse, deer, ox, wolf, cave bear, cave hyena, and mammoth occurred in lesser degree. The deposits covered an area of more than one hectare (2.5 acres), and in some places attained a thickness of 10 meters (33 feet). That the relic-bearing deposits were of Solutrean age in part only was clearly indicated in a section published by Arcelin in 1890 (Fig. 86).

FIG. 85. GENERAL VIEW OF THE STATION OF SOLUTRÉ.

This is the type station of the Solutrean Epoch. It covers an area of more than a hectare (2.5 acres) and is in some places 10 meters (33 feet) thick. This site was occupied in turn by Aurignacians, Solutreans, and Magdalenians for a period of about 15,000 years. It has yielded remains of their industry, art, and bones of the animals on which they fed, as well as their own fossil remains. Photograph by Fouju.

Beneath the Solutrean deposits was a uniform magma composed of bones of the horse, the most conspicuous among them being jaws and leg and foot bones; many of the bones were burnt. This magma varied from one-half to two meters in thickness (6 feet) and covered an area of 3,800 square meters (4,256 square yards). In it were represented the remains of at least 100,000 horses of a small but rugged breed, now extinct with the possible exception of a remnant still living on the Isle d'Yeu, off the west coast of France, and a type of horse (*E. przewalskii*) found in the desert of Gobi (Fig. 87). In addition to horse bones, the magma contained a few bones of the reindeer, cave bear, ox, and mammoth. The flint chips from the magma resembled Aurignacian, rather than Solutrean, technique; Gravette blades and Font-Robert points, characteristic of the Aurignacian, were also found in the magma.

Beneath the horse magma were two Aurignacian deposits, marked by distinct hearths. In addition to heaps of ashes and numerous hearthstones, there were flint implements and bones of

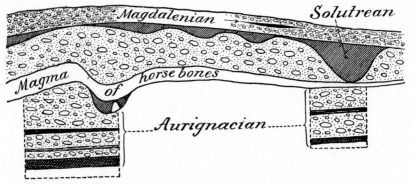

FIG. 86. SECTION SHOWING THE CULTURE SEQUENCE AT SOLUTRÉ.

The Solutrean horizon yielded many flint points of the laurel-leaf type, punches, polishers of horn and bone, perforated animal teeth, and figures carved in bone and stone. Bones of the reindeer were especially plentiful.

The magma of horse bones varied in thickness from 0.5 to 2 meters (1.5 to 6 feet) and covered an area of 3,800 square meters (4,256 square yards); it included the remains of at least 100,000 horses of a small but rugged breed (akin to the horse of Przewalski) upon which the Aurignacians fed. Three Aurignacian human skeletons were found beneath the horse magma in 1923. Adapted from Breuil.

animals, chiefly the horse, on which the Aurignacians fed. Next in point of favor came the reindeer; remains of the mammoth, cave bear, cave lion, cave hyena, marmot, antelope, deer, and elk were less numerous. From a single hearth, 19 meters (62.4 feet) long by 9 meters (29.5 feet) wide, Ducrost gathered more than 35,000 flints. Industry in bone and reindeer horn was represented by punches, polishers, spatulae, perforated antlers, and pendants of bone, ivory, and stone.

Ducrost made a number of soundings in the hillock of the Crot-du-Charnier. Toward the left, at a depth of 1.4 meters (4.6

feet), after passing through friable earth which contained a me-
lange of bones of the horse, ox, and reindeer, he found a large
flagstone; back of the flagstone, and somewhat deeper, he struck
the beginnings of a hearth which formed a deposit 30 centimeters
(11.8 inches) thick, composed largely of blackish ashes, flints, implements of stone and bone, and burnt bones, those of the reindeer
predominating. Further excavations brought out the fact that the

FIG. 87. THE HORSE OF PRZEWALSKI.

This species of horse, now living in the desert of Gobi, is related to one of the species
pictured by Upper Paleolithic artists. The steak of this horse, not our indispensable
beefsteak, was the *pièce de résistance* at all well regulated Upper Paleolithic feasts. Photograph from the New York Zoölogical Park.

flagstone belonged to a series, the others of which, still in place,
formed an oval hearth. This ellipse, bounded by the flagstones
except for a single break in the series, measured 4.5 meters (14.8
feet) long by 3 meters (10 feet) wide. Within it was the skeleton
of a male, buried at full length with the feet toward the west in
the direction of the opening caused by the misplaced flagstone.
Under the bones of the right hand were two carefully chipped
laurel-leaf points; near by were several smaller points of similar
shape, one valve of *Pecten jacobaeus,* and a stone carving repre-

senting a reindeer. Another relief figure had been broken before completion and rejected.

Outside the oval of flagstones, quantities of animal bones were found: remains of nearly one hundred reindeer, including one head practically entire; a molar, lower jaw, pelvic bone, and tibia of the mammoth; bones of the wolf, bear, fox, otter, hyena, and of divers birds. Three other large hearths were excavated by Ducrost, and conditions similar to the foregoing were found, except that only one of the three contained a sepulture.

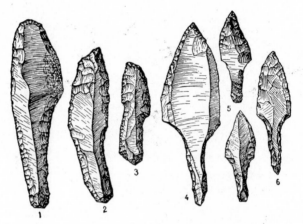

FIG. 88. FLINT POINTS OF THE FONT–ROBERT TYPE.

The pedunculate (stemlike) base of each of these implements might easily have been fixed to a shaft. Upper Aurignacian and Lower Solutrean Epochs. Scale, ½. After Breuil.

According to Cartailhac, the number of sepultures found at Solutré by Ferry and Arcelin, and by Ducrost, amounted in all to at least sixty. It seems that a vast cemetery is spread over the Crot-du-Charnier; as yet only a portion of the burials have been excavated. It is probable that the one described above, as well as most of the others, may be intrusive. The associated animal bones and artifacts might well have been encountered while the grave was being dug, and deposited in the sepulture by reason of superstition. Some of the sepultures certainly date from the Neolithic and even from the Gallo-Roman and Merovingian Epochs. On the other hand, some may belong to the Solutrean and even to the Aurignacian Epoch. As yet no authentic case of a superposition of sepultures has been noted at Solutré.

The Aurignacians made great advances in the preparation of a flint nucleus from which to strike off comparatively long blades. The special contribution of the Solutreans was pressure flaking; they were able to reduce a whole surface by means of well directed retouches through pression, a technique which reappeared in a high degree of perfection during late Neolithic times, especially in Egypt, Denmark, and the New World. This new technique was foreshadowed in the Upper Aurignacian by the expansion of the field of retouches at the tip and the base of the pedunculate points

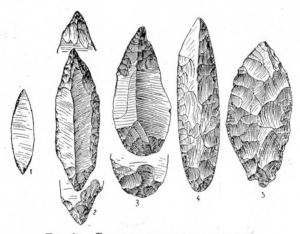

FIG. 89. PROTO-SOLUTREAN FLINT POINTS.

The Solutreans learned to reduce the entire surface of one or both faces of an implement by means of a new technique known as pressure flaking. This was foreshadowed in the Upper Aurignacian by the expansion of the retouches at the tip and base of the Font–Robert points (see Fig. 88). This series illustrates the steps in the evolution of the proto-Solutrean point. Nos. 1 and 2 are from Font–Robert (Corrèze); Nos. 3, 4, and 5 are from the Trilobite cave (Yonne). Scale, ⅓. After Breuil.

found at Font-Robert (Fig. 88), La Ferrassie, and Spy. Then suddenly the proto-Solutrean (Fig. 89) appears, pressure flaking covering practically the whole of one or both faces of the implement. It is well illustrated in the caves of Trilobite (Yonne), Le Figuier (Ardèche), and Le Ruth and Laussel (Dordogne).

The lower half of the Solutrean Epoch was characterized by a lanceolate implement resembling in outline a laurel leaf (*pointe en feuille de laurier*) and chipped on both sides to a remarkable thinness (Fig. 90). The finest and largest examples of this type were found, not at Solutré, but in a cache, or hoard, at Volgu,

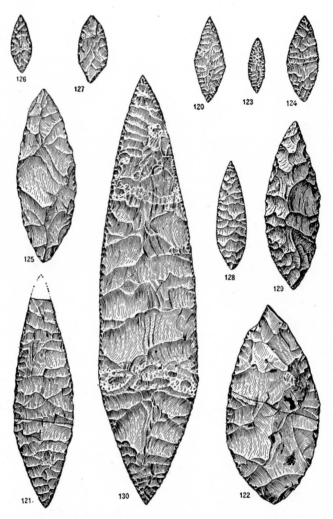

FIG. 90. FLINT POINTS OF THE LAUREL-LEAF TYPE. LOWER SOLUTREAN EPOCH.

These implements were remarkably thin and beautifully chipped on both faces. The one in the center, the largest of its kind ever found, was one of fourteen similar points found at Volgu (Saône-et-Loire); its length is 35 centimeters (13.8 inches) and its thickness 9 millimeters (0.4 inch). All fourteen had been carefully set on edge and pressed together in order to prevent breakage. Nos. 120–123 are from Solutré. No. 124 is from Laugerie-Haute (Dordogne), Nos. 125 and 126 are from the cave of Saint-Martin d'Excideuil (Dordogne), and No. 127, from Gargas (Hautes–Pyrénées). Scale, ⅓, with the exception of 129, which is about ⅔. After de Mortillet.

commune of Rigny (Saône-et-Loire), where, in 1874, workmen engaged in digging a small canal came upon fourteen of these laurel-leaf points at a depth of 1 meter (39.4 inches). The points lay horizontally, not flat but on one edge, and pressed together as if to avoid breakage through pressure from above. They varied

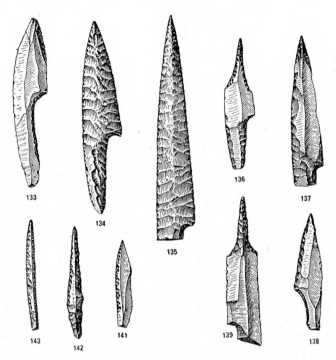

FIG. 91. FLINT POINTS OF THE WILLOW-LEAF TYPE. UPPER SOLUTREAN EPOCH.

The points in the upper row and that in the lower right-hand corner have a lateral notch at the base for purposes of hafting. The largest point in the lower row and the one above it were retouched to form drills after having served as javelin points. The smallest three points have no lateral basal notch; sharp slender points like these were needed to bore holes in animal teeth and shells and to produce the delicate eyeholes in bone and ivory needles. No. 133 is from Laugerie–Haute (Dordogne); Nos. 134–136, 138, and 139 from the Grotte de l'Eglise at Saint-Martin d'Excideuil (Dordogne); No. 137 from Le Placard (Charente); Nos. 141–143 from Grimaldi (Italy). Scale, ca. ⅔. After de Mortillet.

in length from 232 to 350 millimeters (9.1 to 13.8 inches), in breadth from 50 to 88 millimeters (2 to 3.4 inches), and in thickness from 6 to 9 millimeters (0.2 to 0.4 inch). Three were broken by the mattock which uncovered them; eleven are at present in the Museum of Châlon-sur-Saône.

The Upper Solutrean is characterized by the flint point in shape like a willow leaf, in some cases with a lateral notch at the base (*pointe à cran en feuille de saule*) (Fig. 91). Until recently this type of implement had been found only in stations between the Loire and the Cantabrian Pyrenees, being especially abundant (over 5,000 examples) in the cavern of Le Placard (Charente). Toward the end of the Solutrean Epoch, the laurel-leaf point became smaller and thicker, and finally disappeared altogether. Both Lower and Upper Solutrean horizons contain implements which appeared first in the Aurignacian and were derived principally from bladelike flint flakes—scratchers, gravers, drills (often double). Bone needles and batons, which appeared timidly in the Aurignacian, are met with more frequently in Solutrean stations. To this epoch also belongs some of the cave art, both portable and stationary.

The stratigraphic position of the Solutrean industry is in the upper, weathered portion of the recent loess. It is often so near the surface as to be mixed with, and mistaken for, the industry of the Neolithic Period, especially in the absence of faunal remains. Fossil animal remains characteristic of the Solutrean Epoch are as follows: *Elephas primigenius, Rhinoceros tichorhinus, Equus caballus, Bison priscus, Capra ibex, Rangifer tarandus, Cervus elaphus, C. alces, Ovibos moschatus, Canis lupus, C. lagopus, C. vulpes, Ursus spelaeus, Felis spelaea, F. pardus, Hyaena spelaea, Gulo borealis, Castor fiber, Lutra vulgaris, Sorex vulgaris, Myodes torquatus, Lagopus albus, L. alpinus, Lepus variabilis.*

Geographic Distribution of Solutrean Culture

The geographic distribution of Solutrean culture is more restricted than that of either the Aurignacian or the Magdalenian. Like the Aurignacian it does not occur in Switzerland, a fact which has a significant bearing on the relation of these two epochs to the last glacial epoch; both of them evidently coincide with stages of glaciation adverse to Alpine habitation. The best known Solutrean stations are found in middle latitudes from Czechoslovakia, Poland, and Hungary in the east to northern Spain in the west (Fig. 92).

The epoch was not of long duration, and does not seem to have had a very profound effect on the general trend of the cultural (or organic) evolution which began with the Aurignacian and completed its cycle with the close of the Magdalenian Epoch. It is the opinion of both Breuil and Obermaier that Solutrean culture originated in eastern Europe, whence it spread to central and western

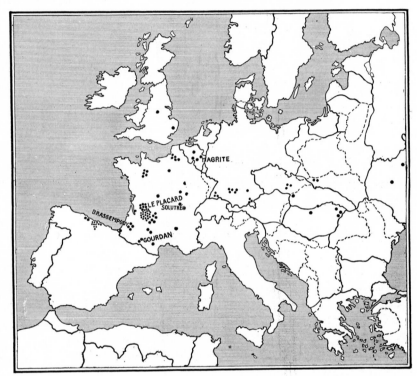

Fig. 92.   Map showing the geographic distribution of Solutrean culture.

Europe. It may be that the early Solutrean of the east is synchronous with advanced Aurignacian in France and that the Solutrean of the west was due to an invasion which, however, did not remain long in the ascendancy; for out of the clash between these two civilizations there arose the Magdalenian culture, to whose further development the east, and not the Mediterranean, contributed.

GEOGRAPHIC DISTRIBUTION OF SOLUTREAN CULTURE

EUROPE

Belgium

Brabant.—Commont reports finds in the brick earth, near Waterloo.
Liège.—Le Docteur.
Namur.—Magrite ; Pont-à-Lesse.

Czechoslovakia

Little Carpathian Mountains.—Pallfy.
Moravia.—Loess stations at Brünn, Millowitz, and Zeltsch-Ondratitz.

England

Derbyshire.—Creswell Crags.
Suffolk.—Mildenhall (brick earth).

France

Ardèche.—Cave of Chaûmadou, at Vallon.
Aube.—Bize.
Aude.—Saint-Benoit-sur-Vannes.
Basses-Pyrénées.—Isturitz.
Charente.—La Chaise ; Combe-à-Roland, commune of La Couronne ; Les Fadets, at Vilhonneur ; Gavechou ; Le Placard ; Grotte du Roc, commune of Sers ; Les Vachons.
Corrèze.—Combe à Negre ; Planchetorte (Pré-Aubert) ; Thevenard cave.
Côte-d'Or.—Cernois, commune of Vice-de-Chassinay.
Dordogne.—Grotte de l'Ane, Les Bernoux, and Fourneau du Diable, at Bourdeilles ; Badegoule ; La Balutie ; Bassilac (rock shelter near Périgueux) ; Bergerac ; Champs-Blancs (or Jean-Blanc) ; Les Combarelles ; Combe-Capelle ; Cro-Magnon ; Église à Saint-Martin d'Excideuil ; Les Eyzies ; Gorge d'Enfer (Abri Pasquet and Grotte de l'Oreille), Laugerie-Haute, Laussel, Liveyre, and Rey, all near the village of Les Eyzies ; Madrazès ; Le Ruth ; Tourtoirac.
Gironde.—Pair-non-Pair.
Haute-Garonne.—Gourdan ; Grotte des Harpons, at Lespugue.
Indre.—Rock shelter of Monthaud.

Landes.—Montaut, a station in the open near Saint-Sever; Grotte du Pape, at Brassempouy; Rivière; Saussaye, at Tercis.

Lot.—Cavart; Lacave; Roussignol.

Mayenne.—Cave de la Bigotte, Grotte à la Chèvre, Grotte du Four, Grotte du Moulin de Rochebrault, and Grotte à Margot, all in the commune of Thorigné-en-Charnie.

Saône-et-Loire.—Solutré; Volgu, commune of Rigny (cache of 14 splendid laurel-leaf points).

Seine-et-Oise.—L'Isle Adam.

Somme.—Belloy-sur-Somme; Boutillerie-les-Amiens; Bultel and Tellier pits at Saint-Acheul; Conty, valley of the Selle; Étouvy pit, at Montières.

Tarn.—Roset.

Tarn-et-Garonne.—Montastruc.

Yonne.—Arcy-sur-Cure (Grotte des Fées).

Germany

Bavaria.—Klause; Ofnet (Grosse and Kleine).

Württemberg.—Bockstein; Cannstatt loess station; Sirgenstein.

Hungary

Budapest.—Kiskevély cave.

Bükk Mountains.—Balla cave; Jankovics at Bajot; Puskaporos.

Szinva Valley.—Szeleta cave; Czobanka cave.

Poland

Galicia.—Kozarina; Wierzchow (lower cave).

Russia

Chernigov.—Mezine.

Ukraine.—Cyrill Street, Kief.

Spain

Asturias.—Cave of Covacha de la Peña; rock shelter of Cueto de la Mina.

Santander.—Altamira; Camargo; Castillo; Cobalejos; La Fuente del Francés; Hornos de la Peña; Peña de Carranceja; El Pendo; Quintanal; Riera; Salitré.

AFRICA

Region of Gafsa.—Ain-Brik.

The Magdalenian Epoch

This last of the epochs that might strictly be called Paleolithic corresponds to the closing episode of Pleistocene times. The type station, a rock shelter, deriving its name from the ruins of the monastery at La Madeleine, commune of Tursac (Dordogne), is one of several stations in the valley of the Vézère which were

Fig. 93.   General view of the rock shelter La Madeleine, Dordogne.

This is the type station of the Magdalenian Epoch.   The rock shelter is at the left; the slope of the Vézère River bank at the right; and the ruins of the monastery of La Madeleine in the background above.

explored by Lartet and Christy in 1862 and 1863.   La Madeleine is but a short distance below, and on the same side of the river as Le Moustier, the type station of the Mousterian Epoch.   It is only 25 meters (82 feet) from, and 6 meters (19.17 feet) above, the level of the river.   Its general character and nearness to the river are seen in Figures 93 and 94.   It is especially notable for the number and importance of the portable art objects found in the deposits.

The skill in chipping flint that made possible the laurel-leaf blade and the point with lateral notch at the base disappeared with the Solutreans. The Magdalenian Epoch was a period of regression in so far as the stone industry was concerned; a marked change is to be noted in the lowest Magdalenian layer at such stations as Badegoule, the two Jean-Blancs, Laugerie-Haute, Laussel, and Le Placard. This leads to the inference that the Magdalenians might have been a different race, with less skill in chipping flint t h a n their predecessors had; certainly they were not so particular as were the S o l u t r e a n s in the choice of raw material for the manufacture of their flint implements, w h i c h consisted for the most part of scratchers, microliths, drills, saws, gravers, etc. (Fig. 95). But, on the other hand, undoubted progress is to be noted in the use of bone and reindeer horn, as exemplified in the delicate bone needle with eyehole (Fig. 96), the *sagaie* or javelin point (Fig. 97), and the har-

FIG. 94. DETAIL VIEW OF THE ROCK SHELTER
OF LA MADELEINE.

This station, first explored by Lartet and Christy in 1863, is one of the richest in portable art objects. The deposits consist of three relic-bearing levels—Lower, Middle, and Upper Magdalenian.

poon of bone and reindeer horn. The arts of engraving and fresco reached their apogee during this epoch.

Six phases or levels have been noted in the Magdalenian Epoch. The first four phases may be combined to form the Lower Magdalenian; the fifth and sixth represent the Middle and Upper Magdalenian respectively. The evolution of the javelin point, or *sagaie,* during the first half of the Magdalenian Epoch is especially instructive. The lowest level is characterized by heavy, thick-based, lanceolate forms made of reindeer horn; at the second level, the base becomes flattened; at the third, the base is conical or beveled, and the shaft grooved, the bevel at first being single, and later double; the fourth witnesses a lengthening of the grooves and the appearance of tubercles near the base of the *sagaie*. At this level there appear flat sculpture with contours cut away, and prototypes of the harpoon, the evolution of which was completed during the fifth and sixth phases of the Magdalenian Epoch.

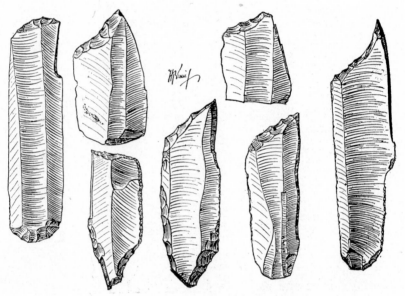

FIG. 95.   FLINT IMPLEMENTS FROM THE LOWER DEPOSITS IN THE CAVE OF LA MAIRIE
AT TEYJAT, DORDOGNE.   LOWER MAGDALENIAN EPOCH.

These are various types of the lateral graver with transverse, oblique, convex, or concave retouching.   They were used by the artists as well as by the makers of tools of reindeer horn, bone, and ivory.   Scale, ⅔.   After Breuil.

In the early stages of its evolution, the teeth, or barbs, of the harpoon were small and numerous, suggesting the jaw of a fish; the material employed was generally bone. This prototype of the harpoon appeared rather early in the Magdalenian of the Pyrenees. If made of reindeer horn, the barbs, reduced in number, were not detached from the shaft; in some cases there was a single row of barbs, in others, a double row (Fig. 98).

Beginning with the fifth phase of the Magdalenian Epoch, a distinct change is to be noted. Harpoons with a single lateral row of

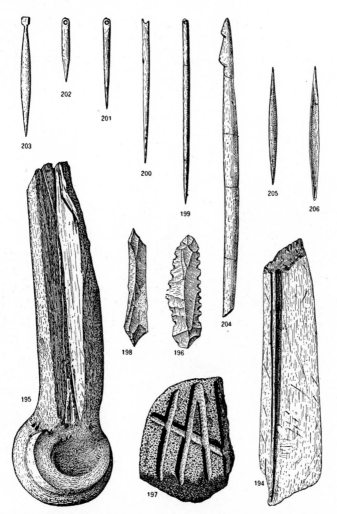

FIG. 96. MAGDALENIAN BONE NEEDLES AND AWLS OR FISHHOOKS WITH MATERIAL
AND TOOLS FOR THEIR MANUFACTURE.

Crude bone needles first appeared in Aurignacian deposits, were rare in Solutrean,
and became abundant in Magdalenian deposits. No. 194, bone from which splinters
have been removed (Laugerie–Basse); No. 195, canon bone of a horse (Laugerie–Basse);
No. 196, dentate flint blade for shaping the cylindrical shaft of bone needles (Bruniquel);
No. 197, gritstone employed in the polishing of needles (Massat); No. 198, flint flake
pointed at both ends, employed to produce the eyehole in needles (La Madeleine);
Nos. 199, 201, 202, bone needles from La Madeleine; No. 200, bone needle from the cave
of Les Eyzies; No. 203, bone needle with head (Massat); No. 204, needle of reindeer
horn with notches in which to fasten a tendon (La Madeleine); No. 205, bone awl or
fishhook from La Madeleine; No. 206, bone awl or fishhook from the cave of Les Eyzies.
Scale, ⅔. After de Mortillet.

barbs and with an enlargement near the base predominate. The
barbs are well defined, although at first rather small and closely
serrated; as the number decreases, the barbs become longer and

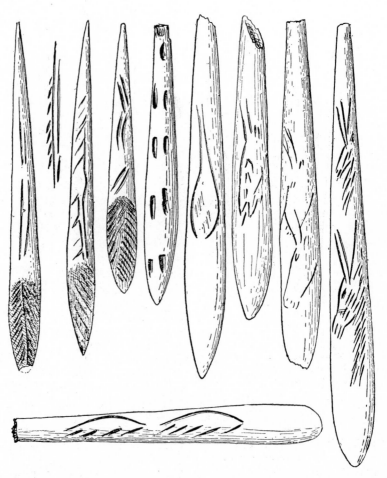

Fig. 97. Early Magdalenian Javelin Points from the Cave of Le Placard,
Charente.

The Upper Paleolithic races decorated many of their weapons with engravings of
the animals they wished to kill for food. The beveled base of the point was attached to a
shaft. After Breuil.

more hooked (Fig. 99). During the sixth phase, harpoons of reindeer horn with a double row of lateral barbs take the lead; there
are two rather distinct and probably successive types. The more

elegant of these, with long, sharp, recurrent barbs not unlike those
of the preceding epoch, are believed by Breuil to be the earlier

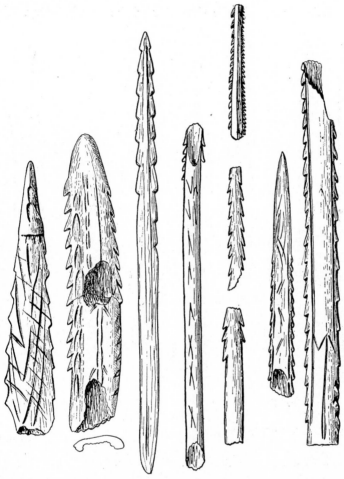

FIG. 98. FIRST STEP IN THE EVOLUTION OF THE HARPOON: PROTOTYPES OF THE
HARPOON WITH TWO ROWS OF LATERAL BARBS. LOWER MAGDALENIAN EPOCH,
FOURTH PHASE.

The Magdalenian Epoch has been divided into six phases; the first four are included
in the Lower, the fifth is the equivalent of the Middle, and the sixth is coextensive with
the Upper Magdalenian. The evolution of the harpoon took place during the last three
phases. Scale, ½ (with the exception of No. 1). After Breuil.

(Fig. 100). The later type, with more or less quadrangular trape-
zoidal barbs, is less artistic and often overloaded with incised

ornament. With this type appeared the flat harpoons, the first indication of the oncoming of the Azilian Epoch.

Synchronous with the progressive development of the harpoon were certain changes in the character of the flint industry. Asso-

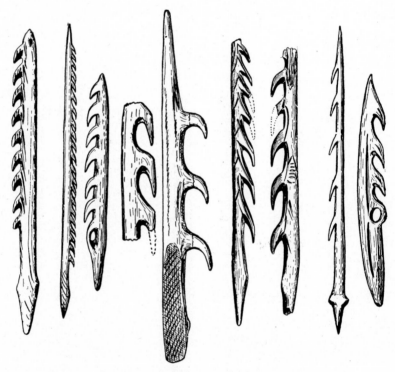

FIG. 99. SECOND STEP IN THE EVOLUTION OF THE HARPOON.
MIDDLE MAGDALENIAN EPOCH.

Beginning with the fifth phase of the Magdalenian Epoch, harpoons of reindeer horn with a single row of lateral barbs and an enlargement near the base predominated. At first the barbs were small and closely serrated; as the number decreased, the barbs became longer and more hooked. Scale, ½. After Breuil.

ciated with the harpoon with a single row of barbs were flint implements recalling those of Upper Aurignacian age; with the harpoon with a double row of well developed barbs were the parrot-beaks,[5] a kind of hooked graver, which, so far as appearance goes, might have been derived from the lateral or the busked graver of the early Aurignacian (Fig. 101).

---

[5] Named by Salmon, *bec de perroquet.*

Although the earliest dart throwers and perforated batons [6] may have antedated the Magdalenian, many of the best examples are undoubtedly of Magdalenian age. Since they are, for the most part, ornamented, they will be discussed in the chapter on Paleolithic art. The purpose served by the baton has been the subject of much conjecture—skull crusher, straightener for the shafts of arrows, bridle, scepter, magic wand, the prototype of the safety

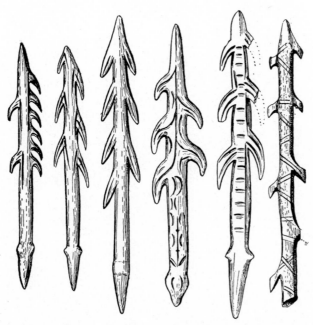

FIG. 100.   THIRD STEP IN THE EVOLUTION OF THE HARPOON.
UPPER MAGDALENIAN EPOCH.

During the sixth phase of the Magdalenian Epoch, harpoons with a double row of long, sharp lateral barbs appeared. They were made of reindeer horn. Scale, ½. After Breuil.

pin, etc. The distribution of the baton is almost as wide-spread as Magdalenian culture; it is found from the Pyrenees northward to Belgium and eastward to Moravia. The age of the baton from Lacave cannot be determined with certainty as it was not in stratigraphic position; it may be either Solutrean or early Magdalenian.

There is no longer any doubt concerning the identity and use of

---

[6] First called *bâton de commandement* by Edouard Lartet.

the so-called dart thrower, since implements of a similar nature
are in use to-day among certain tribes of South America, the
Eskimo, and the Australians.   The dart thrower was also known
to the ancient races of Mexico, Florida, the pueblo region of
southern Utah and northern Arizona, and Central America.   It
consists of a shaft with a handle at one end and a crochet at the
other; it serves to lengthen the arm and to give greater speed to the
dart in its flight.

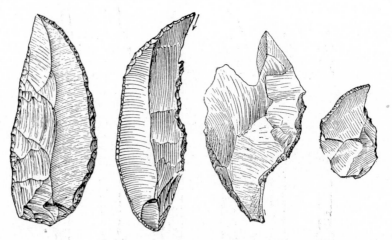

FIG. 101.   FLINT GRAVERS OF THE PARROT-BEAK TYPE FROM LAUGERIE-HAUTE,
DORDOGNE, AND SORDES, LANDES.   UPPER MAGDALENIAN EPOCH.

The parrot-beak graver (Nos. 2 and 4) might easily have been derived from the
Aurignacian Châtelperron blade (No. 1), by the removal of a single flake. Scale, ⅔.
After Breuil.

One of the first dart throwers to be found was mistaken for a
poniard handle with the blade missing; later, it was recognized as
a dart thrower with the handle missing.   The base of the spear,
or projectile, rested against the tip of the upturned tail of the
mammoth (Fig. 102).   This specimen came from the rock shelter
of Montastruc at Bruniquel (Tarn-et-Garonne); it is now in the
British Museum.   Another early discovery, from Laugerie-Basse,
was figured by Lartet and Christy as a harpoon head with a single
barb; later, one from the same station was figured and described
by them as a curved sticklike implement of reindeer horn.   They
made no mention of the short, but distinct, hook or crochet, but
did call attention, and no doubt rightly, to the twenty-four parallel

oblique notches arranged in two series immediately below the hook; these tally marks are a mute witness to the efficiency of the dart thrower in the hands of a skillful hunter (Fig. 103). A splendid example of the dart thrower, carved in ivory and ornamented with a figure in the round of a hyena, was discovered at La Madeleine in 1912 (Fig. 104). It is now in the Saint-Germain Museum. A complete list of stations in which dart throwers

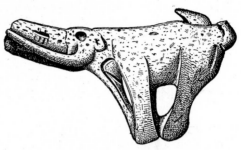

FIG. 102.   MAGDALENIAN DART THROWER OF REINDEER HORN REPRESENTING A MAMMOTH IN THE ROUND.

From the rock shelter of Montastruc at Bruniquel (Tarn-et-Garonne).   The trunk, now broken, served as the handle; the base of the dart or javelin shaft rested against the crochet just above the root of the mammoth's tail.   To the Magdalenian hunters, effigy dart throwers might have possessed efficacious, even magic, qualities.   Scale, ½. After de Mortillet.

have been found is given at the end of this chapter.

The bone needle occurred for the first time, though but rarely, in the Upper Aurignacian Epoch. Somewhat better, but still rare, examples are found in a few Solutrean deposits, as at Lacave. The Magdalenian was the epoch *par excellence* of the bone needle. Needles are suggestive of clothing, as are also buttons of bone and ivory. These buttons vary in shape from flat disks to spool-shaped forms; some are decorated with incised lines and even engraved animal figures (Fig. 105). Objects of similar shape and size might also have been used as toggles; there can be but little doubt that large perforated animal teeth were used as such. Buttons have been reported from

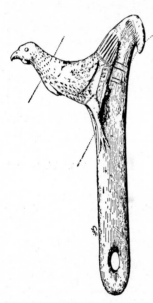

FIG. 103.   MAGDALENIAN DART THROWER OF REINDEER HORN FROM MAS D'AZIL, ARIÈGE.

The portions outside of the dotted lines have been restored by the Abbé Breuil, who believes the figure was intended to be a grouse or ptarmigan.   Scale, ca. ⅓. After Breuil.

Laugerie-Basse, Raymonden, and Combarelles in Dordogne; Gourdan (Haute-Garonne); Enlène (Ariège); and Brünn (Moravia).

Magdalenian fauna is represented in the art of the epoch as well as in the faunal remains. The reindeer and the horse are dominant. One finds, to a lesser degree, the saiga antelope, chamois, musk ox, glutton, arctic fox, and lemming (*Myodes obensis* and *M. torquatus*). The red deer (*Cervus elaphus*) is plentiful, especially in the Cantabrian region. Remains of the mammoth and woolly

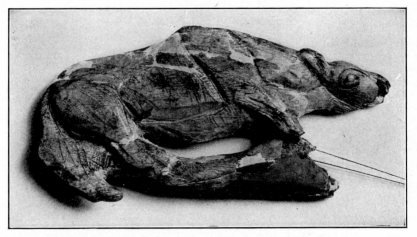

FIG. 104. MAGDALENIAN DART THROWER OF IVORY REPRESENTING A HYENA, FROM LA MADELEINE, DORDOGNE.

The crochet is at the tip of the tail; the end of the handle has been lost. Photograph from the museum at Saint-Germain.

rhinoceros occur in the Lower Magdalenian but disappear completely before the close of the epoch.

To the foregoing may be added: *Bison priscus, Castor fiber; Cervus capreolus, Capra ibex, Canis lupus, C. vulpes; Arvicola amphibius, A. ratticeps, A. terrestris, A. nivalis, Felis leo, F. lynx, F. catus, Meles taxus, Mustela martes, M. vulgaris, Lutra vulgaris, Lepus variabilis, Lagomys pusillus, Lagopus albus, L. alpinus, Mus musculus, M. sylvaticus, Arctomys marmotta, Sciuris vulgaris, Spermophilus rufescens, Sorex vulgaris, Talpa europaea,* etc.

GEOGRAPHIC DISTRIBUTION OF MAGDALENIAN CULTURE

The Magdalenian is the only Upper Paleolithic culture that has been reported from Switzerland and Siberia. The Magdalenians were fortunate in that they were able not only to hold the ground previously occupied by the Aurignacians and Solutreans, but were ready to take advantage of the retreating ice of the Würm glaciation by penetrating to higher altitudes and higher latitudes. Magdalenian stations stretch all the way from Spain to Irkutsk (Fig. 106). Naturally, a primitive culture distributed over such a wide area must undergo regional variation. In northern Africa one encounters a Getulian facies, and in the Yenisei valley, a distinct Siberian facies, while in Italy there is a Magdalenian equivalent [7] in which the reindeer plays no part.

It is the opinion of Breuil that Magdalenian culture might have originated in the French Pyrenees as a more or less direct outgrowth of the Aurignacian culture still persisting there, while the Solutrean wave passed to the northward. It should be recalled that the Solutrean is lacking in the French Pyrenees with the exception of the Grotte des Harpons and Gourdan (Haute-Garonne). It is also probable that some of the elements of Magdalenian culture may have come from central Siberia. It is, however, fairly safe to assume that the strength of the cultural threads which hold together the successive Upper Paleolithic epochs, is accounted for, in a measure at least, by a certain degree of racial contiguity.

FIG. 105.   MAGDALENIAN BONE BUTTON AND TOGGLE FROM LAUGERIE-BASSE, DORDOGNE.

Scale, ⅗. After de Mortillet.

---

[7] Pigorini is of the opinion that the Magdalenian Epoch is not represented in Italy because objects of reindeer horn and ivory are lacking. Mocchi, on the contrary, recognizes an atypic Magdalenian industry which he would class as the equivalent of the final stage of the Magdalenian as at Remouchamps (Belgium), for example.

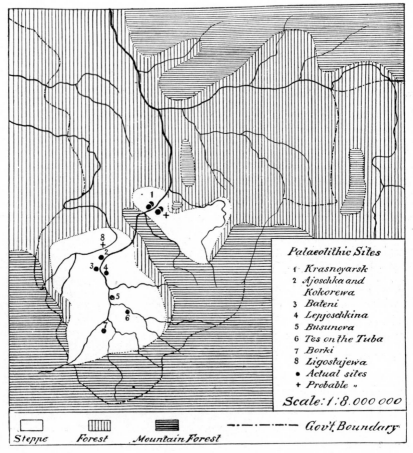

FIG. 106. MAP SHOWING THE LOCATION OF MAGDALENIAN LOESS STATIONS IN THE YENISEI VALLEY, SIBERIA.

GEOGRAPHIC DISTRIBUTION OF MAGDALENIAN CULTURE

EUROPE

Austria

Lower Austria.—Gudenus cave.

Belgium

Liège.—Le Docteur; Engis; Fond-de-Fôret; Remouchamps, near Spa.

Limbourg.—Zonhofen (a curious industry consisting largely of small chipped pebbles, which has been referred to the Upper Magdalenian by Breuil and Cartailhac).

Luxembourg.—Le Coléoptère; Martinrive.

Namur.—Le Chêne; Furfooz (Chaleux, Frontal, Nutons, and Reauviau); Goyet; Naulette.

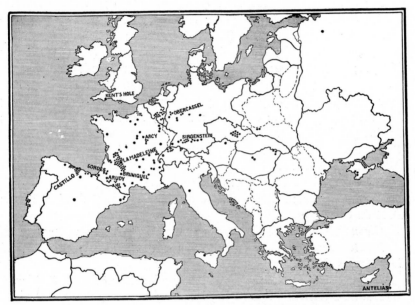

Fig. 107. Map showing the geographic distribution of Magdalenian culture in Europe and the Mediterranean basin.

Czechoslovakia

Moravia.—Caves of Balcarovaskala, Býčískála, Čertova-dira, Kostelik, Kříž, Kulna, Lautsch, Pekarna near Mokrau, Šipka, and Zitny.

England

Devonshire.—Kent's Hole.

France

Ain.—La Colombière; Les Hoteaux.

Ardèche.—Le Figuier.

Ariège.—Enlène; Mas d'Azil (right bank); Massat; Montfort; Trois-Frères; Tuc d'Audoubert; cave of La Vache.

Aude.—Bize; La Crouzade.

Basses-Pyrénées.—Isturitz; Saint-Michel d'Arudy.

Charente.—Ammonite; Bois du Roc; La Chaise; Gavechou; Papeterie; Le Placard.

Cher.—Bellon.

Côte d'Or.—Le Poron des Cuèches.

Dordogne.—Badegoule; Bout-du-Monde, near La Madeleine; Cap-Blanc; Champs-Blanc; Le Cheval, near Brantôme; Combarelles; Les Eyzies (cave of Les Eyzies and rock shelter of the Château); Font-de-Gaume; Gorge d'Enfer (Grotte d'Abzac); Laugerie-Basse (La Grange, Marseilles); Laugerie-Haute; Limeuil (both stations); Liveyre; Longueroche; Le Luc (rock shelter between Verzac and La Roque-Gajlac); La Madeleine; La Mouthe; Pageyral (rock shelter, near Saint-Cyprien); Peyrille; Raymonden; Les Rebières (Recourbie); Les Roches (Delage); Le Ruth; Soucy; La Source rock shelter; Teyjat (La Mairie and Mège).

Drôme.—Bobache.

Eure.—Metreville.

Finistère.—Roc'h Toul, at Guiclan.

Gard.—Chabot; La Saltpetrière on the Gardon, at the foot of the Pont-du-Gard.

Gironde.—Grotte des Fées, commune of Marcamps; Fontarnaud.

Haute-Garonne.—Gourdan; Lespugue (caves of Les Boeufs and Les Harpons); Spugo.

Haute-Loire.—Rock shelter of Le Rond, at Saint-Arcon-d'Allier.

Hautes-Pyrénées.—Aurensan; Lorthet; Lourdes.

Haute-Saône.—Cavern of La Zouzette, at Farincour⁺

Haute-Savoie.—Le Veyrier.

Hérault.—Cave of Laroque.

Indre.—Saint-Marcel (rock shelter of La Garenne).

Isère.—Caves of La Balme and Bethenas.

Jura.—Arlay; La Mère Clochette; La Vieille-Grand'Mère at Mesnay, near d'Arbois.

Landes.—Rivière; Sordes (Dufaure and Duruthy).

Loire.—Saut-du-Perron, at Villerest.

Loire-Inférieure.—Bégrol, at La Haye-Fouassière.

Loiret.—Marnière de Vilette; Le Muids.

Lot.—Batie; Les Cambous; Conduché; Crozo de Gentillo; Lacave; Murat; Pis de la Vache; Rivière de Tulle; Roussignol.

La Manche.—Bretteville.

Meuse.—La Grosse Roche; La Roche Plate, at Saint-Mihiel; cave of Table du Diable.

Oise.—Saint-Just-des-Marais.

Puy-de-Dôme.—Neschers.

Saône-et-Loire.—Solutré (Crot-du-Charnier and Terre Sève); La Goulaine, commune of La Motte-Saint-Jean (a workshop with a cache of several hundred flint implements).

Savoie.—Grande Gave.

Seine-et-Marne.—Beauregard.

Seine-et-Oise.—Cergy.

Somme.—Abbeville (Mautort); Amiens (Boves, Étouvy, and La Voirie); Long, Mareuil, and Porte Mercade (under the peat).

Tarn-et-Garonne.—Bruniquel (Château rock shelters: Lafaye and Plantade; Les Forges, and Montastruc).

Vienne.—Chaffaud; Les Cottés; Les Fadets.

Yonne.—Arcy-sur-Cure (Grottes des Fées, de l'Homme, and du Trilobite; Trou de l'Hyène); Le Mammouth; La Marmotte; La Roche-au-Loup.

Germany

Baden.—Munzingen.

Bavaria.—Grosse and Kleine Ofnet; Hohlenstein cave, near Nördlingen; Kästlhäng; Klause.

Hesse-Nassau.—Wildweiberlei rock shelter.

Hohenzollern.—Propstfels.

Nassau.—Wildhaus; Wildsheuer.

Rhine.—Andernach; Buchenloch; Kartstein; Obercassel, near Bonn.

Thüringen.—Lindental cave, at Gera.

Württemberg.—Bockstein, near Langenau; Ganserfelsen and Hohlefels, near Schelklingen; Niedernau rock shelter; Rosenstein; Schmiechenfels; Schussenquelle; Sirgenstein; Winterlingen, on the Rauhen Alb.

Hungary

Budapest.—Kiskevély; Pillisszanto cave on the Danube above Budapest.

Bükk Mountains.—Balla cave; Jankovics; Peskö cave; Puskaporos.

Italy

Liguria.—Grimaldi (Barma Grande and Les Enfants).

Palermo.—Cave of Natale.

Tuscany.—Cave of Golino, at Talamone.

Poland

Galicia.—Maszycka; Nowa Alexandria (a station in the open on a terrace of the Vistula).

Russia

Crimea.—Caves at Simferopol.
Novgorod.—Bologoie.

Spain

Asturias.—Balmori (or Quintanal); El Buxu; Cangas de Onis; Collubil; Cueto de la Mina; Fonfria, at Barro; La Paloma; Peña de Candamo; La Riera; Sofoxo.
Barcelona.—Abrich Romani.
Burgos.—Cave of Penches.
Gerona.—Cau de las Goyas (or Cova de las Goyas), near San Julián de Renies-Gerona; cave of Serinyá, at Serinyá-Bañolas.
Guipuzcoa.—Cave of Aitzbitarte (or Landarbasso) at Renteria, near San Sebastian (includes remains of reindeer).
Madrid.—Portazgo tileworks.
Oviedo.—Quintanal.
Santander.—Altamira; Camargo; Castillo; Cobalejos (or Puente-Arce); El Cuco, at Ubiarco-Santillana; La Fuente del Frances; Hornos de la Peña; Nuestra Señora de Loreto, near Peña Castillo; Otero, near Secadura; Peña de Carranceja; El Pendo; Rascaño; Salitré; Truchiro, near Rivamontán al Monte; Valle; Villanueva.
Vizcaya.—Cave of Armiña, near Lequeitio-Guernica (with remains of reindeer); Balzola.

Switzerland

Basle.—Arlesheim, near Basle; Birseck; Lausen and Winzernau (loess stations).
Bern.—Bellerive, a loess station 4 kilometers (2.5 miles) from Liesberg; cave of Liesberg, between Delémont (Delsberg) and Laufen.
Olten-Soloturn.—Käsloch, Hard, and Muhleloch (all near Olten).
Schaffhausen.—Freudental; Kesslerloch; Schweizersbild.
Vaud.—Cave of Scé, near Villeneuve.

AFRICA

Northern Africa affords traces of an industry comparable and contemporaneous with the Magdalenian of Europe.

ASIA

China

Father Teilhard de Chardin reports from Shensi an industry with a facies similar to that in the Yenisei valley, but probably belonging to the Mousterian Epoch.

India

Bengal Presidency.—District of Mirzapur (Capsian flints and cavern paintings).

Madras Presidency.—Karnul (caverns); Singapur (rock paintings resembling those at Cogal, Spain).

Siberia

Irkutsk.—Wercholensk mountain, near Irkutsk (reported by B. E. Petri).

Yenisei.—Loess stations of Afontova, Kirpitschnye Sarai, Peresselentscheskij Point, and Woennyi, all near Krasnoyarsk; Bateni; Busunova; Lepjoschkina.

Syria

Phoenicia.—Antelias.

GEOGRAPHIC DISTRIBUTION OF MAGDALENIAN HARPOONS

Belgium

Coléoptère (Luxembourg)   Goyet (Namur)

England

Kent's Cavern (Devon)

Czechoslovakia

Kostelik (Moravia)   Pekarna or Diravica (Moravia)

France

Ammonite (Charente)
Arlay (Jura)
Arudy (Basses-Pyrénées)
Aurensan (Hautes-Pyrénées)
Batie (Lot)
Bruniquel (Tarn-et-Garonne)
Cambous, Les (Lot)
Combarelles (Dordogne)
Coual (Lot)
Conduché (Lot)
Eyzies, Les (Dordogne)
   Abri du Château
   Grotte des Eyzies
Gourdan (Haute-Garonne)
Harpons, Les (Hautes-Garonne)
Isturitz (Basses-Pyrénées)
Lacave (Lot)
Laugerie-Basse (Dordogne)
Laugerie-Haute (Dordogne)

Longueroche (Dordogne)
Lorthet (Hautes-Pyrénées)
Madeleine, La (Dordogne)
Mas d'Azil (Ariège)
Massat (Ariège)
Montfort (Ariège)
Montgaudier (Charente)
Mouthiers (Charente)
Murat (Lot)
Raymonden (Dordogne)
Rivière-de-Tulle (Lot)
Roussignol (Lot)
Sainte-Eulalie (Lot)
Saint Marcel (Indre)
Sordes (Landes) rock shelter of Dufaure
Souci (Dordogne)
Teyjat (Dordogne)
Vache, La (Ariège)

Germany

Andernach (Rhine)
Kästlhäng (Bavaria)

Klause (Bavaria)
Schussenquelle (Schussenried)

Siberia

Wercholensk Mountain (Irkutsk)

Spain

Castillo (Santander)
Mina, La (Asturias) (stag horn)
Morin (Santander)

Rascaño (Santander)
Riera, La (Asturias)

Switzerland

Kesslerloch (Schaffhausen)

Schweizersbild (Schaffhausen)

GEOGRAPHIC DISTRIBUTION OF BATONS (UPPER PALEOLITHIC)

Austria

Gudenus (Lower Austria)

Belgium

Goyet (Namur)

Czechoslovakia

Kulna (Moravia)   Předmost (Moravia)

England

Gough's Cave (Somerset)   Kent's Cavern (Devon)

France

Arudy (Basses-Pyrénées)
   Cave of Saint-Michel
Aurignac (Haute-Garonne)
Badegoule (Dordogne)
Bout-du-Monde, La (Dordogne)
Bruniquel (Tarn-et-Garonne)
Cap-Blanc (Dordogne)
Chaffaud (Vienne)
Conduché (Lot)
Crozc de Gentillo (Lot)
Eyzies, Les (Dordogne), Abri du
   Château
Fées, Les (Gironde)
Ferrassie, La (Dordogne)
Figuier, Le (Ardèche)
Fontaine, La (Meurthe-et-Moselle)
Gargas (Hautes-Pyrénées)
Gorge d'Enfer (Dordogne), Abri du
   Poisson
Gourdan (Haute-Garonne)
Hoteaux, Les (Ain)
Isturitz (Basses-Pyrénées)
Lacave (Lot)
Laugerie-Basse (Dordogne)
Laugerie-Haute (Dordogne)

Limeuil (Dordogne)
Liveyre (Dordogne)
Lourdes (Hautes-Pyrénées)
Madeleine, La (Dordogne)
Mas d'Azil (Ariège)
Massat (Ariège)
Montgaudier (Charente)
Placard, Le (Charente)
Pis de la Vache (Lot)
Raymonden (Dordogne)
Roc, Le (Charente)
Roches, Les (Dordogne), Abri Blan-
   chard
Roussignol (Lot)
Solutré (Saône-et-Loire)
Sordes (Landes), rock shelter of
   Dufaure
Soucy (Dordogne)
Spugo (Haute-Garonne)
Teyjat (Dordogne), rock shelter of
   Mège
Trois-Frère (Ariège)
Tuc d'Audoubert (Ariège)
Veyrier, Le (Haute-Savoie)

Germany

Klause (Bavaria)   Schussenquelle (Schussenried)
Munzingen (Baden)

Italy

Grimaldi (Liguria), Cavillon cave

Poland

Maszycka (Galicia)

Spain

Altamira (Santander)
Camargo (Santander) (staghorn)
Castillo (Santander) (staghorn)

Mina, La (Asturias) (staghorn)
Paloma, La (Asturias) (staghorn)
Valle (Santander) (staghorn)

Switzerland

Kesslerloch (Schaffhausen)

Schweizersbild (Schaffhausen)

Siberia

Afontova (Yenisei Government)

Geographic Distribution of Bone and Ivory Needles
(Magdalenian for the most Part)

Austria

Gudenus (Lower Austria)

Belgium

Blaireaus, Les (Namur)
Chaleux (Namur)
Coléoptère (Luxembourg)

Furfooz (Namur)
Goyet (Namur)

Czechoslovakia

Býčískála (Moravia)
Kostelik (Moravia)
Křiž (Moravia)

Kulna (Moravia)
Pekarna, or Diravica (Moravia)
Zitny (Moravia)

England

Creswell Crags (Derby), Church Hole   Kent's Hole (Devon)

France

Ammonite (Charente)
Arcy-sur-Cure (Yonne), Grotte du
    Trilobite
Badegoule (Dordogne)

Balutie, La (Dordogne)
Batie (Lot)
Bethnas (Isère)
Bize (Aude)

Boeufs, Les (Haute-Garonne)
Brassempouy (Landes) (Upper Solu-
trean)
Bruniquel (Tarn-et-Garonne)
Cambous (Lot)
Cap-Blanc (Dordogne)
Chaffaud (Vienne)
Conduché (Lot)
Coual (Lot)
Crozo de Gentillo (Lot)
Combe-Cullier
Enlène (Ariège)
Eyzies, Les (Dordogne)
Fées, Les (Gironde)
Font-de-Gaume (Dordogne)
Fontaine, Trou de la (Meurthe-et-
Moselle)
Gourdan (Haute-Garonne)
Hoteaux, Les (Ain)
Lacave (Lot) (Upper Solutrean)
Laforge (Lot)
Laugerie-Basse (Dordogne)
Laugerie-Haute (Dordogne)

Liveyre (Dordogne)
Lorthet (Hautes-Pyrénées)
Madeleine, La (Dordogne)
Marsoulas (Haute-Garonne)
Mas d'Azil (Ariège)
Massat (lower cave) (Ariège) (needle
with head instead of eye)
Montgaudier (Charente)
Mouthiers (Charente)
Murat (Lot)
Placard, Le (Charente)
Pis de la Vache (Lot)
Raymonden (Dordogne)
Roches, Les (Dordogne)
Rivère-de-Tulle (Lot)
Saint-Marcel (Indre)
Soucy (Dordogne)
Spugo (Haute-Garonne)
Teyjat (Dordogne)
    Grotte de la Mairie
    Rock shelter of Mège
Trois-Frères (Ariège)
Veyrier, Le (Haute-Savoie)

Germany

Andernach (Rhine)
Kästlhäng (Bavaria)
Klause (Bavaria)

Obercassel (Rhine)
Propstfels (Hohenzollern)

Hungary

Jankovics (Bükk Mountains)

Russia

Simferopol (Crimea)

Mezine (Ukraine)

Spain

Altamira (Santander)
Castillo (Santander)

Paloma, La (Asturias)
Valle (Santander)

Switzerland

Freudental (Schaffhausen)
Kesslerloch (Schaffhausen)

Schweizersbild (Schaffhausen)

Geographic Distribution of Dart Throwers (Upper Paleolithic)

France

Arudy (Basses-Pyrénées)
Bruniquel (Tarn-et-Garonne)
  Abri du Château
  Abri de Montastruc
  Grotte des Forges
Conduché (Lot)
Enlène (Ariège)
Gourdan (Haute-Garonne)

Laugerie-Basse (Dordogne)
Laugerie-Haute (Dordogne)
Lorthet (Hautes-Pyrénées)
Lourdes (Hautes-Pyrénées)
Madeleine, La (Dordogne)
Mas d'Azil (Ariège), station on right
  bank

Switzerland

Kesslerloch (Schaffhausen)

CHAPTER VII

PALEOLITHIC ART

The Evolution of Paleolithic Art

No field in the whole domain of prehistoric archeology has attracted more attention than that of Quaternary art; and rightly so, since its appearance marks a distinct epoch in mental evolution. No history of art can be complete without having for a basis the prehistory of art; and that which is true of art is also true of all the primal lines of human achievement. Greece had for its immediate background Egypt; Rome in turn had Greece; and western Europe, in historic time, has profited by the example of all three. In art as well as in all things else we are largely the children of the past, inheriting from a long line of ancestral ages. In addition, thanks to modern discovery and invention, we are in more or less intimate contact with collateral lines of art development.

Progress, orderly development, evolution in any branch of human endeavor imply a starting point. The light of day is preceded by a period of dawn. It is difficult to conceive of an artist so thoroughly isolated in time and space as to be absolutely devoid of a background of art inheritance. One who could be an artist under such circumstances is certainly worthy of our consideration. It was a cold and unresponsive world on which he looked, with no one to help or understand—man alone with nature, nature untamed, unconquered, unaltered by those ameliorating influences which we are accustomed to think of as cultural environment.

Who this dawn artist was, where and when he lived, and how he solved, one by one, the riddles of art, are questions which admit of approximately correct answers, thanks to the imperishable nature of the records, and to accidental discoveries as well as diligent, well directed search in valley deposits, caves, and rock shelters. According to the records as at present laid bare, the dawn man, and per-

haps also the dawn artist, lived not in Egypt or Greece, but in
western Europe. His personal appearance and anatomical characters can be judged of only from the skeletal parts that by good fortune have escaped the ravages of time. The oldest known skeletal
remains are never found associated with works of art, hence the
races they represent were not artists in the strict sense of the word.

It is true that a careful search in the gravel beds dating from
that early period will often bring to light flints that fortuitously
or otherwise resemble animal forms familiar to man. One may

find, for example, a flint
nodule in the shape of a
bird with the nodular crust
intact; such a piece is certainly a *lusus naturae*.
When, however, one finds
a flint nodule representing
a bird's head with the
nodular crust intact except
for the little chip taken out
e x a c t l y where the eye
should be, there are those

FIG. 108. FIGURE STONE SUGGESTING THE
HEAD OF A BIRD.

Effigy figures of this kind have been found in
sand and gravel deposits at various places; whether
early man was attracted by them and made attempts
to heighten the zoömorphic effect by chipping is still
an open question. From the Dharvent collection.

who would see in it the element of human intent.
They would say that the
specimen had attracted the
attention of some one who
detected the resemblance to
a bird's head and who knew how to supplement nature's unfinished
work by detaching the small chip to represent the eye, thus heightening the zoömorphic effect (Fig. 108).

Examples of this kind might be multiplied indefinitely; they
even seem to receive confirmation in a measure through numerous
authentic examples dating from subsequent periods, to which we
may have occasion to refer later. The reasonableness of the contention is likewise supported by our own experience; for who, by
gazing into the fire burning on his own hearthstone, or into the
clouds, has not in fancy clothed even evanescent forms with life?
Early man might well have been attracted by suggestive natural

forms just as we are to-day. Imagination plays an important rôle in art, and it must have been already active at the very beginning of human culture. On the other hand, it is the part of wisdom to keep in the background these so-called figure stones (*pierres-figures*) until their authenticity shall have been established. If these are set aside for the present, at least, as questionable, we have to pass over a long lapse of time before coming to indubitable works of art. It would hardly be fair, however, to dismiss the problem of these figure stones without mention of two finds, one of which dates from the Paleolithic and the other from the Neolithic Period.

A flint nodule retouched to resemble the head of a monkey was found by Isaie Dharvent *in situ* associated with two large Mousterian flint scrapers in a gravel pit at Roellecourt (Pas-de-Calais) and exhibited before the Congrès préhistorique de France at Vannes in 1906.

In 1919 Baudouin reported his discovery of about a dozen figure stones in a station dating from the lower Neolithic Period at Saint-Gilles-sur-Vie (Vendée). These instances of figure stones associated *in situ* with human industrial remains have done much to strengthen the contention of such enthusiastic collectors as Thieullen, Dharvent, Bertin, Perrot, Newton, *et al.* Mention should also be made of the Egyptian examples noted by Jean Capart and Flinders Petrie. Associated with a number of ape figurines made of ivory and terra cotta, and blocks of stone roughly worked to represent the baboon, Petrie found in the prehistoric Egyptian temple at Abydos an unworked flint nodule which appeared to him to have been saved because of its likeness to a baboon. No other flint nodules were found in the temple area, and this one must have been brought (a mile or more) from the desert. Since it was placed with the rude stone figures of baboons, Petrie believed it to have been a primitive fetish stone picked up because of its resemblance to the animal in question. Capitan and Peyrony report a figure stone from the Upper Aurignacian level of the small cave at La Ferrassie (Dordogne).

During ages long subsequent to the time when the races of Piltdown and of Heidelberg lived, there spread over the greater part of Europe the primitive Neandertal race, of coarse mental and physical fiber, in whose breast the artistic impulse throbbed but

feebly.   This race contributed nothing, in fact, save utilitarian artifacts, the so-called Mousterian industry.

The first appearance in Europe of what we are accustomed to
call the decorative arts, and even the fine arts, is coincident with
the appearance of a new race, the Aurignacian, which supplanted
completely the archaic Neandertal race of Mousterian times.
Physically and mentally the Aurignacians, as represented by the
skeletons of Cro-Magnon and Combe-Capelle, were more nearly
akin to modern European races than to the old Mousterians.   Culturally, the differences were at once as great as to make it very
difficult to conceive of the Aurignacian as having been a direct
outgrowth of the Mousterian Epoch.

The thickness of the Aurignacian deposits from caves and rock
shelters and the evolution of the culture there portrayed prove the
epoch to have been a long one.   Many Aurignacian loess stations
have recently come to light, making it possible to determine approximately, at least, the relation of the Aurignacian Epoch to
glacial chronology.   Aurignacian remains occur in the upper part of
the recent loess, which is assigned to the Würm, or last, Glacial
Epoch.   Moreover, in the cave deposits at Sirgenstein (southern
Germany) and elsewhere, R. R. Schmidt has found immediately
below the oldest Aurignacian layers an arctic fauna characterized
by *Myodes obensis,* a species of lemming.   The Aurignacian began,
therefore, very near the maximum of the last glacial epoch.

We can thus picture the climatic conditions that attended the
birth of Quaternary art in central and western Europe; and climate
is no mean factor in the environment of primitive man.   The great
continental ice sheet covered practically the whole of Scandinavia
and extended into northern Germany.   The northern part of Great
Britain was also under ice, and there was a considerable extension
of the Alpine and Pyrenean glaciers.

Upper Quaternary fauna may be reconstructed from the fossil
remains associated with human cultural remains; it is also reflected
in the art of the time.   Judging from both these sources, one
arrives at the conclusion that the Aurignacians, and the Solutreans
who succeeded them, were contemporaries of an *Equus* fauna with
the horse predominating, the mammoth still abundant, the bison
plentiful, and the reindeer gaining in prominence.   The horse and

reindeer were dominant in the third and last great art epoch of Paleolithic times, the Magdalenian. *Bos primigenius,* representing an ancestral race of cattle, played an important rôle in the art of the time and is also not inconspicuous for its fossil remains. On the other hand, the one station of Solutré (Saône-et-Loire) has furnished skeletal remains of no less than 100,000 horses. Moreover, in the inventory of Quaternary art, representations of the horse are the most numerous. We are therefore justified in assuming that the steak of the horse, and not our indispensable beefsteak, was the *pièce de résistance* at all well regulated Upper Paleolithic feasts.

The fine arts and the love of ornament seem to have developed at the same time; for both in graves and elsewhere are found bone and ivory pendants as well as perforated shells and animal teeth

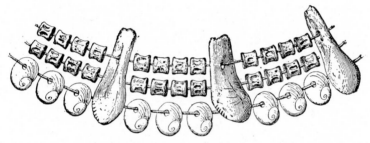

FIG. 109.   UPPER PALEOLITHIC NECKLACE FROM BARMA GRANDE, ITALY, ONE OF THE GRIMALDI CAVES.

This necklace is composed of a happy combination of canines of the stag, fish vertebrae, and shells of *Nassa neritea;* it was found with the skeleton of a young man.   Scale, $\frac{4}{5}$. After Verneau.

that were evidently used as necklaces and otherwise (Fig. 109). In the cave of La Combe (Dordogne), excavated by the Peabody Museum of Yale University during the summer of 1912, was found a human lower molar tooth perforated for suspension as an ornament, the first example of its kind to be reported. At the same level (Aurignacian), we also found perforated animal teeth and one that was grooved to serve as a pendant (see Figs. 69 to 72). Certain female figurines dating from the Aurignacian Epoch are represented as wearing bracelets. The practice of painting or tatooing the body was no doubt common among the cave dwellers.

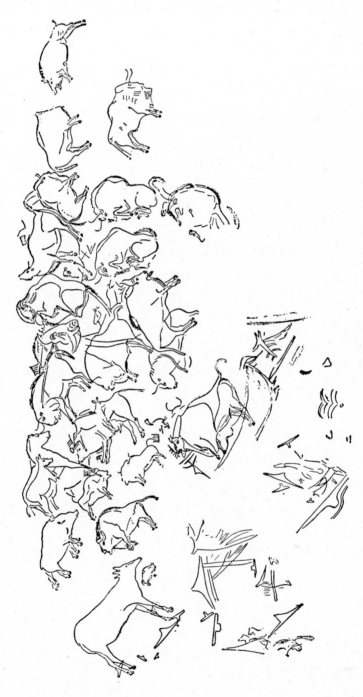

FIG. 110. OUTLINE FIGURES OF THE FRESCOES ON THE CEILING OF THE CAVERN OF ALTAMIRA, SANTANDER, SPAIN.

These frescoes, discovered by Sautuola in 1879, were the first examples of Paleolithic mural art to be recognized and reported as the work of primitive man. Length 14 meters (46 feet). After Breuil.

FIG. 111. ONE OF THE BISONS IN FRESCO ON THE CEILING OF THE CAVERN OF ALTAMIRA, SANTANDER, SPAIN.

The long, slender, sharp horns of the extinct European bison are faithfully portrayed. In this, one of the finest examples of Paleolithic polychrome art, four shades of color are used. Length from front to hips, 1.50 meters (5 feet). After Cartailhac and Breuil.

Paleolithic art objects may be classed under two heads, portable and stationary. The portable class is found in the floor accumulations of caves and rock shelters as well as in valley deposits. It consists in part of decorated tools, weapons, and ceremonial objects, art playing perhaps a supplementary rôle to utility. It also includes engraved pebbles as well as carved fragments of stone, ivory, bone, and the horn of reindeer and stag—in fact, almost anything that could be seized upon to satisfy the exuberant demands of the cave man's artistic impulse. The stationary works of art are those which embellish the walls and ceilings of caverns and rock shelters; in rare instances the clay of the cavern floor was utilized for modeling and sketching purposes.

The scientific world has been more or less familiar with the portable class of Paleolithic art objects since 1863.[1] Our acquaintance with Quaternary mural art is, however, of more recent date. The first discovery was by Sautuola at Altamira, northern Spain, in 1879. Inspired by what he had seen at the Paris Exposition of 1878, Sautuola was searching in the floor deposits of Altamira for relics of ancient man. His small daughter, who had accompanied him, scanned by chance the low ceiling over her head. In the dim candlelight, her eye caught the unmistakable outlines of a strange beast painted in fresco. A cry of surprise brought her father, who soon discovered the other figures on this now celebrated ceiling (Figs. 7 and 110). Sautuola divined from the first the true significance of this remarkable artistic display and published privately a pamphlet on the subject the following year. Not prepared for such a startling innovation, the scientific world remained skeptical. Nearly twenty years later similar discoveries by Chiron, Daleau, and Rivière in France brought to the Spanish savant tardy, even posthumous, but nevertheless complete vindication. One of the reasons for the delay in accepting Sautuola's contention was the failure of the specialists to discern that the bisons (Fig. 111) on the Altamira ceiling have the characters of a fossil species, with long sharp horns and a tail hairy near the root. According to Düerst, the paleontologist, the bison horn has gradually diminished

---

[1] The results of the work of Lartet and Christy were published in the *Revue Archéologique* for April, 1864. The previous discoveries of cave art at Chaffaud and Le Veyrier had not been understood (until 1861).

in length since the time of *Bison priscus*. Abundant in the Lower Magdalenian, the bison grew gradually more and more scarce during the Middle and Upper Magdalenian.

It has been possible to trace the evolution of Quaternary mural art chiefly for two reasons, namely, its relation to the floor deposits and the superposition of mural figures. The age of relic-bearing floor deposits is determinable by the relics themselves. It often happens that mural art is found to be covered by accumulations on

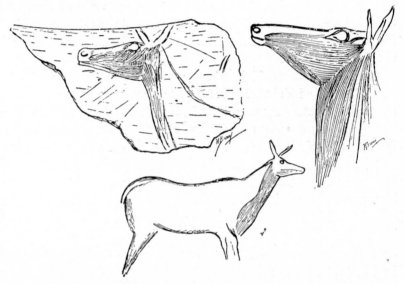

Fig. 112. Engraving of a hind on bone from the cavern of Altamira compared with two mural engravings of the hind from the cavern of Castillo, Santander, Spain.

The engraving on bone was found in deposits of the early Magdalenian Epoch; the two mural engravings may be referred to the same epoch since they represent the same technique and stage of evolution in the art of engraving. By comparisons of this sort much of the mural art may be dated. After Breuil.

the floor of the cavern; mural art in such a case is older than the deposit which covers it. Thus at Pair-non-Pair, rude, deeply engraved wall figures were completely lost to view beneath deposits of Upper Aurignacian age. The engravings are, therefore, anterior to the Upper Aurignacian; they represent the first or oldest phase of engraving. At La Grèze a wall engraving was buried beneath a deposit of Solutrean age. It, also, belongs to the first phase. A mural fragment may become detached, fall to the floor and be buried, thus approximately dating that which remains on the wall. Again, the similarity of art objects from the floor deposits to the mural art may serve to date the latter (Fig. 112).

As to the superposition of parietal figures, it is often very difficult to ascertain which is the older and which the younger if both are incised. On the other hand, if one is incised and the other

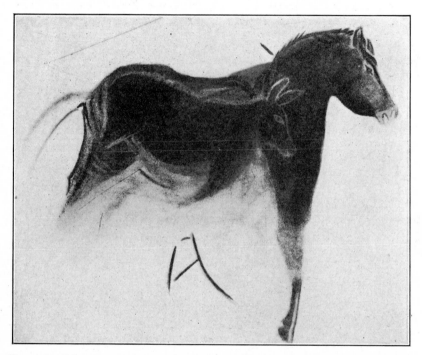

FIG. 113. MURAL FRESCO OF A HORSE SUPERPOSED OVER THAT OF A HIND, FROM
THE CEILING OF ALTAMIRA.

The figure of the horse. a Celtic type, is more recent than that of the hind. Length of horse, 1.60 meters (5.2 feet). After Cartailhac and Breuil.

painted, the problem is simple enough. Either the incised line cuts the painting or is filled by the color. In the first case the engraving is the younger, in the second, the painting. The relation between superposed frescoes is likewise easily established (Fig. 113). Thus Breuil has been able to trace the evolution of Paleolithic mural art through at least four phases.

The first phase includes deeply incised figures generally in absolute profile, that is, with a single foreleg and single hind leg, the outlines being rude and ill proportioned, and details such as hair and hoofs not indicated (Fig. 114). The paintings of this stage are in outline, the color being black or red and drawn with a crayon; there is absolutely no thought of modeling (Figs. 115 and 116).

The incisions of the second phase[2] remain deep and broad, but the outlines are more lifelike, although not always well proportioned. All four of the legs are sometimes represented, likewise the hoofs. As the incisions become less deep, they gain in neatness. In places the effect of bas-relief is produced by means of *champlevé*. The more hairy portions are indicated by incised lines.

FIG. 114. MURAL ENGRAVINGS IN THE CAVE OF PAIR-NON-PAIR, GIRONDE, FRANCE. FIRST PHASE. AURIGNACIAN EPOCH.

These engravings came to light only after the removal of floor deposits containing Upper Aurignacian cultural remains, hence they can safely be dated as not later than Middle Aurignacian. They are examples of the first of four phases in the evolution of Paleolithic mural art. The outlines are crude, deeply incised, and in absolute profile—a single foreleg and a single hind leg—and there is an absence of details, such as hoofs, hair, eyes, and nostrils. After Daleau.

Engravings of this stage are especially well represented at Combarelles. The paintings of the second phase evince the first attempts at modeling by shading at certain points. Toward the close of this phase, engraving is combined with painting, especially for the contours. The use of color continues to develop until one arrives at a well modeled monochrome silhouette, usually in black (Figs. 117, 118 and 119).

The engravings of the third phase are generally of small dimensions but admirable in execution. The entire mural decorations in the cavern of La Mairie at Teyjat are in this style (Fig. 120). In

---

[2] First and second are Lower and Upper Aurignacian, respectively (Breuil).

the domain of painting this phase is characterized by an excessive use of color, completely filling the silhouette and producing a flat effect. The modeling which was such an attractive feature of the

FIG. 115. MURAL DRAWINGS IN BLACK OF THE HORSE IN THE CAVERN OF ALTAMIRA, SANTANDER, SPAIN. FIRST PHASE. AURIGNACIAN EPOCH.

Like the engravings, drawings of the first phase were in simple profile. They were drawn by means of a crayon, the color employed being either black or red. Scale, *ca.* $\frac{1}{6}$. After Cartailhac and Breuil.

previous stage is destroyed. The period is, therefore, one of regression in so far as painting is concerned. Black, red, or brown was used, and the drawing was frequently deplorable. As a rule these paintings are not well preserved; the best work of this period is to be seen at Font-de-Gaume; it is executed in black or brown (Fig.

121).   It is often combined with engraving of a high order which was done before the application of color.

During the fourth phase [3] the engravings lose in importance. The lines are broken and difficult to follow.   The small figures of the mammoth at Font-de-Gaume and of the bison at Marsoulas show this tendency to emphasize detail at the expense of the ensemble (Fig. 122). · Paleolithic painting reaches its zenith in the fourth stage.   The fresco is always accompanied by a foundation of engraving.   The outlines are usually drawn in black, as are the

Fig. 116.   Mural drawing in red of an elephant in the cavern of Castillo, Santander, Spain.   First phase. Aurignacian epoch.

Scale, *ca.* ⅕.   After Alcalde del Rio, Breuil, and Sierra.

eyes, horns, mane, and hoofs.   The modeling is done with various shades produced by the mixing of yellow, red, and black.   These polychrome figures are seen at their best on the famous ceiling at Altamira, as well as at Font-de-Gaume and Marsoulas (Frontispiece, and Figs. 110, 123, and 124).

A fifth phase, discernible at Marsoulas, does not include animal figures, either engraved or painted.   There are banded and branching figures, also dotted surfaces, the art resembling that of the Azilian Epoch.

One of the striking features about Paleolithic art is its realism.

---

[3] Third and fourth phases are Lower and Upper Magdalenian, respectively (Breuil).

This is especially true of the phases leading to the period of its highest development. Recent investigations confirm in the main Piette's views as to the evolution of Quaternary art, although the successive stages overlap more than he had supposed. Sculpture appeared in the Lower Aurignacian, but continued without interruption through the Solutrean and to the middle of the Magdalenian [4]—a much longer period than Piette had in mind. Although beginning but little earlier than engraving, sculpture came to

FIG. 117. WALL ENGRAVING OF A MAMMOTH IN THE CAVERN OF LES COMBARELLES, DORDOGNE, FRANCE. SECOND PHASE. AURIGNACIAN EPOCH.

This is an example of the second phase in the art of Paleolithic engraving. The outline, still deeply incised, is more lifelike although not always well proportioned. All four legs are often represented, as well as hoofs, eyes, and nostrils, and the more hairy portions are indicated. Scale, ⅛. After Capitan, Breuil, and Peyrony.

fruition first. Engraving, on the other hand, developed more slowly at first, not reaching its zenith till the Middle Magdalenian, when it supplanted sculpture.

The sculptor's problem is in many respects the simpler, his opportunity of success greater. Not confined to a single aspect of his model, he has as many chances of succeeding as there are angles from which to view his work (Figs. 125 and 126). The engraver or

---

[4] The Middle Magdalenian is characterized by figures made of flat bone with contours cut away.

painter, on the other hand, must seize the likeness at the first attempt or else fail. His model was almost always an animal form, generally a quadruped. The most striking, as well as the most complete, single aspect of a quadruped is its profile. This happens to be the view that can be most easily represented on a plane surface.

In dealing with the human form, however, the problem is more complex. So far as the head is concerned, the profile presents fewer difficulties and is at the same time quite as characteristic as the front view. With the body it is just the reverse, the view from the front being the most complete and characteristic, as well as the easiest to manage. This element of complexity in a given aspect of

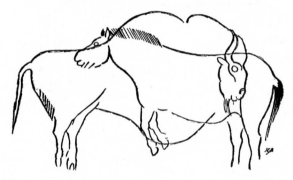

FIG. 118. WALL ENGRAVING OF A BISON AND HORSE FROM LES COMBARELLES. SECOND PHASE. AURIGNACIAN EPOCH.

The proportionate sizes are true to life. After Breuil.

the human form must have confused the primeval engraver and painter not a little, although it was not of such a nature as to disturb the sculptor. Herein may lie the reasons why the latter chose man and four-footed animals as models indifferently, while the former's predilection for quadruped forms was so pronounced. At any rate, the fact is that a large majority of Paleolithic engravings and practically all the paintings are animal profiles. The earliest ones are in absolute profile, thus simplifying the problem of representing the legs without materially detracting from the general effect.

Combined with the artist's skill in handling animal profiles is his skill in executing profile figures that represent the model in a variety of attitudes—running, leaping, walking, standing, browsing, lowing, at rest, chewing the cud, rising from the ground, at bay, etc. By degrees the stiffness of the profile was overcome. The movement of the body and especially of the legs in action is often portrayed with a fidelity that will even stand the test of a comparison with a motion-picture film (Figs. 127, 128, and 129).

The artist seems to have met in a most ingenuous fashion the difficulty of giving to a motionless figure the effect of movement. Objects at rest leave a more distinct image on the retina of the eye than those in motion. Movement in a given direction is likewise more easily followed than movement that changes in direction.

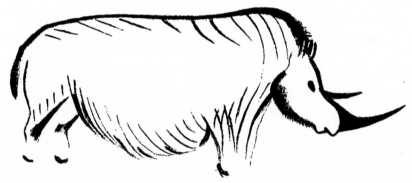

Fig. 119.   Mural drawing in red of the woolly rhinoceros in the cavern of font-de-gaume, dordogne, france.   second phase.   aurignacian epoch.

Drawings of the second phase evince the first attempt at modeling by shading at various points.   In this rare representation of *Rhinoceros tichorhinus*, the crayon was so deftly used as to give the combined effect of hairiness and modeling.   Scale, *ca.* ⅙.   After Capitan, Breuil, and Peyrony.

Confusion is increased in proportion to the number of moving objects viewed at the same time. The eye follows a single spot on the rim of a revolving wheel after the image of the spokes have multiplied by attenuation and finally fuse into one diaphanous disk. The four legs of an animal in motion are especially difficult to follow because of the changes in the direction of motion. The slowing up of the swing in preparation for the change of direction gives the retina a chance to register an image of the member at the two extremes of its trajectory. The number of legs therefore has the appearance of being doubled.

In this way Faure explains the presence of the eight legs given to the wild boar on the ceiling at Altamira; also the fact that several examples of an additional foreleg in motion are quite naturally represented as less distinct than any of the three legs at rest. If this explanation is correct, the Magdalenian artist must be credited with an unprecedented grasp of fundamental principles.

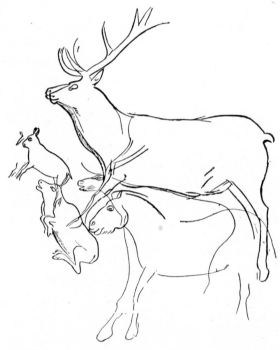

FIG. 120. ENGRAVINGS ON A MASS OF STALAGMITE IN THE CAVE OF LA MAIRIE AT TEYJAT, DORDOGNE, FRANCE. THIRD PHASE. MAGDALENIAN EPOCH.

The stag dominates the group; below are a reindeer and a horse; to the left a female reindeer at rest (chewing her cud) and her fawn. The artist has corrected the stag's muzzle, making it shorter and more pointed, hence more true to nature. The antler, ear, and especially the anatomy of the eye are faithfully rendered. Paleolithic engraving reached its zenith during the third phase. Scale, *ca.* ⅛. After Capitan, Breuil, Peyrony, and Bourrinet.

On the other hand, compositions in the true sense are rare. The placing of figures in proximity often means nothing more than the desire to economize all the suitable space available. In addition, figures are often superposed either unwittingly or otherwise (Fig. 130.) Thus the work of Aurignacian artists was constantly in danger of being injured at the hands of Magdalenian artists, just as the works of both have been mutilated in modern times by the careless and the unscrupulous. The most common way of assembling two or more related forms is the procession, the suggestion of a herd, or a hunting scene.

An excellent example of the herd was recently discovered in the Upper Magdalenian deposit of the cavern of La Mairie at Teyjat

FIG. 121.   BLACK MURAL PAINTING OF AN OX IN THE CAVERN OF FONT-DE-GAUME, DORDOGNE, FRANCE.   THIRD PHASE.   MAGDALENIAN EPOCH.

The third phase marks a period of regression so far as painting is concerned. The drawing was frequently deplorable and the silhouette was so completely filled with color as to produce a flat effect.   Scale, *ca.* $\frac{1}{10}$.   After Capitan, Breuil, and Peyrony.

(Dordogne). It represents a herd of reindeer, the three in the lead being fairly well differentiated, as is also one at the rear; the space between is filled in by hatching similar to that on the bodies of the leaders, representing therefore the undifferentiated bodies of those in the middle of the herd. Above rises a forest of horns, which, being the characteristic feature of the animal, are exaggerated as if to make up for the artist's sacrifice of detail with respect to body and limbs. The entire group is delicately incised on the radius of an eagle. In modern times this piece of work would perhaps be called impressionistic; but it is a good example of the conventionalism that was manifest in cave art even at a rather early period (Fig. 131). Another good example (from Chaffaud) represents a herd of horses engraved on stone (Fig. 132).

After all, many of the processes that lead to conventionalism are but short cuts to the artist's goal, that goal being to convey a given impression. This tendency, however, does not seem to have gained much headway, with the possible exception of certain motives con-

FIG. 122. MURAL ENGRAVING OF A MAMMOTH IN THE CAVERN OF FONT-DE-GAUME.
FOURTH PHASE. MAGDALENIAN EPOCH.

This example of the fourth phase in engraving shows the tendency to emphasize details at the expense of the ensemble. The incised outline is less continuous and the hair plays an exaggerated rôle in the make up of the silhouette. Scale, *ca.* ⅙. After Capitan, Breuil, and Peyrony.

sisting of spirals, circles, and kindred forms that might have been derived from the eye, horns, and other animal features (Fig. 133).

Realism was the essence of Paleolithic art. For an animal figure to be real it should be complete. The animal head, both front and profile views, was, however, sometimes very effectively employed, even being repeated to form a decorative motive. A wand found in the rock shelter of Mège at Teyjat is ornamented with the stag-head motive, viewed from the front. This example comes from deposits of Middle Magdalenian age. Another example of the stag-head motive, viewed from the side, is from the rock shelter of Laugerie-Haute (Fig. 134). To substitute the head for the whole animal is to let down one of the bars to conventionalism, a tendency which is all but universal in art. At the very close of the Magdalenian Epoch we find the horns alone being used as a decorative and symbolic motive to represent not only the stag's

Fig. 123. Reindeer in polychrome in the cavern of Font-de-Gaume. Fourth phase. Magdalenian Epoch.

Compare this with the Frontispiece to this volume. Paleolithic mural art reached its culmination during the fourth phase. The artist first prepared the surface to be painted by scraping and engraving. The outline and such details as eyes, horns, mane, and hoofs were then sketched in color, usually black; the modeling followed with various shades produced by mixing red, black, and yellow (oxides of manganese and iron). Scale, *ca.* $\frac{1}{12}$. After Capitan, Breuil, and Peyrony.

head, but the entire animal. The specimen in question is from La Madeleine (Fig. 135).

Ignorance of the laws of perspective seems to have deterred the troglodyte artist from often attempting the front view of the whole quadruped figure. That such attempts met with indifferent success

is to be seen in the figure of a moose (Fig. 136) engraved on reindeer horn, from the Lower Magdalenian horizon in the cavern of
Gourdan (Haute-Garonne).   In another example from the same
level at Gourdan, representing a bovidian, the engraving seems to
have been signed.   Fairly successful attempts are recorded in which
the head is turned toward the observer, leaving the rest of the
figure in profile.   This is true of the mural figure of a large feline

Fig. 124.   One of the bisons in polychrome in the cavern of Font-de-Gaume.
FOURTH PHASE.   MAGDALENIAN EPOCH.

Compare with Fig. 111.   Scale, *ca.* $\frac{1}{16}$.   After Capitan, Breuil, and Peyrony.

and of the sorcerer in the cavern of Trois-Frères (Fig. 151).   One
frequently finds the horns represented as if seen from the front,
while the rest of the figure is in profile.

That the paleolithic artist did not ignore the matter of scale in
executing a group is seen in even a cursory examination of the
repertory of cave art.   Many examples might be cited in proof of
this: the mother and her young, the male and female of the same
species, a herd of a given kind, the hunter and the animal hunted,
etc.   It goes without saying that figures in accidental juxtaposition,
executed at different times and by different artists, would not con-

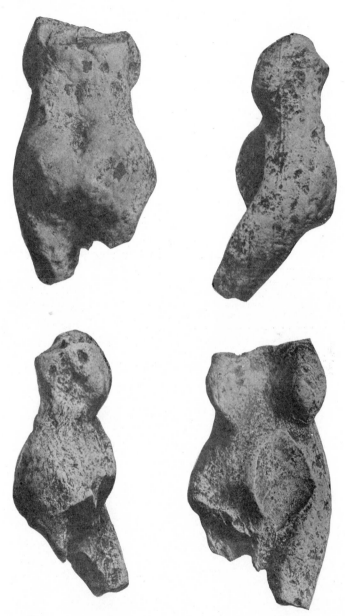

FIG. 125. FOUR ASPECTS OF AN IVORY FEMALE TORSO FROM THE GROTTE DU PAPE AT BRASSEMPOUY, LANDES, FRANCE. AURIGNACIAN EPOCH.

A somewhat weathered but good example of Paleothic sculpture; the dorsal aspect is especially fine. Scale $\frac{4}{5}$. After Piette.

form to the same scale. Again, cases might arise where it would not be expedient to lay stress on the observation of this rule. A case in point is apparently afforded by a decorated bone pendant from Raymonden with an engraved scene depicting a hunters' feast. The dead, dismembered, partially consumed bison rightfully occupies the center of the stage and is drawn to a larger scale than the feasting hunters arranged in two rows, one on each side of the carcass. The position of the forelegs indicates that they had already been disjointed, and the vertebral column had been laid bare. The pendant also served as a hunter's tally (Fig. 137).

FIG. 126. YOUNG FEMALE HEAD IN IVORY FROM THE GROTTE DU PAPE AT BRASSEMPOUY.

Although from the same cave as Figure 125, and of the same material, this piece is inferior both in modeling and proportions. The neck is too long and the features a failure from every angle. Scale, ca. ⅝. After Piette.

In comparing art of the Paleolithic with art of any period that followed, one encounters various difficulties. Paleolithic art differs not only from Neolithic art, but also from the art of modern primitive races. The art of the untutored child is more like Neolithic or modern primitive art than it is like Paleolithic art. The child does not copy the thing itself but rather his

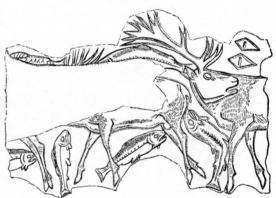

FIG. 127. STAG (RED DEER) AND SALMON ENGRAVED ON REINDEER HORN, FROM THE CAVE OF LORTHET, HAUTES-PYRÉNÉES, FRANCE. MAGDALENIAN EPOCH.

A fine example of Paleolithic composition; the red deer and salmon are not only in motion, but the stag with head turned is calling to the stragglers. The lozenge-shaped figures in the upper right corner are probably the artists' signature. Scale, ½. After Piette.

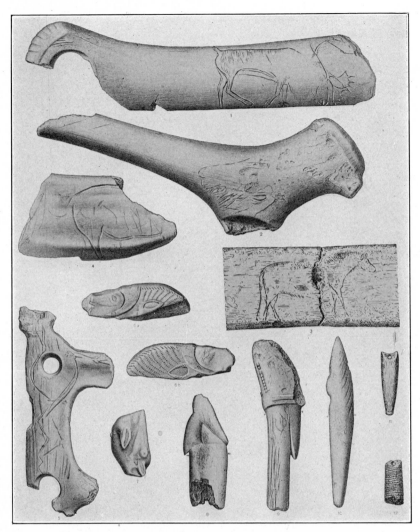

FIG. 128. EXAMPLES OF CAVE ART FROM THE CAVE OF KESSLERLOCH NEAR THAINGEN, SWITZERLAND. MAGDALENIAN EPOCH.

No. 1, browsing reindeer engraved on a baton of reindeer horn; No. 3, horse with shaggy coat (the walking gait is effectively rendered in the horse as well as in the reindeer above); No. 4, posterior half of a woolly rhinoceros (?); Nos. 6a and 6b, head and neck of a musk ox; Nos. 8 and 9, dart throwers. Scale, $\frac{1}{2}$. After R. R. Schmidt.

preconceived notion about the thing. Paleolithic art, free from the influence of preconceived notions, evinces a remarkable familiarity with the object combined with a skilled hand. The artist's models were almost without exception from the animal world, chiefly game animals.

Conditions favoring progress in art are normally just the reverse of those that would make a hunter's paradise. With the increase in density of population there would be a corresponding decrease in game. The animal figures were, no doubt, in a large measure votive offerings for the multiplication of game and success in the chase. The more realistic the figure, the more potent its effect would be as a charm. The mural works of art—figures of male and female, scenes representing animals hunted or wounded—are generally tucked away in some hidden recess which, in itself, is witness to their magic uses.

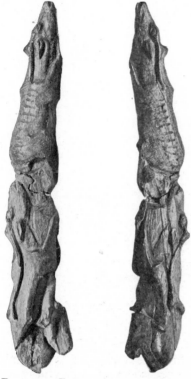

FIG. 129   TWO VIEWS OF A BROKEN IVORY CARVING, REPRESENTING TWO FIGURES IN THE ROUND OF REINDEER, FROM THE ROCK SHELTER OF MONTASTRUC AT BRUNIQUEL, TARN-ET-GARONNE, FRANCE.   MAGDALENIAN EPOCH.

Years after their discovery, the two pieces of the carving, which were broken apart in Paleolithic times, were brought together and found to fit. Scale, *ca.* ½. Photograph from the British Museum.

The mythical representations, so common to modern primitive art and to post-Paleolithic art in general, are foreign to Paleolithic art. There were no gods unless the somewhat rare human figures served as such; no figures with mixed attributes, such as are so well typified in the gold figurines of ancient Chiriqui, on the Isthmus of Panama, or in the Hindu and Egyptian pantheons. The Paleolithic artist left frescoes, engravings, bas-reliefs, and figures in the round of the horse, but there is not a single figure of the centaur, for example.

The cave man's love for the real, the natural, as opposed to the mythical, the artificial, is likewise seen in his representations of

Fig. 130. Superposed mural frescoes in polychrome in the cavern of
Font-de-Gaume. Magdalenian epoch.

The superposition was probably fortuitous. In the order of their execution the
figures are: horse, reindeer, bison, and mammoth. They were not executed at one time:
each was first engraved (upper picture) and then the polychrome fresco was applied (lower
picture). The animals were not drawn to scale, the horse and mammoth being relatively
much too small in proportion to the size of the bison and reindeer; this is added proof that
the panel does not represent a group done at one time. After Capitan, Breuil, and Peyrony.

the human form. A child will draw the figure of a man or woman as clothed but with the legs, for example, showing through the dress. The same thing was done by the artists of ancient Egypt. Not so with the cave artist. That Paleolithic man of the art period wore clothing, the numerous delicate bone and ivory needles afford abundant testimony; but with a single possible exception (Cogul in southeastern Spain, and that, if an exception, dates from the very close of the Paleolithic Period) the human form was represented in the nude (Fig. 138), although some of the figures suggest a more

FIG. 131. HERD OF REINDEER ENGRAVED ON THE WING BONE OF AN EAGLE, FROM THE CAVE OF LA MAIRIE, DORDOGNE. UPPER MAGDALENIAN EPOCH.

This is a good example of the tendency toward conventionalization which became manifest at a rather early period in cave art. Note the stylistic treatment of the middle of the herd as compared with its leaders and the reindeer at the rear. Scale, *ca.* ½. After Capitan, Breuil, Bourrinet, and Peyrony.

pronounced growth of hair over the body than would be considered common at the present time.

Art objects dating from the Paleolithic Period have every appearance of being originals and not copies.[5] Earmarks of the copyist are singularly lacking. The work was done either in the presence of the model or with the image of the latter fresh in the memory. Since the animals almost without exception are repre-

---

[5] Among those who have not seen the originals there is a not unexpected disposition to question the faithfulness of the published reproductions, particularly of the mural art. Practically all of the mural originals have been reproduced by one man, the Abbé Breuil, himself a trained artist. Whenever the condition of the wall would permit, he traced all the figures natural size. These outlines were next reproduced by means of the camera lucida. With the reductions Breuil returned to the cavern and added all the details, completing the final pastel in the presence of the originals. Not a single touch was added after leaving the cavern. The colors were studied with great care and it was often necessary to reinforce them in order to make a readily intelligible copy. Moreover, the reproductions are on white paper, while the cavern walls are more like old and soiled parchment. In view of the limitations imposed, the fidelity of Breuil's reproductions is remarkable.

sented as alive, and since the living wild animal has no inclination
to accommodate the artist by posing either at rest or in action, the
probability is that much of the work was done from memory by
making a composite of the various fleeting glimpses of the model.
Animals were sometimes captured alive; some of them might have
been tethered temporarily for the benefit of the artist. This, however, would be impracticable for an artist whose canvases were not
portable; and the best works are on the walls of dark, narrow, sub-

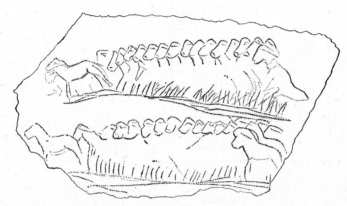

FIG. 132. HERD OF HORSES ENGRAVED ON STONE, FROM THE CAVE OF CHAFFAUD,
VIENNE, FRANCE. UPPER MAGDALENIAN EPOCH.

The problem of representing a herd is met in a fashion similar to that seen in Fig. 131,
but not quite so successfully. Scale, ¾. After Cartailhac.

terranean corridors where the presence of the model would be
impossible.

Such considerations as these lead naturally to the problem of
lighting and the artist's general stock in trade, so far as facilities
were concerned. Nearly all examples of parietal art are far
removed from daylight and the damaging effect of atmospheric
agencies. The picture gallery at Niaux is more than half a mile
from the cavern entrance. The stillness and blackness of darkness
are oppressive, but both have combined to preserve the figures in
their original freshness (Fig. 139). We shall never know how
much of Quaternary mural art has been destroyed by being placed
too near the cavern entrances or in shallow caves and open rock
shelters. In rare instances the mural art of French rock
shelters, Cap Blanc (Dordogne) for example, have been preserved

because covered by subsequently accumulated talus (Fig. 140). In the dry, mild climate of southern Spain, the mural art, even in shallow caves, has been fairly well preserved, especially where the rock is hard.

The cavern artist employed artificial light. A number of stone lamps, similar to the Eskimo lamp, have been found in the floor deposits of certain caves (Fig. 141). The artist's tools were as primitive as his method of lighting. Caverns began as a series of fissures; they were enlarged by subterranean streams that were active in Pliocene times. In their making there was very little consideration for the convenience of the Quaternary artist who came later. Wall space, therefore, had to be selected with care, and the surface was nearly always prepared for the fresco by scraping and by incised contour lines. Embossments of suitable shape and size were often selected so as to give to the figure the effect of relief (Fig. 142). The only tools needed by the engraver were the flint scratcher and graver; they were employed also by those who worked in color (see Figs. 81, 95).

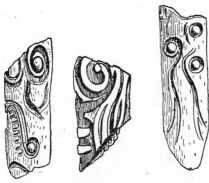

FIG. 133. CONVENTIONALIZED DESIGNS CARVED ON FRAGMENTS OF BONE, FROM THE CAVE OF ESPÉLUGUES AT LOURDES, HAUTES-PRYÉNÉES, FRANCE. MAGDALENIAN EPOCH.

These designs are probably derived from animal features, such as eyes and horns. After Piette.

The colors used by the ancient artists are insoluble in water and contain no organic matter. Ocherous sesquioxide of iron containing a very small quantity of oxide of manganese furnished the warm tints; oxide of manganese with a small percentage of sesquioxide of iron was employed for the darker shades. These minerals were picked up on the surface or in stream beds. Specimens of a uniform tint were chosen, and the material was removed in the form of a powder by means of a flint scratcher. The powder was caught in a stone mortar or other receptacle and reduced to greater fineness. It was mixed with some medium, perhaps grease, and applied by means of an improvised brush. Bivalve shells and even bone tubes (Fig. 143) served as receptacles for the mixed paint; and at least one stone palette with the mixed paint still thick upon it has been reported from Cap Blanc, evidently the one that was employed in

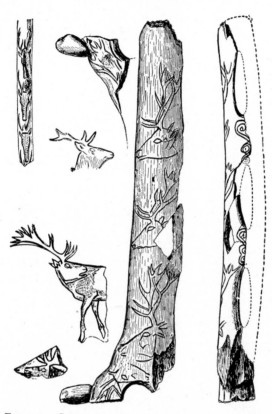

FIG. 134. STAG-HEAD MOTIVES. MAGDALENIAN EPOCH.

The Paleolithic artist excelled in his mastery of the quadruped profile; this is well illustrated in these stag-head motives, in which the profiles are superior to the front views. The front views are on a wand from the rock shelter of Mège at Teyjat; profiles on a baton of reindeer horn from the rock shelter of Laugerie-Haute. Scale ⅓. After Capitan, Breuil, and Peyrony.

painting the relief figures of the horse. The color was applied also by means of crayons whittled from chunks of ocher or oxide of manganese. A number of such crayons have been encountered in the floor deposits, especially at the cave of Les Eyzies near Font-de-Gaume. In some cases the crayons are grooved or perforated for

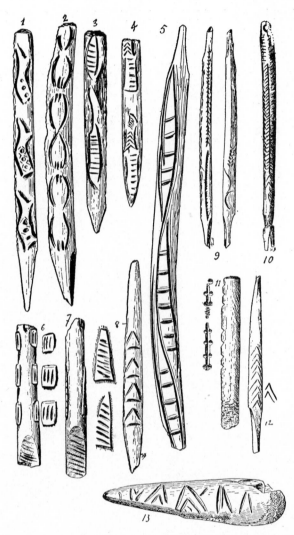

FIG. 135.   DECORATED JAVELIN POINTS AND OTHER OBJECTS OF BONE AND REINDEER
HORN.   UPPER MAGDALENIAN EPOCH. (SIXTH PHASE.)

No. 1, from Le Soucy;   Nos. 2, 6–8, from La Madeleine;   No. 4, from Raymonden;
Nos. 5 and 10 from Lorthet;   Nos. 9 and 13 from Gourdan.   In No. 2 stag horns are em-
ployed as a decorative motive.   Scale, ⅓ (with the exception of No. 5 which is ½).   After
Breuil.

suspension, this affording greater safety and ease of carrying (Fig.
144).

It is worthy of remark that so much in the way of artists'

materials (crayons, mortars, scratchers, gravers, and powdered ocher) should have been found at the shallow, sunny cave of Les Eyzies, while so little has been found in the dark, subterranean cavern of Font-de-Gaume, so rich in mural art. Les Eyzies is in plain sight of the entrance to Font-de-Gaume and only a few hundred yards away. Font-de-Gaume might well have been a studio for an esoteric guild of artists living at Les Eyzies.

With such a wonderful record of achievement in sculpture, bas-relief, engraving, and painting, one might expect to find at least a beginning in the field of ceramic art. Practically all modern primitive peoples are familiar with the plastic possibilities of clay. Remains of the potter's art are abundant, dating from the Ages of Iron and of Bronze as well as the Neolithic Period. A characteristic of Paleolithic stations, however, has been the complete absence of pottery, and there are but few known examples of Paleolithic modeling in clay.

On July 20, 1912, Count Begouen and his three sons, while exploring a subterranean stream bed, Tuc d'Audoubert, near Saint-Girons (Ariège), discovered a series of connected caverns, on the walls of one of which they found a number of engravings—reindeer, horse (Fig. 147), bison. Five days later the author visited the locality under the guidance of Count Begouen and saw what the latter had previously seen. On this occasion other parietal engravings were dis-

FIG. 136. MOOSE ENGRAVED ON REINDEER HORN, FROM THE CAVE OF GOURDAN, HAUTE-GARONNE, FRANCE. LOWER MAGDALENIAN EPOCH.

One of the few examples in which the Paleolithic artist attempted to represent the front view of a quadruped. Scale, full size. After Piette.

covered, including the figure of a horse and mammoth. Two red spots were pointed out as being the only bits of color as yet noted in the caverns; the author at once recognized them as representing a pair of eyes, the head of the animal being formed by the projecting rock. This is a typical example of the readiness with which the Quaternary artist detected and turned to account resemblances in the rock to some familiar animal form.

In October of the same year Count Begouen, continuing his exploration of Tuc d'Audoubert, discovered a chimneylike opening high on the side of one of the galleries. It proved to be a long, narrow corridor. His progress was eventually stopped by two large stalagmite pillars between which one could see that the passage extended farther. After bringing tools and breaking down the pillars, Count Begouen and his sons followed the passage, which led

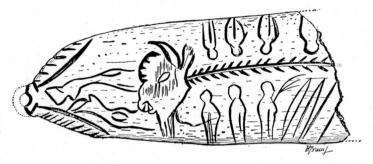

FIG. 137. BONE PENDANT WHICH HAD SERVED AS A HUNTER'S TALLY. FROM THE CAVE OF RAYMONDEN AT CHANCELADE, DORDOGNE, FRANCE.

In this scene, representing a hunters' feast, the question of scale is ignored in order to give greater prominence to the partially dismembered bison. The flesh of the animal has been removed, leaving the spinal column exposed. After Breuil.

to an ample gallery. Traversing this gallery, they came upon two large, lifelike figures of the bison modeled in clay, a female followed by a male (Fig. 148). These figures had never been wholly separated from the clay matrix out of which they were fashioned; they seem to stand out of the sloping clay talus that flanks a fallen rock. The modeling is done in a masterly fashion. The figures are slightly cracked, but otherwise just as the artist had left them. They were not hardened by fire and hence can not be removed. A smaller figure of the bison, found near the first two, has been removed to the museum at Saint-Germain. A fourth figure, also of the bison, is simply outlined in the clay floor back of the stone against which the two large figures recline.

These examples of Paleolithic modeling in clay no doubt owe their preservation to the accidental and fortunate sealing up of the passage to the gallery by stalagmite pillars. In view of their excellence, it is probable that these figures are not unique specimens; many other clay figures, less fortunately situated, have probably been completely destroyed because the modelers were unacquainted with the secret of tempering and firing their products. The foregoing statement, published by the author as a prediction in 1913 and again in 1916, was literally verified in 1923 by Norbert Casteret's discovery of many animal figures modeled in clay in the cavern of Montespan (Haute-Garonne). Chief among these is a large headless bear, 1.10 meters (43.3 inches) long; the presence of a bear's skull between the forepaws leads Casteret to conclude that the model might have been provided with a real bear's head. The body is covered with dart thrusts. Around the bear were some twenty smaller clay models in relief, rendered unrecognizable by the action of dripping water; the three best preserved are horses with ample paunches and abundant mane and beard. Elsewhere there were three large feline figures (1.50 to 1.60 meters) leaning against the wall and much damaged; the breast of one is marked by numerous javelin thrusts. A horse's head in clay about the size of a man's hand was discovered in the same gallery. On a bank of clay, Casteret found half of a woman's body modeled in clay, also several clay balls.

That the artist purposely mutilated some of his models by means of numerous dart stabs is a performance full of magic significance. It seems that the Montespan clay models were never so perfectly formed as were the bisons from Tuc d'Audoubert; the latter were chosen apparently to live for breeding purposes, while those from Montespan were marked for death at the hands of the hunter. No text could be more eloquent of the true meaning of cave art than these mute but by no means inglorious witnesses from Tuc and Montespan, testifying to the cave man's need for clothing and sustenance, and to the magic means invoked in his behalf. In addition to the clay models, Casteret found at Montespan mural engravings of the bison, horse, reindeer, stag, ass, wild goat, mammoth, and hyena; the horse and bison predominate. There were also engravings done in the clay of the cavern floor; those above high water have been preserved, among them the horse reproduced in Fig. 149.

Casteret deserves high praise for the courage and skill exhibited

Fig. 138. Mural painting in the rock shelter of Cogul, Lerida, Spain. Probably final Magdalenian.

Nine partially clothed females surrounding a nude male. This scene represents a dance or some analogous ceremony having to do with initiation or creative ritual. The only instance in which the Paleolithic artist depicted the human form as clothed. Scale, ⅓. After Breuil and Cabre.

in meeting difficulties. The danger of losing one's way, even where there is no water to wade or swim, would suffice to discourage many. Casteret had not only to swim, but also dared to pass through the neck of a siphon under water; for the ceiling was so low at one place as to be completely immersed. The story of how he alone, in bathing attire and carrying a candle and matches in a rubber case, swam a subterranean stream for 1,190 meters (about 3,900 feet) and passed through a siphon to be finally rewarded by his notable discoveries at Montespan, is a striking

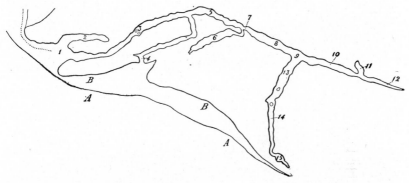

Fig. 139. Plan of the cavern of Font-de-Gaume, Dordogne, France.

This celebrated cavern, which contains far in its interior a wealth of Paleolithic mural art, is over 123 meters in length with three branches. The width varies from 2 to 3 and the height from 7 to 8 meters.

A, base of rock cliff; B, top of rock cliff; 1, entrance archway; 2, parallel grotto with bones of bears; 3, barren entrance gallery; 4, small entrance; 5, narrow passage; 6, lateral gallery before the "Rubicon"; 7, narrow passage called the "Rubicon" where the mural art begins (see Fig. 154); 8, principal gallery of frescoes, first part; 9, fork; 10, principal gallery second part; 11, room with polychromes of small bison; 12, *diverticule*, with walls covered with drawings; 13, lateral gallery; 14, gallery prolonged in narrow pass; 15, small room leading to the narrow end of the branch. After Capitan, Breuil, and Peyrony.

example of the courage and endurance demanded of those who would wrest from the caverns their subterranean secrets. In this class belong the achievements of Count Begouen and his sons at Tuc and Trois-Frères (Casteret is one of Count Begouen's students at Toulouse), and the Abbé Lemozi with the youthful David at the cavern of David in Lot.

In 1922, David, a boy of fourteen, inspired by the discoveries of the Abbé Lemozi at Marcenac, Sainte-Eulalie, and Murat, decided to do some exploring on his own account. Armed with a

candle and matches, the lad began exploring on his father's own
land; he squeezed through an aperture, crept into a small gallery,
and after unusual difficulties eventually found himself in another
gallery of large dimensions. Much excited by his success, the lad
climbed back to sunlight and reported his discovery, first to his
father and then to the Abbé Lemozi. Within a month the Abbé
Lemozi and David found on the walls of the large gallery some
forty figures engraved or painted in black and red: bison, mam-

FIG. 140. HORSE CARVED IN HIGH RELIEF ON THE WALL OF THE ROCK SHELTER OF
CAP-BLANC, DORDOGNE, FRANCE. MAGDALENIAN EPOCH.

The length of the figure is 2.15 meters (7 feet). The lower half, carved in softer rock
than the upper, has been destroyed; when first uncovered, traces of ocher were still per-
ceptible on the head, neck, and shoulders. This is one of six mural equine figures in relief,
forming two groups of three each. After Lalanne.

moths, horse, fish, and the human hand. The great gallery is connected with two smaller galleries, one of which contains a beautiful
mural figure of a bear; the other contains mural engravings and
paintings, bones, and fossilized excrement of the bear. Of special
significance are the engraved figures of men (ithyphallic) followed
by women with prominent pendant breasts. The art in the cavern
of David has been referred in part to the Aurignacian and in part
to the Magdalenian Epoch.

Thanks to their sheltered location, figures of a bison and a trout,
sketched in the fine clay of the cavern floor at Niaux (Ariège), and

similar figures of *Bos* at La Clotilde de Santa Isabel (Santander) have been preserved to us. The latter, which were traced with the finger tips on a coating of clay that adhered to the ceiling, are believed to be of Aurignacian age (Fig. 150). Figures of animals incised in clay have been reported from four other caverns: bison and horse from Gargas (Hautes-Pyrénées); *Bos* and wild goat from

FIG. 141. LAMP OF GRITSTONE, FROM THE CAVERN OF LA MOUTHE, DORDOGNE, FRANCE.

Most of the Paleolithic mural art is so situated that it could have been executed only by the aid of artificial light. Scale, ⅓. After de Mortillet.

Hornos de la Peña (Santander); *Homo* and horse from San Garcia (Burgos); wild boar from Quintanal (Oviedo).

The need of something less difficult to manipulate than stone, ivory, bone, or horn must have been ever present in the experience of the troglodyte artist. That he should finally have chanced upon clay is not surprising; he did not, however, discover t h e ceramic art. This was left for the later and more practical, if less artistic, Neolithic races.

Tuc d'Audoubert is perhaps the most picturesque of all the caverns ornamented by Paleolithic man. From the entrance by means of a boat (Fig. 145) to the immaculate

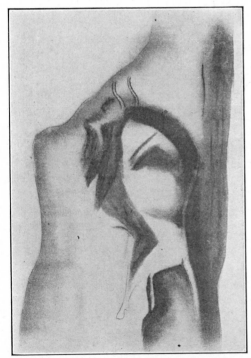

FIG. 142. BISON PAINTED IN BLACK ON STALAGMITE IN THE CAVERN OF CASTILLO, SANTANDER, SPAIN. MAGDALENIAN EPOCH.

This figure illustrates the cleverness of the Paleolithic artist in detecting resemblances to animal forms in the natural configuration of the rock and his skill in heightening them. Scale *ca.* 1/10. After Alcalde del Rio, Breuil, and Sierra.

and marvelous stalactites of the *Salle Cartailhac* and the *Salle des Noces,* and finally to the mysterious Hall of the Bisons requires a long afternoon and no small degree of physical endurance, but in return it nets one a climax of unforgettable experience.

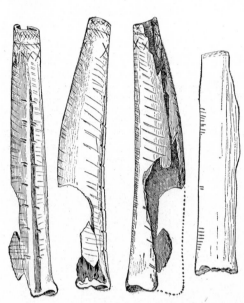

FIG. 143. DECORATED RECEPTACLES FOR PAINT, FROM THE CAVE OF LES COTTÉS, VIENNE, FRANCE. AURIGNACIAN EPOCH.

The receptacle on the left, of which three views are shown, is made from a cannon bone of the reindeer. Bivalve shells and stone mortars were also employed as receptacles for paint. Scale, ⅓. After Breuil.

The cavern of Trois Frères was discovered by the sons of Count Begouen in 1914, through a pit leading there from the summit of the hill, under which also lies Tuc d'Audoubert. This pit had trapped many an unwary beast in Pleistocene times, as witness the several almost complete skeletons of the bison and reindeer in the Begouen collection.

Like Tuc, Trois-Frères is a series of galleries connected by corridors. Pleistocene man had a more convenient means of entering these than by the overhead pit, probably by way of the cave of Enlène which Count Begouen found later to be connected with the Trois-Frères complex; it now serves as the entrance to the latter.

If nature has not done so much for Trois-Frères as for Tuc, the cave artist has done more in that he left there some four hundred mural engravings. If Tuc has its bisons modeled in clay, Trois-Frères has its sorcerer, the most remarkable one of several hundred figures that adorn the walls of the terminal gallery of the lower level. The figure of the sorcerer, about 75 centimeters (29.5 inches) in length, is situated high on the wall at the end and dominates the entire chamber (Fig. 151). It is completely engraved, and the application of black paint further emphasizes certain features of the outline. The figure is that of a man masked, in motion, and not wholly erect. The body is in profile with the head turned to face the beholder, bringing to view a pair of long, pointed ears and cervidian antlers. The transformation is further emphasized by the presence of a long, horselike tail. The arms are diminutive, the legs and feet are typically human, and from between the half-flexed legs the sexual organs are intentionally brought to view.

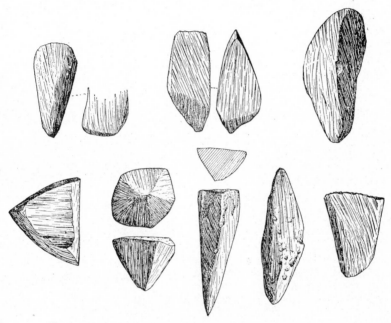

FIG. 144. CRAYONS OF RED OCHER, FROM THE CAVE OF LES EYZIES AND THE ROCK SHELTER OF LAUGERIE-HAUTE, DORDOGNE, FRANCE.

These crayons, showing marks of long use, were employed by Paleolithic artists in mural drawing. Scale, *ca.* ¾. After Capitan, Breuil, and Peyrony.

By the close of the Magdalenian Epoch, the continental ice sheet had retreated far to the north, and the area of Alpine and Pyrenean glaciation was much reduced. Cold-loving animals such as the mammoth, the woolly rhinoceros, and the reindeer had followed the retreating ice toward the north, where, of the three, only the reindeer still survives (Fig. 152). Forever linked with these and other animal forms, Paleolithic art likewise disappeared. The complete story of its taking off has not yet been written; it came unheralded and went in like manner. The time had come for the swing of the culture pendulum in another direction, perhaps toward other and more serviceable, if less artistic, channels of thought expression. The Azilians seem to have turned their attention toward a system of cursive writing, if certain archeologists are correct in their interpretation of the painted pebbles of Mas d'Azil and other stations dating from the same epoch. The first experi-

FIG. 145. ENTRANCE TO THE CAVERN OF TUC D'AUDOUBERT, ARIÈGE, FRANCE, BY MEANS OF THE SMALL SUBTERRANEAN STREAM, THE VOLP.

In addition to its treasures of art, including mural figures and the bisons modeled in clay, Tuc is perhaps the most beautiful of all known Paleolithic caverns. The cavern proper is reached by ascending the stream for a short distance in the little boat and then climbing up through a small hole into the first of a series of galleries. Photograph by the author.

menters in the great domain of agriculture and domestication of animals were yet unborn, since unmistakable traces of their work do not appear until Neolithic times.

From the standpoint of priority of antiquity then, the artist has special reason to be proud. He follows a calling that had its worthy devotees ages before any other method of leaving imperishable records of human thought was known. Man was artist before he was the maker of even hieroglyphs; he tamed his imagination and his hand to produce at will objects of beauty long ages before he tamed the first wild beast or made the humble plant world do his bidding. The farmer, whose calling we are apt to think of as representing the life primeval, is a mere upstart in comparison with one who practices the fine arts.

FIG. 146. A DISTINGUISHED GROUP AT THE ENTRANCE TO THE CAVERN OF TUC D'AUDOUBERT.

In the foreground are Count Begouen (owner of the cavern) his three sons, the Abbé Breuil (center, seated), and Professor Cartailhac (extreme right). Photograph by Count Begouen.

In the history of art there are many bright pages. Certain epochs have shone more resplendent than others; but the age pre-eminent of fundamentals in art dates from the last epoch of the great Ice Age, estimated at from 40,000 to 20,000 years ago. Without a background of art inheritance and beset by insuperable difficulties on every hand, the troglodyte artist left a record, of which any age might well be proud. He was not without his reward. First, there was the sense of satisfaction in the achievement, which must have been keen. Then came oblivion for countless ages; nature and human ignorance combined to weave an enduring and

protective mantle over those primeval art objects, a mantle which
was not lifted till near the close of the nineteenth century. Finally
they came to light, not to be destroyed, but preserved in so far as
this can be done by combined local, governmental, and scientific
agencies.

It is the policy of the governments concerned to set aside as
national monuments all caverns and rock shelters which contain

FIG. 147. MURAL ENGRAVING OF A HORSE WOUNDED BY A DART IN THE SIDE, IN
THE CAVERN OF TUC D'AUDOUBERT. MAGDALENIAN EPOCH.

Note the texture of the mural surface. Photograph by Count Begouen.

good examples of Quaternary art (see Appendix III). Only in one
or two rare instances have parietal engravings and frescoes been
cut from their original places. Such a step is to be taken only when
not to remove the specimens in question would be to invite certain
destruction. Where works of this nature are easily accessible and
can be permanently protected, there is as little call to remove them
as there would be to mutilate the walls of the Sistine Chapel. If
France has her Louvre, she likewise has her Font-de-Gaume (Figs.
153 and 154); and the art student who would visit the Prado

Museum in Madrid should not fail to include the Quaternary gallery at Altamira.

No other part of France, or of Europe for that matter, is so rich in remains of Paleolithic man as the valley of the Vézère in the Dordogne; this is especially true of Paleolithic art. Here one finds within a radius of only a few kilometers from the village of Les Eyzies several dozen sites in about half of which art objects

FIG. 148. BISONS, MALE (LEFT) AND FEMALE, MODELED IN CLAY ON THE CAVERN FLOOR OF TUC D'AUDOUBERT.

The skill with which these figures are executed is proof of the Paleolithic artist's mastery over clay as a medium of art expression, and of his belief in the efficacy of magic in controlling the reproduction of game animals. Scale, *ca.* $\frac{1}{16}$. After Count Begouen.

have been found. As regards portable art, no other station in Europe is so rich as Laugerie-Basse; and for mural art, Font-de-Gaume and Les Combarelles have few equals. The fact that two Paleolithic type stations—Le Moustier and La Madeleine—are located here gives an added interest to this region (Figs. 155 and 156).

THE FIELD OF PALEOLITHIC ART

The Paleolithic artist's range of models included both the animate and inanimate, but was confined almost wholly to the fauna.

Among the fauna, mammals (including man) largely monopolized his attention. Birds and fishes came in for a relatively small share of attention; reptilian representations are practically nonexistent, and the same may be said of invertebrates. Plantlike forms are very rare.

The inanimate field is represented by claviform and tectiform figures; by spirals, circles, chevrons, frets or grecques, volutes, wave

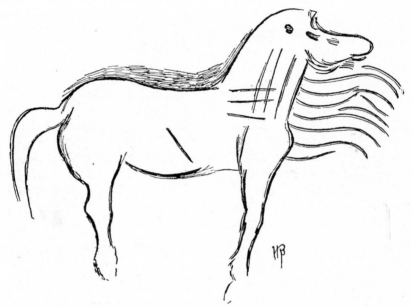

FIG. 149. HORSE DEEPLY INCISED IN CLAY, CAVERN OF MONTESPAN, HAUTE-GARONNE, FRANCE. LOWER MAGDALENIAN.

The clay just above the back of the horse was removed, giving the effect of relief. The head, neck, and belly are marked by long gashes representing wounds. Scale, ¼. After Count Begouen (from a drawing by Breuil).

ornaments, and alphabetiform signs, some of which were derived from animate objects through processes of conventionalization.

In a study of cave art, one is impressed, although not surprised, by the extent to which it reflects the fauna of the times. To the hunter, game animals would naturally loom large on the horizon; that which makes the strongest appeal to the senses is the first to find expression, especially when it happens to be essential to one's existence. Among the animal forms reproduced in Paleolithic art, game animals occur much more frequently than any other kinds;

the horse far outnumbers the hyena, as does the red deer, the lion.
To one animal killed because it was dangerous to man (or to the
animals on which he fed), there would be scores of game animals
captured. Besides, many of the representations are prayers for the
increase of species useful for food, and these would account, in part
at least, for the preponderance noted above.

The cave artist not only had predilection for such species as

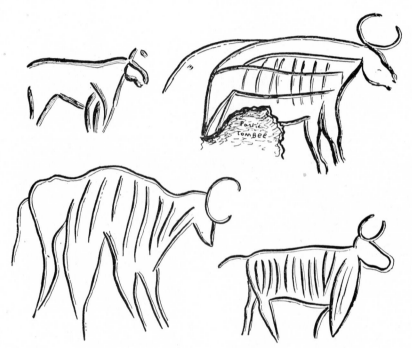

FIG. 150. FIGURES OF "BOS PRIMIGENIUS" TRACED IN CLAY, CAVERN OF LA
CLOTILDE DE SANTA ISABEL, SANTANDER, SPAIN.   PROBABLY AURIGNACIAN.

These figures of the ox were traced in the clay of the cavern floor by means of the finger
tips.   Scale, *ca.* ⅐.   After Alcalde del Rio, Breuil, and Sierra.

the horse and red deer, but he also seems to have had a preference
for the female of the species. In some species sex distinction is much
more pronounced than in others. The stag can be distinguished
from the hind by the presence of antlers; figures of the hind far
outnumber those of the stag, a fact which is significant. The hind
is the symbol of fecundity; the larger the number of hinds, the
greater the increase of the herd. With the horse it is difficult for

the artist to differentiate between the sexes; were it not so, we would probably find a like majority of mares over stallions. The same preference holds good among representations of *Homo,* no doubt for kindred reasons.

FIG. 151. THE SORCERER, CAVERN OF TROIS-FRÈRES, ARIÈGE, FRANCE.

This mural figure, partly engraved and partly in black, represents a masked magician whose aid would insure success in the chase and the multiplication of game animals. The sorcerer dominates a series of several hundred mural engravings of animals which adorn the terminal gallery at the lower level of the cavern. The Trois-Frères cavern complex and that of Tuc are in close proximity and may once have been connected by a corridor now closed. Scale, $\frac{1}{10}$. Photograph by Count Begouen.

The human form played by no means an insignificant rôle as a model for the cave artist; with but few exceptions, however, the artistic treatment of it was not so successful as was that of the lower animals. The failure of the Paleolithic artist in this respect is, no doubt, more apparent than real since, both by inclination and training we are more critical of human representations than of any other. Moreover, the cave artist had what seemed to him more weighty reasons for portraying game animals than man. In any event, he evinced much skill in emphasizing the characteristic features of a given species, such, for example, as the peculiar outline of the mammoth with its high cephalic dome, deep nuchal notch, and broad, sloping dorsal dome; or the short, pointed muzzle of the stag and the finer distinctions between its antlers and those of the reindeer.

FIG. 152.   HERD OF REINDEER, FINNMARKEN, NORWAY.
Similar herds of reindeer roamed over southern France during the cave-art period.

The Human Form.—Forms distinctly human, including the entire figure or a part thereof, and anthropomorphic forms, including masked figures, have been found in France, Belgium, Austria, Germany, Russia, Czechoslovakia, Italy, and Spain. France leads with examples from thirty-two stations; Spain is represented by examples from two dozen stations; three stations have been reported from Czechoslovakia, and one each from Austria, Belgium, Germany, Italy, and Russia (Fig. 158).

In the cave artist's treatment of the human form, the first things to attract the attention are: (1) the pictorial predominance of the female over the male; (2) the recurrence of a female type suggestive of the Hottentot or Bushman; and (3) the wide geographic and chronologic distribution of this type in Europe. To the figures in the round from Brassempouy, Lespugue, Grimaldi, Willendorf, and Mainz there should be added the bas-reliefs from Laussel, all conforming to one type. Those from the Brassempouy have been referred to the Lower Aurignacian, and those from Willendorf,

FIG. 153. PROMONTORY IN WHICH THE CAVERN OF FONT-DE-GAUME IS SITUATED.
The entrance (see also Fig. 171) is at the right and near the top of the escarpment. After Capitan, Breuil, and Peyrony.

Lespugue, and Laussel to the Upper Aurignacian. All are suggestive of a symbolic, rather than a physical, type (Figs. 159 to 164). In 1922, Dr. Otto Schmidtgen found the lower half of two female statuettes of this type in the loess within the city limits of Mainz on the Rhine. A Magdalenian female bust from Mas d'Azil, carved from the incisor of a horse, has long, pendant breasts also suggestive of the Hottentot.

A feature common to this group is the summary treatment of the head and extremities; the chief attention is bestowed upon the

FIG. 154.   INTERIOR VIEW OF THE CAVERN OF FONT-DE-GAUME AT THE POINT CALLED
THE "RUBICON."

The ladder leads to the narrowest corridor in the cavern (see Fig. 139).   After Capitan,
Breuil, and Peyrony.

primary and secondary sex characters.   In a few examples where
the head is present, there is an attempt to suggest hair, or a coiffure,
by means of cross-hatching, as is the case of the *figurine à la*

capuche (see Fig. 126) from Brassempouy and the negroid head from Barma Grande (Grimaldi). Still more remarkable is the suggestive way in which kinkiness is represented in the coiffure of the Venus of Willendorf (see Fig. 160).

One of the most active of the cave explorers in France is Dr. R. de Saint-Périer. For the past dozen years he has explored a group of caves (Grotte des Harpons, Grotte des Boeufs, Grotte

FIG. 155. THE ROCK SHELTER OF LAUGERIE-BASSE NEAR LES EYZIES, DORDOGNE, FRANCE.

Laugerie-Basse faces toward the east and commands an extensive view up and down the Vézère River; its yield of Paleolithic art objects has been exceedingly rich (see Appendices I and II).

des Rideaux) at Lespugue between Saint-Gaudens and Saint-Martory (Haute-Garonne). All three of these caves have yielded portable examples of Paleolithic art as well as artifacts: engravings of the horse on bone, harpoons of reindeer horn, a stone lamp, etc. from the Grotte des Harpons; bone javelin points, bone needles, baton, engravings on bone, and a fish (probably flounder) in bone with the contours cut away from the Grotte des Boeufs; bone javelin points, flint implements, and an ivory statuette of a

human female (Venus) from the Grotte des Rideaux (Fig. 159).
This last is of special importance because practically complete, of
fine workmanship, and of a type especially favored by Aurignacian
artists.

This Lespugue statuette surpasses in length (14.7 c.m.) any
other one of the group, although by no means as large as the reliefs

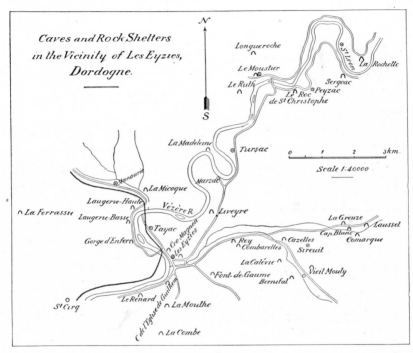

FIG. 156. MAP SHOWING THE LOCATION OF PALEOLITHIC CAVES AND ROCK SHELTERS
IN THE VICINITY OF LES EYZIES, THE CENTER OF PALEOLITHIC CAVEDOM.

from Laussel. It has the characters common to all—undifferentiated features, large pendant breasts and mountain of Venus, enormous hips, large thighs, slender arms, and legs diminishing from
the knee down—but resembles the Venus of Willendorf more
closely than any other. Unfortunately, the pick of the workman
injured the statuette in the region of the breasts; the restoration
serves to emphasize the close resemblance to the Willendorf specimen. The pose is exactly the same even to the resting of the
lower arms on the breasts and the outlines tapering from the hips

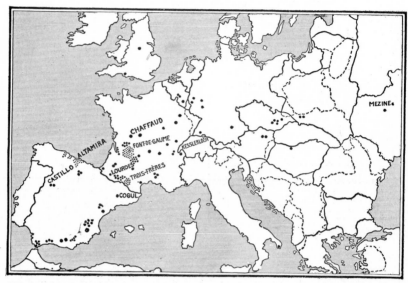

FIG. 157. MAP SHOWING THE GEOGRAPHIC DISTRIBUTION OF PALEOLITHIC ART STATIONS.

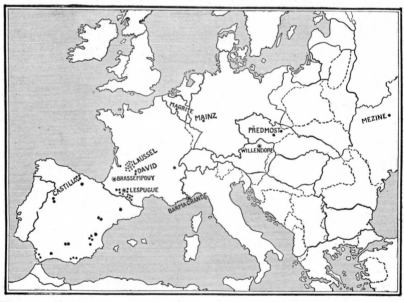

FIG. 158. MAP SHOWING THE GEOGRAPHIC DISTRIBUTION OF PALEOLITHIC ART STATIONS IN WHICH REPRESENTATIONS OF THE HUMAN FORM OCCUR.

in both directions. They are as much alike as a tall, slender figure can be like a short, stocky figure of the same general type. The artists who made them were both masters of the same canons and traditions. The striking similarity of these two figures throws a flood of new light on the homogeneity of Aurignacian culture over

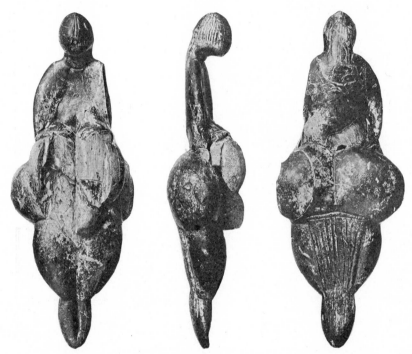

FIG. 159. FRONT, SIDE, AND REAR VIEWS OF A FEMALE FIGURINE IN IVORY FROM THE GROTTE DES RIDEAUX AT LESPUGUE, HAUTE-GARONNE,   FRANCE. AURIGNACIAN EPOCH.

A symbol of fecundity rather than an example of the Aurignacian ideal of feminine beauty.   The features are not differentiated and the footless legs are fused and taper to a point, a tendency in all female statuettes of this type and even in the low relief from Laussel (Fig. 162).   Scale, *ca.* ⅗.   Photograph by Count de Saint-Périer.

a wide geographic area. If specimens of this kind can still be found in caves of the Dordogne and Garonne valleys, the caves of Grimaldi, and in loess of the Danube and Rhine valleys, the probabilities are that they once existed at all the principal Aurignacian settlements in central and western Europe.

The human form is often treated in summary fashion. At Gourdan and Raymonden the figures (engraved on bone) are sketchy, quite small, and arranged in processions; the grouping and general treatment in the two localities are strikingly similar.

Scarcely better defined is the diminutive human from La Madeleine incised on reindeer horn, which is represented as carrying a stick on the right shoulder; it is comparable with a series of small human figures (incised on a rib), each with a staff on the shoulder, from the Abri du Château at Les Eyzies; and with the hunting scene (?) on reindeer horn from Laugerie-Basse, generally

FIG. 160. FEMALE FIGURINE IN LIMESTONE, FROM THE LOESS STATION OF WILLENDORF, LOWER AUSTRIA. AURIGNACIAN EPOCH.

This figurine, known as the "Venus of Willendorf," is as much like the Lespugue Venus as a short, stocky figure can be like a tall slender one. The remarkable similarity of these figures points to a homogeneity of Aurignacian culture over a wide geographic area Scale, ca. ¾. Photograph by Bayer.

FIG. 161. FEMALE TORSO IN IVORY FROM THE GROTTE DU PAPE AT BRASSEM-POUY, LANDES, FRANCE, AURIG-NACIAN EPOCH.

This torso is so much like the figurines from Lespugue, Willendorf, and Grimaldi, that one is justified in assuming for it a head without features, diminutive arms (or perhaps none at all), and feetless legs tapering to a point. Scale, full size. After Piette.

referred to as the *chasse à l'aurochs*.

In addition to a number of masked figures (Fig. 167) which are obviously human, there are a quite a number of

FIG. 162. FEMALE FIGURE HOLDING A BISON HORN, FROM THE ROCK SHELTER OF
LAUSSEL, DORDOGNE. AURIGNACIAN EPOCH.

This figure is cut in low relief in limestone. The features and feet are undifferentiated.
The artist turned the head to the right so as to obtain a profile view but for some reason
left the work unfinished. The same tendency to ignore the features (and feet) is seen
in the female statuettes from Grimaldi, Lespugue, and Willendorf. Scale, *ca.* ¼. Photograph by Lalanne.

examples, chiefly engravings, in which the human form is but vaguely indicated; they may best be referred to as anthropomorphic figures. Representations of this kind have been found at Combarelles, Cro-Magnon, Font-de-Gaume, Gourdan, Laugerie-Basse, Marsoulas, and Mas d'Azil (Fig. 168) in France; and at Altamira and Hornos de la Peña in northern Spain.

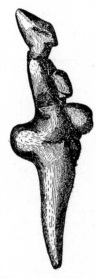

Human (and animal) Paleolithic representations are, as a rule, of the entire figure. Figures of an arm, leg, phallus or vulva are rare. On the other hand, and this is the exception which proves the rule, figures of the human hand are abundant, especially in the Pyrenean and Cantabrian regions. They have been reported from eight caves in France (Gargas, Bedeilhac, Trois-Frères, Les Eyzies, Font-de-Gaume, Beyssac, David, and Blanchard at Sergeac) and from four in Spain (Altamira, Castillo, Santian, and Pretina).

The technique employed in representing the human hand was wholly different from that employed in any other cave art; it was ingenious but not of special artistic merit. Two methods

FIG. 163. FEMALE STATUETTE OF CRYSTALLINE TALC FROM ONE OF THE GRIMALDI CAVES. AURIGNACIAN EPOCH.

After Piette.

were employed. That chiefly used was to press one hand against the cavern wall and with the other to apply coloring matter, in the form of a powder, over an area sufficient to leave a negative imprint on removal of the hand (Fig. 169). The imprint is usually of the left hand, for the simple reason that a majority of mankind are (and were even in Paleolithic times) right-handed. The other method was by means of color transference. The palm of the hand was dampened, covered with dry powdered paint, and then applied to the damp wall of the cave. As one might be led to expect, a majority of the hands stamped or printed on the walls are right hands; this is additional proof of dextral predominance.

At Santian (Santander) mural figures in red, suggestive of the human hand and lower arm, were reproduced by the ordinary free-hand application of color. They are more or less schematic, or stylistic, in treatment, grading off into forms suggesting the bird foot; they may have little to do with the positive and negative hand imprints. The latter are the oldest examples of Paleolithic mural art; the technique involved, therefore, represents an initial stage in the evolution of art in general. At Castillo the hands underlie

FIG. 164. FRONT, SIDE, AND REAR VIEWS OF A FEMALE STATUETTE IN CRYSTALLINE TALC FROM THE CAVE OF BARMA GRANDE NEAR MENTONE. AURIGNACIAN EPOCH.

That the artist was dealing with a symbolic type is evident from the exaggeration of some parts and the reduction or elimination of others; the same parts are always sacrificed—head (especially the features), arms, and feet. Scale, $\frac{2}{3}$. After Reinach.

figures in yellow, and these in turn underlie figures in red. In Spain, figures of the human hand are most abundant at Castillo; in France, at Gargas. Many of the figures at Gargas represent hands that have been mutilated by the removal of one or more joints from one or more fingers.

The custom of finger mutilation, already in existence during the Aurignacian Epoch, has persisted until recent times; it has been recorded from both the Old World and the New. As early as 1812 Burchell observed the custom among the Bushmen of South Africa, who, Stow states, performed the operation with a stone knife. According to Patterson, the Hottentots of the Orange River cut off the first joint of the little finger as a cure for sickness.

FIG. 165. MALE FIGURE FROM THE ROCK SHELTER OF LAUSSEL. UPPER
AURIGNACIAN EPOCH.

This figure, cut in low relief in limestone, is a fine athletic type. Scale, *ca.* ⅓. Photograph by Lalanne.

The following table gives the localities (63), nature, and age (when determinable) of human representations in Paleolithic art:

HUMAN REPRESENTATIONS IN PALEOLITHIC ART

AUSTRIA

| Willendorf | Portable | Female figurine of stone | Aurignacian |

BELGIUM

| Magrite | Portable | Ivory figurine | Aurignacian |

CZECHOSLOVAKIA

| Brünn | Portable | Ivory statuette | Solutrean |
| Předmost | Portable | Six female figurines of bone; stylistic engraving of female | Aurignacian |

FRANCE

Aurensan	Portable	Homo engraved on slate	
Bedeilhac	Mural	Human hand	
Beyssac	Mural	Negative figure of hand	
Brassempouy	Portable	Ivory figurines	Aurignacian
Colombière, La	Portable	Figures of Homo engraved on bone	Aurignacian (or Magdalenian)
Combarelles	Mural	Human leg; anthropomorphic figures	
Cro-Magnon	Portable	Female (full length) engraved on bone	Aurignacian
David	Mural	Engraved figures of men (ithyphallic) followed by women with pendant breasts; human hands in red or black	
Eglises, Les	Mural	Human figure under tectiform	
Eyzies, Les (Abri du Château)	Portable	Human figures engraved on rib; human hands	Magdalenian
Ferrassie, La	Portable	Vulva	Aurignacian
Font-de-Gaume	Mural	Human profile (?); hands	
Gargas	Mural	Human hands	Aurignacian
Gorge d'Enfer	Portable	Human phallus (double)	
Gourdan	Portable	Anthropomorphic figures	
Laugerie-Basse	Portable	Female figurine (Venus impudique); engraving of female with reindeer; hunter with bison	

Laussel	Portable	Five low reliefs of *Homo* on stone	*Aurignacian*
Lespugue	Portable	Female figurine of ivory	*Aurignacian*
Lourdes	Portable	Human leg (?); a sorcerer	
Madeleine, La	Portable	Human figure; human arm	*Magdalenian*
Marcamps	Portable	Human head (?) of reindeer horn	
Marsoulas	Mural	Anthropomorphic figures	
Mas d'Azil	Portable	Female bust carved from horse incisor; anthropomorphic figure on bone	*Magdalenian*
Montespan		Vulva	*Aurignacian*
Portel	Mural	Human figure	
Raymonden	Portable	Engraving of *Homo* on bone	*Magdalenian*
Rivière-de-Tulle	Portable	Anthropomorphic engraving on reindeer bone	
Roches, Les (Blanchard)	Portable	Vulva engraved on stone; phallus carved from bison horn	*Aurignacian*
Terme Pialat	Portable	Relief figure of *Homo* on limestone	*Aurignacian*
Teyjat	Portable	Phallus engraved on reindeer horn; diminutive figures with chamois masks	
Trois-Frères	Mural	Human hand imprints; a sorcerer	
Vache, La	Mural	Human stylistic forms	

GERMANY

Mainz	Portable	Lower half of two female statuettes	*Aurignacian*

ITALY

Baoussé-Roussé	Portable	Six figurines of crystalline talc, five of them female, also a negroid head	*Aurignacian*

RUSSIA

Mezine	Portable	Stylistic human figures in the round	*Aurignacian*

SPAIN

Albarracin	Mural	Paintings of *Homo*	
Alpera	Mural	Paintings of hunters and women	
Altamira	Mural	Anthropomorphic figures	
Arco, El	Mural	Paintings of *Homo*	
Batuecas, Las	Mural	Stylistic human figures in color	*Azilian* (probably)

Carasoles del Bosque,		
Los	Mural	Paintings of men
Castillo	Mural	Paintings of hand
Charco del Agua		
Amargo	Mural	Paintings of females
Chiquita de los Trenta	Mural	Paintings of men
Cogul	Mural	Paintings of men and women
Cortijo de los Treinta	Mural	Paintings of women
Garcibuey	Mural	Paintings of *Homo*
Hornos de la Peña	Mural	Anthropomorphic figures
Jimena	Mural	Anthropomorphic figures
Lavaderos de Tello	Mural	Stylistic figures of men
Minateda	Mural	Stylistic figures in color
Monte Arabi	Mural	Figure of man
Peña	Mural	Stylistic human figures in color
Peñon de la Tabla de	Mural	Stylistic human figures (probably *Neolithic*)
Pochico		
Pileta, La	Mural	Figure of *Homo* in color
San Garcia	Mural	Figure of *Homo* traced in clay
Santian	Mural	Human hand
Tajo de las Figuras	Mural	Stylistic figures of *Homo*
Tortosillas	Mural	Figures of *Homo*

Mammalia.—Ever since man began his omnivorous career, mammals have probably furnished his chief supply of animal food; this is obviously true of the whole Paleolithic Period, especially of

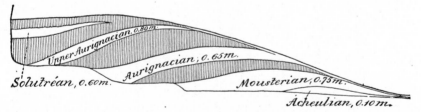

Fig. 166. SECTION OF THE ROCK SHELTER OF LAUSSEL, DORDOGNE, SHOWING THE SEQUENCE OF DEPOSITS.

The human figures in bas-relief (Figs. 162 and 165) were found in the Upper Aurignacian deposit. Collections from Laussel, with the exception of one bas-relief, are in the Lalanne collection at Bordeaux.

its last four epochs. From their kitchen refuse we know which animals the Mousterians fed upon. During the Upper Paleolithic Period, we have two lines of evidence, the kitchen refuse and the animal representations in art; both point to one and the same conclusion.

The relative frequency of a given genus as a model for the artist varies approximately in direct ratio with the frequency with which the bones of that particular genus occur in the kitchen refuse. Other things being equal, both hunter and artist no doubt drew most heavily on the animal that was most abundant. Some animals are caught more easily than others by a hunter limited to primitive means. The horse and red deer were both fleet of foot,

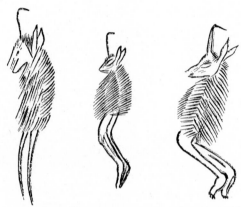

but either was a prize well worthy of special effort. It is surprising that the mammoth should have been sought so extensively as food and that the hare should not have played a more important rôle; the latter is scarcely represented in the art of the period.

The original home of the horse was in North America. Remains of the true horse (*Equus stenonis*) are found in Pliocene and early Pleistocene deposits of Europe. One finds in Upper

FIG. 167. MASKED (CHAMOIS HEAD) FIGURES ENGRAVED ON A BATON, FROM THE ROCK SHELTER OF MÈGE AT TEYJAT, DORDOGNE, FRANCE. MAGDALENIAN EPOCH.

Masks might have been used in stalking game as well as ceremonially (see Fig. 151). A figure with a bear-head mask has been reported from Mas d'Azil and one with a horse-head mask from the cave of Espélugues at Lourdes. Scale, full size. After Capitan, Breuil, Bourrinet, and Peyrony.

Paleolithic deposits remains of several varieties of horse not unlike the modern Celtic or Arab type, the Nordic or forest type, and the steppe type (*E. przewalskii*), also the wild ass or kiang (*E. hemionus*). These varieties are even recognizable in the art of the Upper Paleolithic.

The horse was easily the favorite in France; in Spain it was second, conceding first place to the red deer.[7] Taking Europe as a whole, the horse predominates (Fig. 170); it is followed in turn by the red deer, bison, wild goat, Bovidae, reindeer and Cervidae.

---

[7] Red deer (*Cervus elaphus*) was common during the Aurignacian; became rare in the Solutrean and Lower Magdalenian; was abundant in the Upper Magdalenian; and finally supplanted the reindeer.

By tabulating the occurrence of animal forms represented in art from eight of the principal Paleolithic stations in France, the horse is found to predominate in six, the reindeer in one, and the bison in one. The stations in question are: Bruniquel, Combarelles, Font-de-Gaume, Laugerie-Basse, Lorthet, Lourdes, La Madeleine, and Mas d'Azil. The horse, reindeer, bison, Bovidae (chiefly *Bos*), and mammoth are, in the order mentioned, the most frequently represented. The horse, reindeer, and Bovidae (not including bison) occur in all eight stations; the bison is lacking in one (Lorthet), and mammoth in three (Lorthet, Lourdes, and Mas d'Azil).

At Bruniquel the animal representations in the order of frequency are: Equidae (chiefly *Equus caballus*), reindeer, wild goat, *Bos,* Capridae, chamois, bison, red deer, mammoth, musk ox, and *Homo.*

The order of frequency at Combarelles is: Equidae, mammoth, antelope, Bovidae, reindeer, wild goat, bison, Capridae, cave bear, *Felis,* and wolf.

The bison easily leads at Font-de-Gaume (Fig. 171) and is followed by Equidae, mammoth,

FIG. 168. ANTHROPOMORPHIC FIGURE ENGRAVED ON A RONDELLE OF BONE CUT FROM A SHOULDER BLADE; FROM THE STATION OF MAS D'AZIL, ARIÈGE, FRANCE, RIGHT BANK.
After Piette.

reindeer, *Bos,* Capridae, rhinoceros, *Felis,* cave bear, *Homo,* and wolf.

The reindeer leads at Laugerie-Basse, followed by Equidae, fish, Bovidae, bison, red deer, wild goat, *Cervus, Homo, Felis,* and otter.

At Lorthet the order of frequency is: Equidae, Cervidae, fish, red deer, Bovidae, roebuck, glutton, and reindeer.

The horse is first at Lourdes (Fig. 172), with Bovidae a fairly

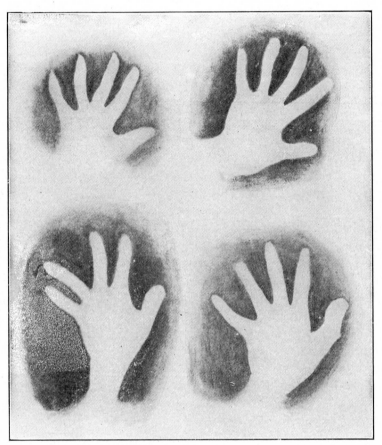

FIG. 169. NEGATIVE IMPRINTS OF THE HUMAN HAND, ON THE WALL OF THE CAVERN OF CASTILLO, SANTANDER, SPAIN.

Imprints are usually of the left hand, the right being used to throw powdered coloring matter against the wall (see also Fig. 327). Scale, ⅛. After Alcalde del Rio, Breuil, and Sierra.

close second; then follow, in their turn, the bison, bird, reindeer, red deer, cave bear, Cervidae, fish, and rhinoceros.

At La Madeleine the horse again leads; the reindeer, a close second, is followed by the red deer, Bovidae, Cervidae, ruminants, bison (Fig. 173), *Felis, Homo,* and mammoth.

The eighth station, Mas d'Azil, concedes first place to Equidae by a wide margin; after the horse there come in turn: reindeer, wild goat, Bovidae, fish, bison, *Bos,* bird, red deer, antelope, Cervidae, anthropomorphic figures, *Homo,* ruminant, and wild boar.

A somewhat different faunal composition is reflected in the Paleolithic art of Spain, even though the field be limited to the region north of the Cantabrian Mountains. The horse concedes

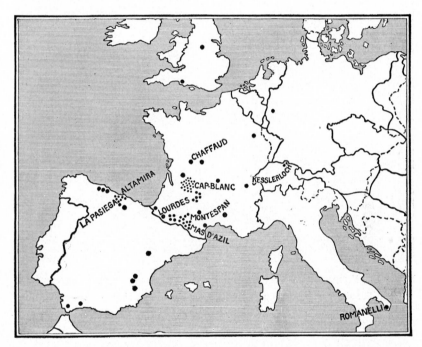

FIG. 170. MAP SHOWING THE GEOGRAPHIC DISTRIBUTION OF REPRESENTATIONS OF THE HORSE IN PALEOLITHIC ART.

first place to the red deer, with the bison a close third. *Elephas* is rare, and the reindeer disappears altogether. The animals most frequently represented are, in the order given, red deer, Equidae, bison, and Bovidae (chiefly *Bos*).

The foregoing is based on the stations of Altamira, Castillo, Hornos de la Peña (Fig. 174), and La Pasiega. The order of frequency in each is:

Altamira.—Bison, horse, red deer, Capridae, wild boar, chamois, Cervidae, *Elephas,* and wild goat.

Castillo.—Red deer (mostly female), bison, horse, *Bos,* wild goat, Capridae, chamois, and *Elephas.*

Hornos de la Peña.—Horse, bison, *Bos,* wild goat, and red deer.

La Pasiega.—Red deer (mostly female), horse, *Bos,* bison, wild goat, chamois, and *Elephas.*

FIG. 171. ENTRANCE, OPENING AT RIGHT, TO THE CAVERN OF FONT-DE-GAUME. See also Figs. 139, 153, and 154. After Capitan, Breuil, and Peyrony.

The mammals occurring in cave art are listed in the following tables, which serve as indices to the relative frequency and geographic distribution of art works dealing with mammalian forms.[8]

REPRESENTATIONS OF MAMMALS IN PALEOLITHIC ART

BADGER

France.—Gourdan.

BISON

France.—Bedeilhac; Bernifal; Bout-du-Monde; Bruniquel; Cap Blanc; Champs-Blancs; La Colombière; Combarelles; La Croze-à-Gon-

---

[8] Stations are arranged alphabetically under the various countries; each station stands for one, or many, representations of the animal in question.

tran; David; Les Eglises; Les Eyzies; Font-de-Gaume; Gargas; La Grèze; Isturitz; Laugerie-Basse; Limeuil; Lourdes; La Madeleine; Marcenac; Marsoulas; Mas d'Azil; Montespan; La Mouthe; Nancy; Niaux; L'Ombrive; Pair-non-Pair; Le Placard; Le Portel; Raymonden; Spugo; Teyjat; Trois-Frères; Tuc d'Audoubert.

Spain.—Las Aguas de Novales; Altamira; El Buxu; Castillo; Cogul; Hornos de la Peña; La Pasiega; La Pileta; Pindal; Venta de la Perra.

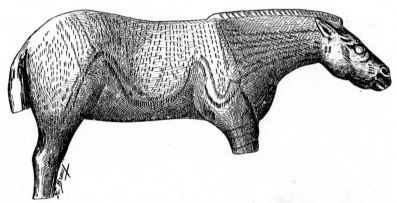

FIG. 172. SCULPTURED FIGURE OF THE HORSE, FROM THE CAVE OF ESPÉLUGUES, LOURDES, HAUTES-PYRÉNÉES, FRANCE. LOWER MAGDALENIAN EPOCH.

Representations of the horse in Paleolithic art are more widely distributed than those of any other animal (see Fig. 170); this fact, taken in connection with the quantities of horse bones in the kitchen refuse, especially at Solutré, points to the horse as the chief game animal during the Upper Paleolithic Period. Scale, full size. After Piette.

BOVIDAE [9] (chiefly *Bos*)

Czechoslovakia.—Kostelik.

France.—Arcy; Bruniquel; Les Combarelles; La Croze-à-Gontran; Enlène; Les Eyzies; Font-de-Gaume; Gourdan; Laugerie-Basse; Limeuil; Lorthet; Lourdes; La Madeleine; Mas d'Azil; Marsoulas; Le Placard; La Vache.

Spain.—Albarracin; Alpera; Altamira; El Arco; Las Batuecas; Calapatá; Castillo; La Clotilde; Cogul; Covalanas; El Charco del Agua Amarga; Hornos de la Peña; La Loja; Minateda; Monte Arabi; La Pasiega; La Pileta.

---

[9] In some cases the execution of the work is such as to make it impossible to determine which genus or species of a family was intended; in such cases the family name is employed.

Canidae

France.—Gourdan; Laugerie-Basse; Lorthet; Mas d'Azil.
Spain.—Alpera.

Canis

Spain.—Alpera; Minateda.

Capridae

France.—Bruniquel; Font-de-Gaume; Massat.
Spain.—Altamira; Castillo.

Fig. 173. Bison carved in reindeer horn, from the rock shelter of La Madeleine. Dordogne, France. Magdalenian epoch.

After the horse, the bison was one of the favorite Upper Paleolithic game animals. Photograph from the museum at Saint-Germain.

Cervidae

France.—Ammonite; La Colombière; Combarelles; Les Eyzies; Fontarnaud; Gourdan; Isturitz; Laugerie-Basse; Lorthet; Lourdes; La Madeleine; Mas d'Azil; Massat; Le Placard; Raymonden; Les Roches; Saint-Mihiel; Soucy; Tuc d'Audoubert.
Spain.—Altamira; El Arco; El Buxu; Cogul; El Charco del Agua Amarga; Minateda; Monte Arabi; El Tajo de las Figuras.

Chamois

France.—Bruniquel; Les Cambous; Gourdan.
Spain.—Altamira; Castillo; La Pasiega; Tortosillas.

ELEPHAS [10] (chiefly mammoth)

France.—Bernifal; Bruniquel; Chabot; Combarelles; La Croze-à-Gontran; David; Le Figuier; Font-de-Gaume; Gargas; Laugerie-Basse; La Madeleine; Montespan; La Mouthe; Pair-non-Pair; Raymonden; Les Roches; Saint-Mihiel; Trois-Frères.
Germany.—Klause.
Spain.—Altamira; Castillo; La Pasiega; Pindal.

FIG. 174.   GENERAL VIEW OF HORNOS DE LA PEÑA.
The cavern entrance is visible in the center, about halfway up the mountain side. After Alcalde del Rio, Breuil, and Sierra.

ELK (or moose)

France.—Gourdan; Les Rebières.
Spain.—Alpera.

---

[10] *Elephas antiquus* is probably represented in cave art at Castillo (Spain).

EQUIDAE[11]

France.—David; Isturitz; Mas d'Azil; Montespan; Les Roches; Soucy; Trois-Frères.
Switzerland.—Schweizersbild.

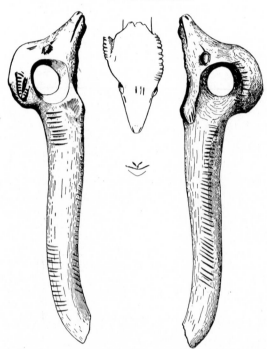

FELIS

France. — Arudy; Bruniquel; La Colombière; Les Combarelles; Font-de-Gaume; Gourdan; Isturitz; Laugerie - Basse; La Madeleine; Montespan; Trois-Frères.

FOX

France. — Arudy; Les Combarelles; Le Placard. (Fig. 175.)

GLUTTON

France.—Laugerie-Haute; Lorthet.

FIG. 175. BATON OF STAGHORN CARVED TO REPRESENT THE HEAD OF A FOX, FROM THE CAVE OF LE PLACARD, CHARENTE, FRANCE. LOWER MAGDALENIAN EPOCH.

The hole in the baton was used as a gauge for javelin shafts. Scale, *ca.* ¼. After Breuil.

HARE

France.—Isturitz.

HORSE

England.—Robin Hood; Sherborne.
France.—Arudy; Aurensan; Batie; Bernifal; Bout-du-Monde; Brassempouy; Bruniquel; Calévie; Cap Blanc; Chaffaud; La Colombière; Comarque; Combarelles; La Crouzade; La Croze-à-Gontran; Enlène; Les Eyzies; La Ferrassie; Gargas; Gourdan; Laugerie-Basse;

---

[11] Most of the equidian figures represent *Equus caballus;* a few, the wild ass.

Laugerie-Haute; Laussel; Lespugue; Limeuil; Lorthet; Lourdes; La Madeleine; Marcenac; Marsoulas; Mas d'Azil; Montespan; La Mouthe; Murat; Nancy; Neschers; Niaux; Le Placard; Pont-du-Gard; La Pépue; Le Portel; Raymonden; Sainte-Eulalie; Saint-Marcel; Saint-Mihiel; Sordes; Spugo; Teyjat; Trois-Frères; Tuc d'Audoubert.

Germany.—Obercassel.

Italy.—Romanelli.

Spain.—Albarracin; Alpera; Altamira; El Buxu; Castillo; Covalanas; Doña Trinidad; La Haza; Hornos de la Peña; Minateda; Monte Arabi; La Paloma; La Pasiega; La Pileta; Pindal; San Antonio; San Garcia; La Sotarriza; Valle.

Switzerland.—Kesslerloch.

HYENA

France.—Laussel; La Madeleine; Montespan.

LYNX

Spain.—Las Batuecas.

MOOSE (see ELK)

MUSK OX

France.—Bruniquel; La Colombière; Laugerie-Haute.
Switzerland.—Kesslerloch.

OTTER

France.—Laugerie-Basse.

RED DEER

France.—Bellet; Bout-du-Monde; Bruniquel; Chaffaud; La Chaise; Les Eyzies; Gourdan; Les Hoteaux; Laugerie-Basse; Laugerie-Haute; Laussel; Limeuil; Lorthet; Lourdes; La Madeleine; Marcenac; Mas d'Azil; Montespan; Montfort; Murat; Pair-non-Pair; Le Placard; Planche-Torte; Le Portel; Le Pouzat; Raymonden; Teyjat; Trois-Frères.

Spain.—Albarracin; Alpera; Altamira; Las Batuecas; El Buxu; Calapatá; Chiquita de los Trenta; Cortijo de los Treinta; Covalanas; Castillo; El Charco del Agua Amarga; Doña Trinidad; Estrecho de Santonje; Hornos de la Peña; Lavaderos de Tello; Minateda; Monte

Arabi; La Paloma; La Pasiega; La Pileta; Pindal; Salitré; Tortosillas; Valle.

REINDEER

France.—Bout-du-Monde; Bruniquel; La Colombière; Combarelles; Corgnac; Crozo de Gentillo; Les Eyzies; Font-de-Gaume; Gourdan; Laugerie-Basse; Limeuil; Liveyre; Lorthet; Lourdes; La Madeleine; Mas d'Azil; Massat; Montespan; La Mouthe; Le Portel; Sainte-Eulalie; Saint-Marcel; Solutré; Soucy; Teyjat; Trois-Frères; Tuc d'Audoubert.
Germany.—Schussenquelle.
Spain.—Minateda.
Switzerland.—Kesslerloch.

RHINOCEROS

France.—Arcy; La Colombière; La Ferrassie; Font-de-Gaume; Gourdan; Lourdes; Le Placard; Trois-Frères.
Spain.—Minateda.
Switzerland.—Kesslerloch.

RODENT

Germany.—Obercassel.

ROEBUCK

France.—Lorthet.

RUMINANT

France.—La Madeleine; Mas d'Azil; Pair-non-Pair; La Pépue.

SAIGA ANTELOPE

France.—Gourdan; Lacave; Laugerie-Haute; Mas d'Azil.
Spain.—Minateda.

SEAL

France.—Brassempouy; Gourdan; Montgaudier; Sordes; Teyjat; La Vache.

URSUS (chiefly cave bear)

France.—La Colombière; Comarque; Combarelles; David; Font-de-Gaume; Isturitz; Lespugue; Lourdes; Massat; Montespan; Teyjat; Trois-Frères.
Spain.—Venta de la Perra.

WILD BOAR

France.—Mas d'Azil.

Spain.—Altamira; Charco del Agua Amarga; Minateda.

WILD GOAT

France.—Arudy; Bruniquel; Combarelles; La Croze-à-Gontran; Les Eglises; Les Eyzies; La Ferrassie; Gourdan; Laugerie-Basse; Limeuil; Liveyre; Lorthet; Marcenac; Marsoulas; Mas d'Azil; Montespan; La Mouthe; Nancy; Niaux; Planche-Torte; Trois-Frères; Le Veyrier.

Spain.—Alpera; Las Batuecas; El Buxu; Calapatá; Cortijo de los Treinta; Coto de la Zarza; Castillo; Las Grajas; Hornos de la Peña; Minateda; Monte Arabi; La Pasiega; La Pileta; El Prado del Azogue; Quintanal.

WILD SHEEP

France.—La Colombière.

WOLF

France.—Bruniquel; Les Combarelles; Font-de-Gaume; Gourdan; Lourdes.

Birds.—Representations of the bird are relatively rare, but they occur both in portable and mural art. They include figures in the round, in low relief, and engravings; they adorn batons as well as dart throwers, and at least two examples of mural art have been noted. Among the figures that can be identified with a fair degree of certainty one finds the crane, duck, goose, grouse, owl, penguin, partridge, and swan (practically all edible forms).

A bird figure from the loess station of Andernach (Rhine) illustrates, perhaps as well as any other example of portable art, the ability of the artist to improvise, to seize upon resemblances, and to make of the imagination a labor-saving device. It is carved from the basal portion of a reindeer horn in which the sculptor saw the head, beak, and eyes of a bird; the wings and tail were added by means of a few incised lines on the shaft which formed the body of the bird (Fig. 176). Another notable figure of a bird, in the round, ornaments a dart thrower of reindeer horn found by Piette at Mas d'Azil (see Fig. 103). The fragmentary specimen has been restored by Breuil, who sees in the figure a grouse. The crochet of a dart thrower of reindeer horn from Raymonden is carved so as to represent the head and beak of a bird.

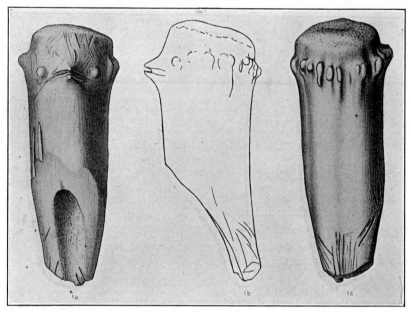

FIG. 176. FIGURE OF A BIRD ON A REINDEER HORN, FROM THE LOESS STATION OF ANDERNACH, GERMANY. UPPER MAGDALENIAN EPOCH.

By means of summary incisions and chopping off obliquely, the artist converted the basal end of the horn into this figure of a bird. Scale, *ca.* ⅔. After R. R. Schmidt.

The bird has been found in cave art at nineteen stations:

REPRESENTATIONS OF BIRDS IN PALEOLITHIC ART

France.—Arudy; Bruniquel; Fontarnaud; Gargas; Gourdan; Isturitz; Lourdes; Mas d'Azil; Raymonden; Soucy; Teyjat; Trois-Frères.
Germany.—Andernach.
Italy.—Romanelli.
Russia.—Mezine.
Spain.—Minateda; Monte Arabi; El Pendo; El Tajo de las Figuras.

Fish.—The fish occurs more frequently than the bird in cave art. It is represented in some two dozen stations in France, but is met with somewhat rarely outside. Among the kinds that figure in art, the carp (?), flounder, pike, plaice (or brill), salmon, Spanish mackerel, and trout have been identified (Fig. 177; see also Fig. 127).

There are some fifty Paleolithic stations in Europe at which harpoons of reindeer horn have been found, all dating from the Magdalenian Epoch. The harpoon may well have served in fishing, especially for salmon and pike. In about half the stations from which figures of the fish have been reported, harpoons also occur, namely at Arlay, Arudy, Bruniquel, Les Cambous, Goyet, Isturitz, Kostelik, Laugerie-Basse, Lespugue, Lorthet, La Madeleine, Mas

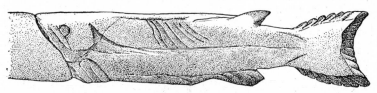

Fig. 177.   Figure of a fish (probably salmon) carved in bone, from the cave
of Rey, Dordogne, France.

The rib bone of a ruminant, split in the plane of its maximum dimensions, was carved at one end to represent this fish.   In a less perfect example from Rey (now in the Yale University Collection), the artist very ingeniously fashioned the split rib so as to represent the split body of a fish, the vertebral column and ribs alone showing on the split surface of the rib.   Scale, full size.   Redrawn from Rivière.

d'Azil, Sordes, and Teyjat. At Teyjat the harpoons and the engraving of a fish came from the same Magdalenian level. Among the remaining stations with representations of the fish, some were not inhabited during the harpoon-making epoch (Upper Magdalenian). In the Piette collection from Mas d'Azil there is the figure of a fish carved from reindeer horn and represented as having been pierced by a harpoon.

The hook was also employed in fishing. Many examples of a primitive fishhook made of a straight sliver of bone or reindeer horn, 3 to 4 centimeters (1 to 1.5 inches) long and pointed at both ends (see Fig. 96), were found at Bruniquel; other examples, including barbed hooks, are reported from some of the caves and rock shelters of the Dordogne. The engraving of a fish found by Labrie at Fontarnaud is of unusual interest because the fish is represented at biting at what appears to be intended for a barbed hook.

In certain figures of the fish the surface is so incised as to resemble the vertebral column and spines or ribs; this is true of the specimen recently found by Passemard at Isturitz and the one found by de Saint-Périer in the Grotte des Boeufs at Lespugue. An even greater stretch of phantasy is to be noted in one (now in the Peabody Museum of Yale University) of the fish figures from the cave of Rey. The body and tail of the fish are cut from a ruminant rib split in the plane of its maximum dimensions. The artist imagined the split rib to represent the split body of the fish and by means of incised lines made the framework of the fish to show on the split side only.

Perforated fish vertebrae were often used as beads during the cave-art period. In one of the Mentone caves Rivière found a Paleolithic necklace composed of twenty-four salmon vertebrae. The remarkable necklace found by Verneau with one of the skeletons (the young man) composing the triple burial at Barma Grande cave (Grimaldi) was made up of a happy combination of canine teeth of the deer, univalve shells (*Nassa neritea*), and fish vertebrae (see Fig. 109). One of the batons of reindeer horn found by Lartet and Christy at La Madeleine has a marginal decoration closely resembling a series of fish vertebrae in their anatomic relation.

Figures of the fish have been found at thirty-one stations and in at least five European countries as follows:

Representations of Fish in Paleolithic Art

Belgium

Goyet (Namur).—Trout engraved on a baton of reindeer horn; discovered by Dupont.

Czechoslovakia

Kostelik (Moravia).—Ramus of the lower jaw of a horse carved to represent the form of a fish; discovered by Křiž.

France

Arlay (Jura).—Engraving reported by Girardot.
Arudy (Basses-Pyrénées).—Fish carved from ivory and one

engraved on reindeer horn; discovered by Mascaraux in the cave of Saint-Michel.

Brassempouy (Landes).—Engraving found by Piette.

Bruniquel (Tarn-et-Garonne).—Engraving on bone from the Château rock shelter; discovered by Cartailhac.

Cambous, Les (Lot).—Tail of a fish engraved on reindeer horn; published by Bergougnoux.

Chaffaud (Vienne).—Fish carved in reindeer horn with contours cut away (Gaillard de la Dionnerie collection at Poitiers).

Croze de Tayac, La (Dordogne).—Engraving on an antler of Cervidae; published by Rivière.

David (Lot).—Pike.

Fontarnaud (Gironde).—Engraving on reindeer horn (Laôrie collection).

Gorge d'Enfer (Dordogne).—Large figure in relief of a salmon on the ceiling of the Grotte du Poisson; discovered in 1912 by Marsan.

Gourdan (Haute-Garonne).—Pike engraved on reindeer horn; a fish tail carved in the round from the palmate portion of a reindeer horn and provided with a hole for suspension as a pendant or charm; both published by Piette.

Isturitz (Basses-Pyrénées).—Salmon (or trout) engraved on a baton; discovered by Passemard.

Laugerie-Basse (Dordogne).—Pike engraved on a fragment of reindeer lower jaw; some sixteen other examples engraved on bone and reindeer horn, one of which (probably salmon) is associated with the figure of an otter.

Lespugue (Haute-Garonne).—Fine figure of the flounder made of bone with the contours cut away; recently discovered by de Saint-Périer in Magdalenian deposits of the Grotte des Boeufs. Engraving of fish on bone from the Grotte des Harpons.

Lorthet (Hautes-Pyrénées).—Several salmon in juxtaposition with two stags and a hind, engraved on a baton of reindeer horn with the artist's mark affixed (two lozenge-shaped signs); figure made of flat bone with contours cut away, Piette collection (see Fig. 127).

Lourdes (Hautes-Pyrénées).—Figure made of flat bone with contours cut away; Piette collection.

Madeleine, La (Dordogne).—Carp (?) engraved on both sides of a section of reindeer horn; figures engraved on batons of reindeer horn. Marginal decoration resembling a series of fish vertebrae in their anatomic relation engraved on a baton. Published by Lartet and Christy.

Mas d'Azil (Ariège).—Fish carved from reindeer horn and represented as having been pierced by a harpoon; two other figures on bone and one on reindeer horn (one of these was apparently used as a dart thrower, a fin forming the crochet). Piette collection.

Niaux (Ariège).—Two figures of the trout incised in the compact clay of the cavern floor of the diverticulum at the entrance to the *salon noir*. Published by Cartailhac and Breuil.

Placard, Le (Charente).—Javelin points decorated with stylistic figures of the fish, one of which resembles the flounder (eyes not indicated). Published by Breuil.

Pont-du-Gard (Gard).—Figures of the fish and horse on bone; discovered by Cazalis de Fondouce in the cave of Salpêtrière.

Rey (Dordogne).—Two sculptured figures, probably salmon, each carved from a ruminant rib split in the plane of its maximum dimensions. In one, the entire fish is represented with a high degree of artistic skill (see Fig. 177); in the other, the body and tail only are represented (Yale collection).

Sordes (Landes).—Engraved figure of a pike on a perforated canine tooth of the cave bear; discovered by Lartet and Duparc in the cave of Duruthy.

Teyjat (Dordogne).—Engraved figure of a fish on a spatulate bone implement from the cavern of La Mairie. Published by Capitan, Breuil, Bourrinet, and Peyrony.

Trois-Frères (Ariège).—Figure of a fish on bone; discovered by Count Begouen.

Serpentiform figures, probably of the eel, or snake, have been reported from Gourdan, Lorthet, La Madeleine, Montgaudier, and Teyjat.

POLAND

Wierzchow (Galicia).—Engraving of a fish on the rib of a reindeer; discovered by Zavisza.

SPAIN

Batuecas, Las (Salamanca).—Mural painting of the fish, probably of Azilian age; discovered by Breuil.

Minateda (Albacete).—Mural figures of the fish; reported by Breuil.

Pileta, La (Malaga).—Several engravings of the fish, one of which, 1.5 meters (4.9 feet) long, represents a marine fish, probably a plaice or brill. Published by Breuil, Obermaier, and Verner.

INVERTEBRATES IN PALEOLITHIC ART

The invertebrate world does not seem to have appealed to the Paleolithic artist. Invertebrates did not serve the needs of man to any appreciable extent; when wanted, they could be obtained without resort to magic. Five examples only have been reported—four from France and one from Belgium:

1. An ivory beetle of Magdalenian age from Cap Blanc.
2. An ivory ladybug from Laugerie-Basse.
3. The facsimile of a *Cypraea* shell carved from ivory, with a large loop for suspension, from Pair-non-Pair (Middle Aurignacian).
4. A lignite coleopter from Arcy.
5. An ivory coleopter from the Grotte du Coléoptère.

FLORA IN PALEOLITHIC ART

As was the case with invertebrates, and perhaps for similar reasons, the plant world was of little concern to the cave artist. Figures that might be construed as plant representations have been found at only a few stations, including Arcy (Grotte du Trilobite), Gaubert, Gourdan, Lourdes, Marsoulas (?), Mas d'Azil, and Le Veyrier (Fig. 178).

THE INANIMATE IN PALEOLITHIC ART

Portrayal of animal forms was the cave artist's chief concern, with decided predilection for vertebrates, and among the ver-

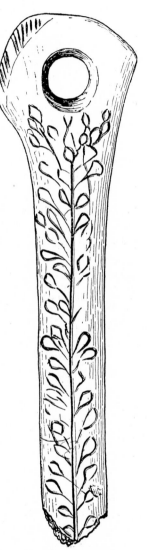

FIG. 178. PLANTLIKE FIGURE INCISED ON A BATON, FROM THE CAVE OF LE VEYRIER, HAUTE-SAVOIE, FRANCE. UPPER MAGDALENIAN EPOCH.

This is one of the few known examples of plantlike figures in Paleolithic art. Scale, ½. After Breuil.

tebrates Mammalia easily ranked first. There is but a faint pictorial reminder of the cave man's contact with the invertebrate life about him; the same is true so far as the plant world is concerned. Inanimate objects fared somewhat better; among portrayals of these are classed decorative motives that might well have been derived from animal forms through processes of conventionalization, such as chevrons, frets or grecques, spirals, volutes, etc.

Alphabetiform Signs.—Before the close of the Magdalenian Epoch symbolism began to play an important rôle in Paleolithic art. Piette believed these Magdalenian symbols to be figures or images employed as signs of objects, hence representative of words. One of the early Paleolithic symbols is the dotted circle, supposed to be a sun symbol (found at Gourdan and Lourdes); it reappears as an Egyptian hieroglyph, also on dolmens and menhirs, on

FIG. 179. BIRD BONE INCISED WITH ALPHABETIFORM SYMBOLS, FROM THE CAVE OF LE PLACARD, CHARENTE, FRANCE. MAGDALENIAN EPOCH.

Marks such as these have been interpreted as alphabetiform signs. Piette even went so far as to distinguish two successive systems of writing in the Magdalenian Epoch, hieroglyphic and cursive. After Piette.

Bronze Age funerary urns, and on ornaments of the Iron Age. The circle without the dot passed into the ancient alphabets and from them into modern alphabets. The lozenge was employed as an artist's signature at Lorthet.

Piette distinguished two successive systems of writing in the Magdalenian Epoch, the hieroglyphic and the cursive; he believed the latter to be derived from the former, but admitted that since symbols are the results of convention, they may from the beginning have been figures formed by geometric lines instead of being simplified images.

Alphabetiform symbols which have been interpreted as proof of the existence of primitive writing during the Magdalenian Epoch have been found at a number of stations in France, including Crozo de Gentillo (Lot), Gourdan (Haute-Garonne), Lorthet (Hautes-Pyrénées), La Madeleine (Dordogne), Mas d'Azil (Ariège), and Le Placard (Charente); and in Spain at La Pasiega (Fig. 179).

Chevrons, Frets, Spirals, Volutes, and Wave Ornaments.—As has been previously stated, ornamental motives may be derived from realistic originals. To what extent this is true of Paleolithic chevrons, frets (or grecques), spirals, volutes, sigmoids, etc., it would be difficult to say (Fig. 180). They are no doubt, in a measure the result of a tendency to conventionalize, to standardize, to symbolize, after the tide of realism had begun to ebb. The chevron is employed at a number of stations, including Mas d'Azil, Sordes, and Teyjat in France, and Wildscheuer in Germany.

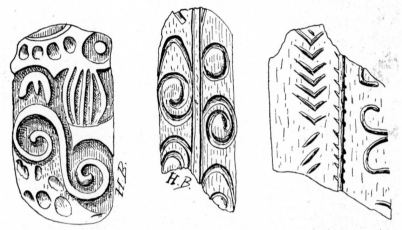

FIG. 180. CONVENTIONALIZED FIGURES ENGRAVED ON REINDEER HORN, FROM LOURDES, AND ARUDY, FRANCE. MAGDALENIAN EPOCH.

The one on the left (sigmoid) is from Lourdes; the others (volutes and chevrons), from the cave of Espélungues at Arudy. Such figures have been found at a number of stations. They were probably derived from such animal features as eyes, horns, etc. After Piette.

Spirals, sigmoids, and volutes, identical in pattern and technique, have been found at Arudy (Basses-Pyrénées), Lespugue (Haute-Garonne), and Lourdes (Hautes-Pyrénées) in France; and at Hornos de la Peña (Santander), Spain. A wave ornament incised on a large rib found in Solutrean deposits at Předmost (Moravia) is unique so far as the author's knowledge goes. The fret, or grecque is developed to an unexpected degree in the loess station of Mezine (Ukraine). The patterns, incised on ivory, were reported by Volkov as belonging to the Upper Aurignacian; but Tchikalenko believes them to date from near the close of the Magdalenian Epoch.

Fig. 181. MURAL ENGRAVING OF THE HEAD AND NECK OF A REINDEER (FACING TO THE LEFT) IN THE CAVERN OF TUC D'AUDOUBERT, ARIÈGE, FRANCE.

Across the head, between the eye and ear, there is an incised claviform or club-shaped figure indicating that the animal is marked for slaughter. An additional claviform sign may be seen just above the antlers. Scale, ca. $\frac{3}{10}$. After Count Begouen.

Claviform Signs and Darts.—Club-shaped figures, both engraved and in color, and, so far as the author can recall, found only on cavern walls, have been discovered in a number of French and Spanish stations. They are associated for the most part with animal figures and are no doubt of magic import. Examples occur

FIG. 182. GENERAL VIEW FROM THE CAVERN OF PINDAL, ASTURIAS, SPAIN, OVER-
LOOKING THE BAY OF BISCAY.

After Alcalde del Rio, Breuil, and Sierra.

in three important caverns of Ariège—Niaux, Trois-Frères, and
Tuc d'Audoubert. In the latter a claviform is incised across the
head of a reindeer engraved on the cavern wall (Fig. 181). The
claviforms in Niaux and Trois-Frères, in color, are not directly

associated with animal figures. At Combarelles (Dordogne) a figure that admits of being interpreted as a claviform is incised longitudinally on the head of a horse.

The claviform is found in northern Spain at Altamira and La Pasiega (Santander) and Pindal (Asturias). More than a dozen claviforms are painted on the famous ceiling at Altamira, some of

FIG. 183. WOUNDED BISON AND CLAVIFORM SIGNS IN THE CAVERN OF PINDAL.

This depicts the magical means invoked by the hunter for success in the chase. The figures are partly engraved and partly painted in red. At the right of the bison are the outlines in red of a horse's head and neck. Scale, ⅕. After Alcalde del Rio, Breuil, and Sierra.

them encroaching upon the space occupied by the legs of the large figure of the hind. Beneath a wounded bison at Pindal are six club-shaped figures in color, all oriented in the same manner (Fig. 183). The original bludgeons for which these stand were obvi-- ously of wood since none has been preserved; they were about 1 meter (39.4 inches) in length if the artist drew them and the accompanying bison to the same scale. Some of the clay statues and figures in high relief of various animals in the newly discovered cavern of Montespan (Haute-Garonne) are riddled with punctures representing dart wounds. Darts are often depicted as dangling from the sides of wounded animals; the best known examples are from Niaux and Tuc d'Audoubert in Ariège and Pindal in northern Spain. The figure of a harpoon, incised on a pebble, was found at Gourdan in Haute-Garonne.

Tectiforms.—Tent-shaped figures, incised and in color, have been found on the cavern walls of both France and Spain. They are often placed directly on animal figures, as if to imply a direct,

FIG. 184. TENT-SHAPED SIGNS PAINTED OVER THE POLYCHROME MURAL FIGURE OF A BISON IN THE CAVERN OF FONT-DE-GAUME.

These tectiforms may represent the temporary shelters in use among the Upper Paleolithic races. Magdalenian Epoch. Scale, *ca.* ⅛. After Capitan, Breuil, and Peyrony.

or desired, association. Two of these tectiforms resembling a front view of two overlapping tents without sides are incised on the figure of a mammoth at Bernifal (Dordogne). Two tectiforms of more elaborate form, including sides and openings, are painted on the body of the mammoth at Font-de-Gaume (Fig. 184). Two other similar figures are painted on the body of a reindeer at Font-de-Gaume, where a total of nineteen tectiforms has been listed. Other French stations in which tectiforms have been found include: Les Combarelles and La Mouthe (Dordogne), Les Eglises and Trois-Frères (Ariège), and Marsoulas (Haute-Garonne). Tectiform signs also occur in northern and southern

Spain: Bolao (Oviedo), Castillo and La Pasiega (Santander), and La Pileta (Malaga).

The term tectiform is well chosen because of the resemblance of the figures to primitive dwellings, and because of the probability that they actually represent the kind of temporary abode employed by the Upper Paleolithic races. Man of that period dwelt in caves and rock shelters wherever they were found, and if habitable; elsewhere he had recourse to artificial and more or less temporary shelters. The tectiforms reproduce in a remarkable manner the simple shelters, tents, and huts in use to-day among primitive and nomadic races in various parts of the world.

Materials Employed in Paleolithic Art.—Artists of the historic period have a wide range of materials to serve as a vehicle of art expression,—chiefly canvas, paper, textiles, plaster, the various metals, pottery, porcelain, glass, skins, clay, bone, horn, ivory, wood, stone, and coloring matter. The first seven of these were certainly unknown to the Paleolithic artist, who was limited to what he found ready to hand in nature. This limitation, together with the paucity of utensils at his command, was a serious handicap to the artist of the Old Stone Age. It is known from the record, that in portable art he made use of stone, bone, ivory, reindeer horn, and staghorn, the chief vehicles being bone and stone. Bones of both mammals and birds were freely employed, usually without any previous preparation. Various kinds of stone were pressed into service, some of them quite hard, others soft. The list includes: crystalline talc, lignite, schist, gritstone, slate and several varieties of limestone. Sometimes pebbles were employed.

All these ready-to-hand materials are of a more or less non-perishable nature; but it would not be logical to assume that such were the only materials employed. Wood is well adapted as a medium of art expression; it was available and the cave artist had tools well adapted for working it. Wood is widely used for art purposes by primitive living races; there is every reason, therefore, to assume that the paleolithic artist made free use of wood from which, no doubt, many of his ornamented dart throwers were made. The skins which served for clothing were probably decorated, and the practice of painting the body might also have been in vogue.

Stationary art, which adorns the walls, ceilings, and sometimes even floors of caverns, and the walls of rock shelters, if engraving or fresco, is done in stone and in rare cases, stalagmite; if painted, ocher and oxide of manganese furnished the coloring matter. In the field of stationary art, clay was used perhaps much more extensively than the record would indicate; for only under exceptionally favorable conditions could one expect figures of untempered and unbaked clay to survive the ravages of time.

CHAPTER VIII

FOSSIL MAN

Physically, man is a vertebrate and belongs to the great class of Mammalia. We may differentiate further and place man in one of the families composing the order Primates, which includes the Simiidae and the tailed monkeys of the Old and the New World. In physical structure man most nearly approaches the Simiidae, the family to which the chimpanzee, gorilla, orang-utan, and gibbon belong. These are all Old-World forms, the gorilla and chimpanzee being found in Africa, and the orang and gibbon in the Far East.

The relation between man and the anthropoids seems to be as close physiologically as it is structurally. Friedenthal, Uhlenhuth, and others have made tests of the blood of man which go to show that chemically it resembles that of anthropoids quite as closely as the blood of the dog resembles that of the wolf, for example; or that of the horse, the donkey.

It is hardly necessary in this place, however, to insist upon man's similarity in structure to the anthropoids, and even to more lowly forms. Neither shall we discuss the bearing of embryology and vestigial structures on the nature of man's organic evolution. It is enough to say that both classes of phenomena lead to the conclusion that humanity is an integral part of the animate world.

The principal anatomic differences between man and the higher apes are those arising from adjustment to posture. The erect posture requires certain anatomic adjustments and facilitates others. The sum total of these adjustments roughly represents the differences between the human and the simian organism. The chimpanzee head, perched upon a vertical column, would find itself in unstable equilibrium; increase in the size of the brain case and decrease in the size of the face restores the equilibrium. The three changes are interdependent; it would be difficult, illogical, to imagine any one of them as taking place independently of the other two. In other words, if a human head supported by a chimpanzee spinal

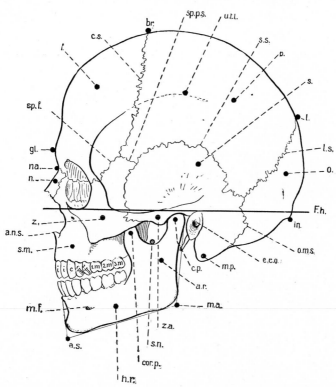

FIG. 185. SIDE VIEW OF A MODERN SKULL, WITH KEY TO CRANIAL STRUCTURE.

s.m., superior maxillary (upper jaw); *a.n.s.*, anterior nasal spine; *z.*, zygoma (cheek bone); *n.*, nasal bone; *n.a.*, nasion (point where the internasal suture meets the naso-frontal suture); *gl.*, glabella (median point between the brow ridges or eyebrows); *sp.f.*, spheno-frontal suture; *f.*, frontal bone; *c.s.*, coronal suture; *br.*, bregma (point where the sagittal suture meets the coronal suture); *sp.p.s.*, spheno-parietal suture marking the region of the pterion (in the apes this suture is normally absent and in its place there is a fronto-squamosal articulation); *u.t.l.*, upper temporal line; *s.s.*, squamosal suture; *p.*, parietal bone; *s.*, squamosal or temporal bone; *l.*, lambda (point where the sagittal suture meets the lambdoid suture); *l.s.*, lambdoid suture; *o.*, occipital bone; *F.h.*, Frankfort horizontal (plane passing through the lower margin of the orbits and the upper margin of the external ear opening); *in.*, inion (median point on the external occipital protuberance); *o.m.s.*, occipito-mastoid suture; *e.e.o.*, external ear opening; *m.p.*, mastoid process; *c.p.*, condyloid process; *a.r.*, ascending portion of the ramus of the lower jaw (ramus mandibulae); *m.a.*, mandibular angle; *z.a.*, zygomatic arch; *s.n.*, sigmoid notch; *cor.p.*, coronoid process; *h.r.*, horizontal portion of the lower jaw (corpus mandibulae); *a.s.*, angle of symphysis; *m.f.*, mental foramen.

column is unthinkable, so are a human spinal column and brain case in connection with a chimpanzee face.

The spinal column must carry a different load, and must carry it in a different way. Instead of being supported at both limb girdles, it is supported at only one—the posterior. Its former anterior support becomes a load in addition to the original load —the head. Obviously the only recourse for the spinal column, weakened by the loss of one support to carry an additional burden,

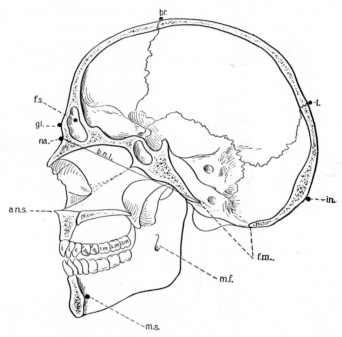

FIG. 186. SAGITTAL SECTION OF A MODERN SKULL.

a.n.s., anterior nasal spine (akanthion, nasospinale); *na.*, nasion; *gl.*, glabella; *f.s.*, frontal sinus; *br.*, bregma; *l.*, lambda; *in.*, inion; *f.m.*, foramen magnum (occipital foramen; the median point on the anterior margin of the foramen magnum is known as the basion and marks one end of the basi-nasal line); *m.f.*, mandibular foramen (inferior dental foramen); *m.s.*, mental spine (genial tubercles) serving for the attachment of muscles which move the tongue in speech; *b.n.l.*, basi-nasal line, a straight line connecting the nasion with the base of the cranium (basion).

is to get directly under its load, that is, to assume a vertical position. Assumption of the vertical position means a change in the angle formed between the axis of vision and the axis of the spinal column from an obtuse angle to a right angle. The increase in the size of the brain case and the decrease in the size of the face not only permit such an adjustment, but are inconsistent with any other state of affairs (Figs. 185 and 186).

Meanwhile the spinal column undergoes change to meet the new conditions. Instead of presenting a single long curve, concave in front, from the atlas to the last lumbar vertebra, it presents a triple curve: a cervical curve, convex in front; a thoracic curve, concave in front; and a lumbar curve, convex in front. The sacral curve not only remains concave in front, but the concavity is largely increased. The sacrum also becomes broader and shorter under the additional load (see Figs. 233 and 234).

With the change to erect posture, the thorax gains in transverse diameter, but loses correspondingly in antero-posterior diameter. This change is reflected in the greater curvature of the ribs in man. The spinous processes of the cervical vertebrae, set in a position more oblique to the axis of the spinal column, are bifid and relatively small.

The pelvis, as a whole, undergoes profound changes due to the fact that it must support not only the entire body but the anterior limbs as well. In addition, it must serve as a direct support for the abdominal viscera. Coincident with the broadening and shortening of the sacrum, the ilia increase in breadth and decrease in length; they also become concave, forming a real basin. The angle of the great sciatic notch becomes smaller and the sciatic spine more pronounced.

The erect posture also profoundly affects the legs and feet. The femur straightens and is provided with a posterior longitudinal crest or ridge. The straightening process changes the angle of the tibial articular facets at the knee. Terrestrial habits and the erect posture combine to release the foot from all save a single function —that of support. The new conditions imposed require many adjustments affecting the heel and the toes, especially the great toe, which gains in size and strength what it loses in flexibility and opposability.

Morton contends that the primate must pass through the arboreal period before the first steps leading to an erect posture can be taken. Arboreal life encouraged the development of the foot and caused the shifting of the axis of support to a line between the great toe and the rest of the foot. At present, man shares the constant employment of this inner axis of support only with the chimpanzee, gibbon, and gorilla (Fig. 187). All four groups are made up of relatively large individuals. With increase in size, the use of the tree tops as avenues of travel was no longer possible; arboreal life, therefore, was restricted to tree trunks and large branches, a condition which would encourage the tendency toward the assumption of an erect posture.

Terrestrial conditions do not encourage such a change of posture, but they are such as to welcome the change once it had been initiated. The modification of the human frame for the erect pos-

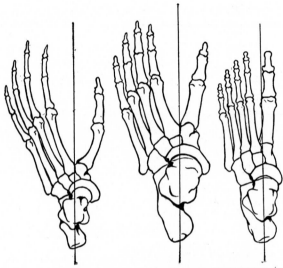

Fig. 187.　Axis of support of the left foot of the chimpanzee (left) and the gorilla (center) compared with that of man (right).

After Morton.

ture could take place only after an extended period of arboreal existence prior to the ultimate adoption of terrestrial life. The apes, halting on their way from the trees to the ground, have never completely gained erect posture; the complete break with arboreal life, successfully made only by the human precursor, has been rewarded to an unusual degree. In giving up the arboreal domain, the human precursor won the far richer and vaster terrestrial domain with its unlimited possibilities for cultural evolution.

DISTRIBUTION OF THE PRIMATES

The primates appear in the early Tertiary and recur in every geologic epoch since that time, as seen in the tabulation opposite.

GEOGRAPHIC DISTRIBUTION OF THE PRIMATES

	AFRICA	EUROPE	ASIA	NORTH AMERICA	SOUTH AMERICA
RECENT	Chimpanzee (*Pan* or *Troglodytes*) Gorilla (*Gorilla*) Cynopithecidae Lemuridae Cheiromyidae Galaginae	*Macacus*	Orang (*Pongo* or *Simia*) Gibbon (*Hylobates*) Cynopithecidae Lorisinae Tarsiidae		Cebidae Hapalidae
PLEISTO-CENE	*Macacus* *Cynocephalus* *Megaladapis* *Lemur* *Paleopropithecus* *Archeolemur* *Bradylemur* *Hadropithecus*	*Macacus*	*Pithecanthropus* *Semnopithecus* *Cynocephalus*		*Cebus* *Mycetes* *Eriodes* *Callithrix* *Hapale*
PLIO-CENE		*Anthropodus* *Dryopithecus* *Dolichopithecus* *Macacus* *Semnopithecus* *Mesopithecus* *Oreopithecus*	*Paleopithecus* *Macacus* *Semnopithecus* *Cynocephalus*		
MIO-CENE		*Pliopithecus*			*Homunculus*
OLIGO-CENE	*Propliopithecus* *Parapithecus* *Moeripithecus*				
EOCENE		*Caenopithecus* *Adapis* *Cryptopithecus* *Necrolemur* *Pronycticebus*		*Pelycodus* *Notharctus* *Anaptomorphus* *Trogolemur* *Omomys*	

The lemurs, now most abundant in Madagascar, represent the Eocene cycle of primate dispersal; a few are still found in central and western Africa, southern India, and the East Indies. Forms closely related to modern lemurs occur in the Eocene of Europe and North America (United States), but disappear from these continents at the end of that epoch. In South America they seem

FIG. 188. ADULT RUFFED LEMUR FROM MADAGASCAR.
Photograph from the New York Zoölogical Park.

to have evolved into the so-called New-World monkeys, while in the Old World they gave rise to the higher primates. Matthew believes the center of lemur dispersal to have been Asia, north of the Himalayas, followed by the dispersal of Old-World monkeys and baboons from a central Asiatic center, dating back at least to the beginning of the Oligocene. The anthropoid apes, surviving to-day in the forests of western Africa and the East Indies, represent a late Tertiary cycle of dispersal, probably from a central Asiatic source.

Fossil Primates.—The primate stem took root at the beginning of the Tertiary (Eocene) in North America. Its earliest representatives were closely related to the insectivores. In respect to skull, brain, and dentition, they resembled certain living lemurs

FIG. 189. WHITE-THROATED SAPAJOU (CEBUS) FROM SOUTH AMERICA.
Photograph from the New York Zoölogical Park.

of the Malay archipelago. The species that Cope would have made the common ancestor of man and the apes is the small primate known as *Anaptomorphus homunculus.* Its muzzle is short and its brain relatively large. The North American Lemuridae become more highly differentiated in the Miocene with a tendency toward types that are still living.

From North America the earliest primates migrated to South

America, becoming the ancestors of the living *Platyrrhinians.*
Ameghino invented a whole series of genera to connect the earliest
forms with *Homo,* such as *Homunculus patagonicus, Anthropops
perfectus, Tetraprothomo, Triprothomo, Diprothomo,* and finally
Prothomo. As these never existed save in Ameghino's imagination,
the sooner the terms are forgotten the better.

FIG. 190.   GIBBON (HYLOBATES) FROM SUMATRA; ADULT FEMALE.
Photograph from the New York Zoölogical Park.

The earliest Lemuridae of western Europe belong to the middle
Eocene of France and Switzerland. The phosphoritic deposits of Lot
and Lot-et-Garonne have yielded complete crania of several interesting genera—*Adapis, Necrolemur,* and *Pronycticebus.*   Somewhat related types still live in Asia and Africa, especially Madagascar.

The Catarrhinian apes of the Old World and the Simiidae

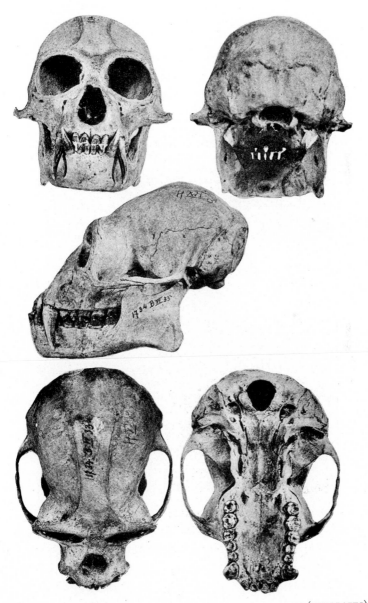

FIG. 191. FIVE ASPECTS OF THE SKULL OF AN ADULT GIBBON (HYLOBATES).
Scale, ½. After Oppenheim.

appeared in the Miocene. An interesting link in the chain connecting the diminutive primitive *Anaptomorphus* with *Pliopithecus,* to which he gave the name *Propliopithecus haeckelii,* was discovered by Schlosser in 1910 in the fluvio-marine Oligocene of Fâyum, Egypt.

There is a distinct affiliation between the Miocene apes of Europe and Asia and living forms. In 1837 Lartet announced the

FIG. 192. ORANG FROM BORNEO; MALE, NOT FULL GROWN.
Photograph from the New York Zoölogical Park.

discovery at Sansan of *Pliopithecus antiquus,* a type related to the gibbon; it was thought by Schlosser to be descended from *Propliopithecus* of Egypt. *Pliopithecus* has since been found in other parts of Europe.

In 1856 Lartet discovered in southern France, near Saint-Gaudens, *Dryopithecus fontani,* a form considered by him to be even more closely related to man. A more complete lower jaw discovered in 1890, however, proved to the satisfaction of Gaudry that *Dryopithecus* was a form inferior to the living anthropoids. Pilgrim has found three species of *Dryopithecus* in the Siwalik Hills of India. The genus is now so well represented as to make plain its ancestral relationship with the living chimpanzee and gorilla.

The fossil ape found at Monte Bamboli in Tuscany and described by Gervais under the name *Oreopithecus bambolii* partakes of the characters of both Cynomorphs and Anthropomorphs.

One of the best known species of the fossil ape is *Mesopithecus pentelici*, found at Pikermi in Greece by Gaudry, who discovered the remains of twenty-five individuals and was able to reconstitute a complete skeleton. According to Gaudry, *Mesopithecus* had two limbs suggestive of *Macacus* and a dentition related to that of *Semnopithecus*.

Pilgrim has once more turned the attention of paleontologists and anthropologists to the Siwalik Hills of northern India because of his recent discovery there of a late Miocene ape to which he has given the name *Sivapithecus indicus*. As yet the species is known only from fragments—a few isolated teeth and two pieces of lower jaw. It appears that the molars resemble human molars more closely than do those even of living anthropoids. For this reason Pilgrim looks upon *Sivapithecus* as a transition form between anthropoids and man. In his critique of Pilgrim's view, Gregory points out that the lower jaw of *Sivapithecus* bears a closer resemblance to that of *Dryopithecus* or to the lower jaw of a female orang than it does to the most primitive human lower jaw.

The fossil apes discovered in the Pliocene as well as the Pleistocene of Europe are more or less closely related to the living Cynomorphs. Both *Semnopithecus* and *Macacus* are found in the Pliocene of France and Italy. A fossil *Macacus* closely related to living species of the same genus has been found in the Quaternary of several European countries, including England, France, and Germany. As yet the Pliocene and Pleistocene of Europe have yielded no remains of Simiidae unless the lower jaw of Piltdown and, perchance, a tooth from Taubach may be counted as such.

Living Primates.—The present home of the Lemuridae or Prosimians is restricted to Madagascar (where ninety per cent

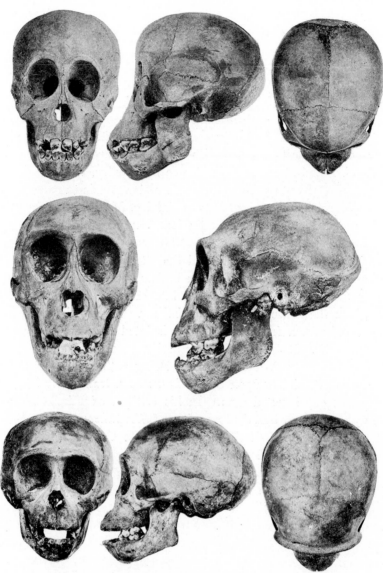

Fig. 193. Various aspects of the skulls of a young orang-utan (above), young gorilla (center), and young chimpanzee (below).

Scale, ⅓. After Oppenheim.

of them live), Africa, and Indo-Malaysia. Lemurs are arboreal in their habits. Their legs are large in comparison with their arms; the phalanges are armed with claws. The skull differs from that of other primates in that the muzzle is more pronounced, the orbits are turned more laterally and communicate with the temporal fossae, and the position of the foramen magnum is farther back.

FIG. 194. GORILLA FROM THE FRENCH CONGO; FEMALE, NOT FULL GROWN.
Photograph from the New York Zoölogical Park.

Lemurs represent the lowest level of intelligence among Primates (Fig. 188).

Monkeys and anthropoids form a more or less graduated series from the lowly Hapalidae to the chimpanzee, which makes the nearest approach to man, both physically and mentally. All have certain characters in common which distinguish them from the lemurs and from the ranks of the lower mammals. The brain case is large in comparison with the face; with the shortening of the muzzle the face gains in mobility and in ability to reflect a higher degree of intelligence. The foramen magnum opens at the base rather than at the back of the cranium, and the cerebrum overhangs the cerebellum. The orbits no longer communicate with the temporal fossae. The arms are longer than the legs; the forearm is capable of pronation and supination and serves as an excellent

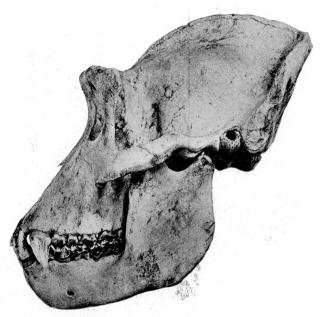

FIG. 195.   SIDE VIEW OF THE SKULL OF AN ADULT GORILLA.
Scale, ⅓.   After Oppenheim.

organ of prehension. The food of apes consists chiefly of tender branches, fruits, eggs, and insects.

Apes are divided into two groups, the Platyrrhinians of the New World and the Catarrhinians of the Old World. The Platyrrhinians are so named because of their wide nasal septum and hence widely separated nostrils. In the Catarrhinian group the dentition is the same as in *Homo*—two incisors, one canine, two premolars, and three molars. In the Platyrrhinian group there are family differences as to dentition: the Hapalidae have thirty-two teeth, the same number as in *Homo,* but instead of two premolars and three molars, we find three premolars and two molars; the Cebidae have not only three premolars, but also three molars, making thirty-six teeth in all.

The Hapalidae are the smallest of monkeys. They resemble lemurs in that their phalanges are armed with claws instead of nails; the thumb is not opposable. Among the Cebidae the thumb is slightly opposable. The tail is long and generally prehensile. The present home of both Hapalidae and Cebidae is in South America (Fig. 189).

The Catarrhinians comprise two great groups, the Cercopithecidae or Cynomorphs and the Simidae or Anthropomorphs (also known as anthropoids). The Cercopithecidae have tails; the Simiidae are tailless, but both groups agree in having the same dental formula as *Homo*. The Cercopithecidae are found both in Asia and in Africa and include a number of genera—*Cercopithecus, Semnopithecus, Cynocephalus,* and *Macacus.* Among the Cercopithecidae the tail is never prehensile; it is sometimes rudimentary.

Anthropomorph or anthropoid rather than simian more nearly expresses the true position of the great tailless apes in the primate scale. They belong to four genera, *Hylobates, Simia, Gorilla,* and *Troglodytes* (also known as *Anthropopithecus* and *Pan*).

To the genus *Hylobates* belong the gibbons of southeastern Asia and the Malay archipelago (Figs. 190 and 191). They are the smallest of anthropoids and, in certain characters, the most lowly. In cranial conformation and in posture, however, they resemble *Homo.* Their arms are very long; in the largest species, the siamang, the stature rarely exceeds one meter.

The genus *Simia* is better known by the names orang and pongo. It is found in Borneo and Sumatra. The orang is larger and more powerful than the gibbon (Fig. 192). Its legs are short, its head relatively short or brachycephalic, and its canines powerful (Fig. 193).

The genus *Gorilla* is the largest of anthropoids, surpassing even *Homo* in bulk and physical strength (Fig. 194). Its home is in central west Africa. Its skull is dolichocephalic and provided with enormous orbital, temporal, and occipital crests, especially in the adult male (Figs. 193 and 195); its canines are very large.

Western Africa is also the home of the chimpanzee, the most

intelligent and manlike of all the apes (Fig. 196). The cranial
contours are not so rugged as in the gorilla; neither are the sexual
differences so pronounced. Like the gorilla and gibbon, the chimpanzee is dolichocephalic (Fig. 193).

In passing from Simiidae to Hominidae, one is struck by the
wide differences in respect to geographic distribution at the present time. Whereas all
four genera of anthro-
poids are confined to
relatively small por-
tions of only two con-
tinents, both in the
Old World, Homini-
dae a r e found in
every continent and
the islands of every
sea. It is customary
to speak of the liv-
ing representatives of
Hominidae as belong-
ing to only one genus
and species — *Homo
sapiens.* W h e n it
comes to dividing the
species *sapiens* i n t o
races or varieties, one
is struck by the di-
vergences of opinion
depending u p o n the
individual   classifier
and the p e r i o d in

FIG. 196. CHIMPANZEE FROM THE FRENCH CONGO;
MALE, NOT FULL GROWN.

Photograph from the New York Zoölogical Park.

which he lived. Leaving out the classifications of antiquity, the
system of Linnaeus is the first to deserve serious notice. He distinguished four or five races, as Blumenbach and Cuvier did in
their turn.

THE BIRTHPLACE OF HOMO

Since the members of the Simiidae, the family nearest akin to man, living as well as fossil, are all found in the Old World,[1] the presumption is that the Old World, probably the central plateau of Asia north of the Himalayas, was the cradle of the human race, a presumption which finds confirmation in the nature of the human skeletal and cultural remains which have thus far come to light. No one who has compared the finds from the two hemispheres can fail to be impressed by the cumulative evidence pointing to a much greater antiquity for man in the Old World than in the New. The climatic conditions attending the birth of *Homo* were probably temperate and more or less arid, since his loss of body hair was apparently due more to the clothes-wearing habit than to tropical climate.

The beginnings of things human, so far as we have been able to discover them, have their fullest exemplification in Europe. The cradle of the human race has not yet been definitely located. When found, it will no doubt prove to be at least within easy reach of Europe, which, structurally, is the keystone of the Old World arch —still planted firmly against Asia and once in more intimate contact with Africa than at present. Asia and Africa have not been so thoroughly explored, prehistorically speaking, as Europe has; both are full of archeological possibilities. It should be recalled that Darwin looked with favor on Africa as the place where man might have originated, because it is still the home of the gorilla and chimpanzee, the simian forms most nearly approaching man. Moreover, *Propliopithecus,* a fossil ape recently discovered by Schlosser in the fluvio-marine Oligocene of Egypt, seems to be the ancestor not only of all Simiidae, but of Hominidae also.

The superficial resemblances between the archaic Neandertal race and the modern Australians led Schoetensack to the belief that Australia was the first home of man. Such a view is untenable for several reasons. In the first place, Australia, like Pa-

---

[1] In 1920 Osborn described two molars from the Pliocene of Nebraska; he attributed these to an anthropoid primate to which he has given the name *Hesperopithecus*. The teeth are not well preserved, so that the validity of Osborn's determination has not yet been generally accepted.

tagonia and Madagascar, belongs to an early Tertiary southern land mass, where, beginning with the Miocene, mammalian evolution came almost to a standstill; again, the resemblance of modern Australians to *Homo neandertalensis* is more fancied than real. The architecture of the skull is quite different; the Neandertal skull is long, while the Australian face is short. The Australian limb bones are long and light; those of the Neandertal race are

FIG. 197.   THE TRINIL STATION ON THE BENGAWAN OR SOLO RIVER IN CENTRAL JAVA.

The white cross in the center marks the spot where the remains of *Pithecanthropus erectus* were found in 1891–1892.   After Selenka and Blanckenhorn.

short and stocky. Other minor differences might be noted (Figs. 233 and 234).

In discussing Asia's claims, one naturally thinks first of the fact that it is not only the home of the orang-utan and the gibbon, but also of *Pithecanthropus erectus*. In this connection it is well to recall the main features of Klaatsch's theory. He compared certain types of man with certain anthropoids and found that the differences between the gorilla and the orang-utan are in a measure parallel to those between the primitive human type of Neandertal and the later type of Aurignac. On this as a basis, two lines of human descent are postulated. One goes back to an ancestor in Africa, common to the gorilla and *Homo neandertalensis;* the other to an ancestor in Asia, common to the orang-utan and *Homo aurignacensis*. These two types met and mingled in Europe, producing a new type which was dominant at the close of the Paleolithic Period. The Klaatsch hypothesis has met with a rather cold reception, especially at the hands of Keith, who is especially fitted to expose its weaknesses on account of his familiarity with the anatomy of the orang.

The data bearing on human origins come primarily under two heads, human skeletal remains and cultural remains. Human skeletal remains are the rarest and at the same time the most convincing of documents. They bring us nearest to the root of human origins; we may reasonably expect to find ancestral forms even antedating the oldest recognizable cultural remains. The Hominidae are generally looked upon as a family consisting of but one genus. Even the monogenist must admit that *Homo neandertalensis* is a separate species from *Homo sapiens*. Such differences in any other group of the animal kingdom would be looked upon as even generic. The differences between modern man and the man of Heidelberg are still greater. It will thus be seen that the definition of what constitutes man, anatomically speaking, has not yet been clearly formulated. It is, however, safe to assume that when we find skeletal remains approximating those of man, associated with distinct cultural remains, we have not the precursor of man but man himself.

PITHECANTHROPUS ERECTUS

Let us first examine the skeletal remains that might be calculated to throw some light on the physical characters of early man and his precursor. Beginning with the most remote, anatomically and geologically speaking, we have *Pithecanthropus erectus,* found near Trinil, Java, by Dubois in 1891 and 1892, in what were then supposed to be Pliocene deposits (Figs. 197 and 198). The parts found then include the upper part of the cranium (Fig. 199), two teeth, and a left thigh bone. Later, part of a lower jaw and a tooth were discovered. The specimens were all found at the same level, but the femur and cranial cap were some 15 meters (49.2 feet) apart. The teeth were found near the cranial cap; they are the third right (Fig. 201), and the second left, upper molar, and the second left lower pre-molar (not found by Dubois and apparently too small to belong with the molars).

The geology of the site has been very thoroughly explored in recent years, with the result that the age of the *Pithecanthropus* remains is now considered by some authors to be Lower Quaternary instead of Upper Tertiary, the evidence being furnished by the nature of the associated fauna and flora. According to the flora,

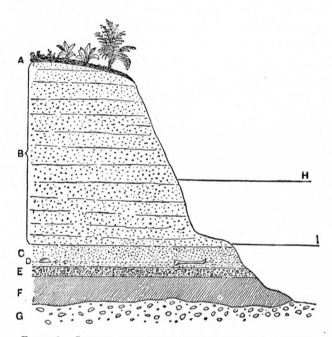

FIG. 198. SECTION OF THE DEPOSITS AT THE TRINIL STATION.

A, vegetal earth; *B*, sand-rock; *C*, bed of lapilli-rock; *D*, level in which the remains of *Pithecanthropus* were found; *E*, conglomerate; *F*, clay-rock; *G*, marine breccia; *H*, rainy-season level of the river; *I*, dry-season level of the river. After Dubois.

the climate was cooler and damper than now, suggesting a stage nearer to a glacial epoch than we are to-day. The faunal remains of the *Pithecanthropus* layer include three species of elephant, all referred to the genus *Stegodon* and one closely related to *Elephas antiquus* of Europe; hippopotamus, rhinoceros, boar, tapir, the Indian buffalo, axis and rusa deer, porcupine, several felines, hyena, otter, and a macaque. The Trinil fauna, therefore, bears a striking resemblance to that of the Siwalik Hills of India, which have been referred to the late Pliocene. The fossil flora of this layer includes several species of *Ficus, Eugenia jambolana* (plentiful), *E. decipiens, Magnolia, Dillenia,* and *Michelia.*

Dubois considered that the *Pithecanthropus* remains all belong to one individual or species, but opinions still differ on this point. On account of the pithecoid character of the skull and the straightness of the femur, he gave it the name *Pithecanthropus erectus.*

The femur, being complete, admits of exact measurement and indicates a stature of about 1.69 meters (5 feet, 6.5 inches); the cranial capacity was at first estimated at 850 cubic centimeters but subsequent study of the endocranial cast and surface points to a capacity of about 985 cubic centimeters, a figure slightly larger than the minimum (960 cc.) among the Veddahs and much larger than the cranial capacity of the adult male gorilla (550 cc.). The cranial capacity of the male Bushman averages 1,240 cubic centimeters, which is higher than the capacity of certain human fossil skulls. The endocranial surface is perfectly preserved and shows that the convolution of Broca, the seat of articulate speech, is different from the same region in any known ape's brain.

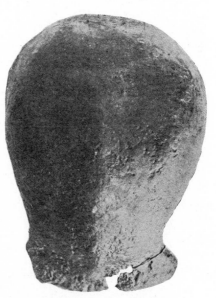

FIG. 199. THE CRANIAL CAP OF PITHE-CANTHROPUS ERECTUS.

Original in the Teyler Museum at Haarlem. Scale, ⅖. After Dubois.

The Java skull is remarkable for its prominent brow ridges, narrowness in the frontal region, and its feeble height. There are other simian features, such as the practical fusion of the supramastoid with the occipital crest and the fact that the skull reaches its extreme height at the bregma (Fig. 200).

Although larger than those of man, the teeth are more nearly human than simian; the roots are large and spreading (Fig. 201).

The Selenka Expedition of 1907–08, one of whose results was to reduce the age of *Pithecanthropus* remains to Lower Quaternary, secured a tooth which is said by Walkoff to be definitely human. It is a third lower molar from a neighboring stream bed and from deposits older (Pliocene) than those in which *Pithecanthropus erectus* was found. Should this tooth prove to be human, *Pithecanthropus* could no longer be regarded as a precursor of man. Instead it would simply give us the cross section of a different

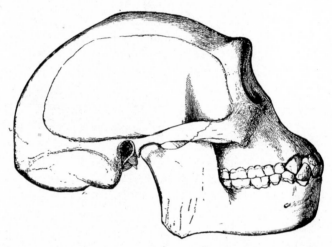

Fig. 200.   Restoration of the skull of Pithecanthropus erectus.

Although many features of this skull, such as the prominent brow-ridges and the lack of height, resemble those of an ape, the teeth and the convolution of Broca are more human than simian. Scale, $\frac{2}{5}$. After Dubois.

limb of the primate tree from the limb whose branches now represent the various types of Hominidae.

The superficial resemblance of the *Pithecanthropus* femur to the modern human femur is striking (Fig. 201). A cross section in the poplitic region, 4 centimeters from the margin of the condyles, shows, however, a striking difference, as seen in the illustration (Fig. 202). It is practically cylindrical. If beginning at an anterior point *m,* we take two antero-posterior diameters, one ending at the median point *p,* the other at the point *n* situated on the external branch of the bifurcation of the linea aspera, we find *mn > mp.* Manouvrier found only one human femur in one thousand that presented this character to the same

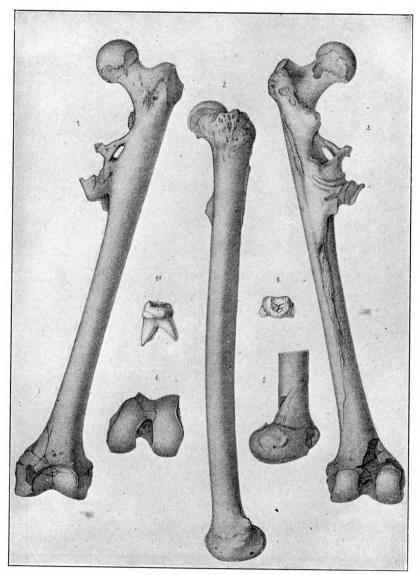

FIG. 201. VARIOUS ASPECTS OF THE FEMUR AND TOOTH OF PITHECANTHROPUS
ERECTUS.

Nos. 1–5 are views of the femur; the exostosis (jagged protuberance) just below the
small trochanter is due to a wound. No. 6 is the third upper right molar from above and
from the back (6a). Originals in the Teyler Museum at Haarlem. Scale, Nos. 1–5, *ca.* ⅓;
Nos. 6, 6a, *ca.* ⅔. After Dubois.

degree as the Java femur, a pathologic Parisian femur of the Middle Ages. The Java femur is likewise pathologic; the lesion near the small trochanter might well have caused the shaft to be cylindrical in the poplitic region. Had the other femur been found, it might have been very different in this respect.

Owing to the fragmentary nature, not only of the *Pithecanthropus* remains, but also of available collateral evidence, three rather distinct views are still held as to the status of the Java specimen. In the first place, there are those who, with Dubois,

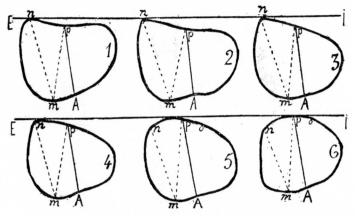

FIG. 202. TRANSVERSE SECTION OF THE FEMUR OF PITHECANTHROPUS ERECTUS COMPARED WITH SEVERAL VARIATIONS OF THE HUMAN TYPE.

The sections are made 4 centimeters above the condyles. They show the transition from the common human type (1) through several variations of the human type (2–5) to the femur of *Pithecanthropus erectus* (6).   *EI*, transverse axis; *Ap*, antero-posterior axis. After Manourrier.

believe it to represent a transition form between man and the higher apes (especially the gibbon), and to be in a sense the precursor of man—in other words, a creature which had won for itself the erect posture, but whose brain was still too primitive to be called the brain of *Homo*. Others, like Keith, believe the name given by Dubois to be justified in a zoölogical sense, but, in view of the many human characters, would go a step farther and call it *Homo javenensis*. For them the human line of descent would lead directly to, and through, *Pithecanthropus* on its way back to the parent trunk (Fig. 203). Both Hrdlička and McGregor would place *Pithecanthropus* much nearer to man than to any known ape.

Lastly, there are those who believe that *Homo* and *Pithecanthropus* represent different limbs of the parent trunk but go back to a common ancestor. Viewed in this light, *Pithecanthropus* would have no living lineal descendant. The fragment of a lower jaw, with the first (or second) premolar and the socket for the tooth immediately in front, was found in the same horizon as *Pithecanthropus,* but at a distance of some 40 kilometers (25 miles) southeast of Trinil; it might well have belonged to another individual o f t h e s a m e genus.

The remains of *Pithecanthropus erectus* are preserved in the Teyler Museum at Haarlem.

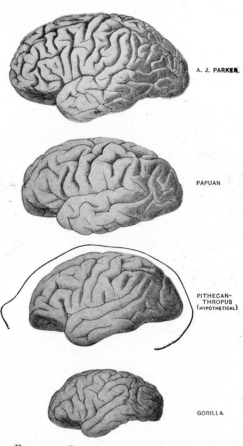

A. J. PARKER,

PAPUAN

PITHECAN-
THROPUS
(HYPOTHETICAL)

GORILLA

FIG. 203.   STAGES IN BRAIN DEVELOPMENT.

The brain of a Papuan and the hypothetical contour of the brain of *Pithecanthropus* (drawn into the cranial outline) interposed between the brain of a highly intelligent man and that of a gorilla. Modified from Spitzka.

HEIDELBERG MAN

On October 21, 1907, there was discovered in a sand pit near the village of Mauer, 10 kilometers (6.25 miles) southeast of Heidelberg, Germany, a fossil human lower jaw, remarkable alike for its completeness, stratigraphic position, and anatomical characters. The announcement of the discovery was duly made by Schoetensack of the University of Heidelberg. Mauer lies in the valley of the Elsenz, a tributary of the Necker. The human lower jaw was found *in situ* in the so-called Mauer sands, at a depth of 25.04 meters (82 feet) and 0.87 meter (2.8 feet) from the bottom of the deposit. The first 10.92 meters (35.8 feet) at the top of the section are composed of loess which is classed as Upper Quaternary, while the Mauer sands forming the rest of the section are Lower Quaternary. The loess itself represents two distinct periods, an older and a younger (Fig. 204).

The horizon from which the human lower jaw came has furnished other mammalian remains, including *Felis spelaea, F. catus,*

FIG. 204.   SAND PIT AT MAUER, NEAR HEIDELBERG, GERMANY.

The white cross (*x*) in the Mauer sands, at a depth of 25.04 meters (82 feet), marks the spot where the Heidelberg jaw was found.   1, Lower Quaternary, Mauer sands;   2, Upper Quaternary, ancient loess;   3, Upper Quaternary, recent loess.   After Schoetensack.

Canis, Ursus arvernensis, Sus scrofa var. *priscus, Cervus latifrons, Bison, Castor fiber, Equus, Rhinoceros etruscus,* and *Elephas antiquus.* Schoetensack likens this fossil mammalian fauna to that of the preglacial Forest Bed of Norfolk and the Upper Pliocene of southern Europe. This is particularly true of *Rhinoceros etruscus* and of the horse of Mauer, which is a transition form between *Equus stenonis cocchi* and the horse of Taubach, both of which may be referred definitely to the Pliocene. The rest of the mammalian fauna belongs to the Lower Quaternary.

The coexistence of man with *Elephas antiquus* at Taubach, near Weimar, gave Schoetensack special reasons for expecting to find human remains also at Mauer. The possibility of such a discovery had kept him in close touch for twenty years with Rösch, the owner of the sand pit. The discovery was made by one of the workmen, with whom at the time were another workman and a boy. Schoetensack was immediately informed and arrived the following day. The lower jaw was intact, but the stroke of the workman's shovel had caused the two halves to separate along the line of symphysis. It was discolored, and marked by incrustations of sand exactly as are all fossil bones from the Mauer sands. A limestone pebble

FIG. 205. LOWER JAW OF HOMO HEIDELBERGENSIS FROM THE MAUER SAND PIT. PRE-CHELLEAN EPOCH.

The jaw is massive and of a primitive chinless type, but the teeth, all of which were intact when found, are distinctly human. Original in Heidelberg. Scale, $\frac{4}{5}$. After Schoetensack.

was so firmly cemented to the left half of the jaw, covering the premolars and first two molars, that the crowns of all four stuck to the pebble when the latter was removed. Both the jaw and the pebble were marked by dendritic formations (Fig. 205).

Perhaps the first thing to attract one's attention is the absence of a chin, giving to the specimen a gorilloid aspect, much more so than in any Neandertal lower jaw. The genial pit is also deeper and larger than in the Mousterian race. The corpus mandibulae is massive and relatively long in proportion to the bicondylar breadth. The ramus is characterized by unusual breadth, 60 milli-

meters (2.3 inches), as opposed to an average of 37 millimeters
(1.4 inches) for recent man. The angle formed by the lines
tangent to the basis and the posterior border of the ramus is 107
—smaller than the average. The processus coronoideus is exceedingly blunt, and the incisura mandibulae correspondingly shallow.
The condyloid process is noteworthy on account of the extent of
articular surface due to an increased antero-posterior diameter
(13 to 16 millimeters, .5 to .6 inch), since the transverse diameter
is relatively short. The neck constriction is very slight, approaching in this respect the anthropoid forms.

The teeth have a distinctly human stamp, not only in their
general appearance but also in point of size—larger than the average, but smaller than in exceptional cases. They are, in fact, just
what one would expect to find in an ancestor of the Neandertal
man. One is impressed with the relative smallness of the teeth as
compared with the massive jaw in *Homo heidelbergensis.* The
alveolar arch is almost long enough to allow space for a fourth
molar. The crowns of the teeth are worn enough to show the
dentine, proof that the individual had reached the fully adult stage.
All the molars except the third left have five cusps. The tendency
in modern man is toward a four-cusp type for the third molar, if
indeed there be a third molar. The breaking away of the crowns
of four teeth on the left side, although to be deplored, nevertheless tended to facilitate the study of the relatively large pulp cavities and the walls.

One of the even more surprising features about the dentition
is that the canines are less apelike than are some modern examples,
and in no wise interfere with the side-to-side movement of the
teeth, which distinguishes man from the anthropoids. Had they
been missing, they would no doubt have been reconstructed along
very different lines, which is one of the many arguments in favor
of not taking reconstructions too seriously.

The position of the Mauer lower jaw near the bottom of the
old diluvium, and its association with the remains of *Elephas
antiquus* and *Rhinoceros etruscus,* suggest for it a place below the
Middle Quaternary. According to most authors, it represents Pre-
Chellean man.

Piltdown Man

The discoveries at Piltdown were announced at a meeting of the Geological Society in London on December 18, 1912. Briefly the facts are these. Several years before, in passing up the Ouse valley from his home in Lewes (Sussex) into the Weald, Charles Dawson, a lawyer by profession and a fellow of the Society of

FIG. 206. FLINT-BEARING GRAVEL BED AT PILTDOWN, SUSSEX, ENGLAND.

The darkest stratum resting on the bedrock in the section is the one from which the famous Piltdown cranium, lower jaw, and canine tooth were taken. After Dawson and Smith Woodward.

Antiquaries and of the Geological Society, noted that the roadway had been recently mended with flints of a kind that he had not noticed before in that vicinity. These were traced to their source, which proved to be a pit near Piltdown Common, Fletching (Sussex). Nothing was found that day, but on a subsequent visit to the pit one of the men handed to Dawson a part of an "unusually thick parietal bone." A portion of the frontal bone of the same skull, including part of the left brow ridge, was picked up by Dawson himself in 1911 from one of the refuse heaps. He submitted this piece to Smith Woodward of the Natural History Museum, London, who thereafter took part in the search. Other fragments of the skull were recovered from the refuse heaps; the right half of a lower jaw with first and second molars *in situ* was dug out of the undisturbed gravel by Dawson. At precisely the same level, 1.2 meters (some 4 feet) below the surface, and within 0.92 meter (1 yard) of the point where the jaw was found, Woodward dug up a piece of the occipital bone of the cranium. By reason of their proximity, as well as in point of size, the cranium and lower jaw were "referred to the same individual without any hesitation." The bones are mineralized and stained to a ruddy-brown color, as are the sands and flints among which they were found.

The finding of human and fossil animal remains in the same pit and stratum, both associated with rudely worked flints (see Fig. 31), makes Piltdown one of the most extraordinary prehistoric stations ever uncovered. The fossils include broken pieces of a molar of *Mastodon, Stegodon,* and teeth of *Hippopotamus* and *Castor*. On the surface of an adjacent field were found the tooth of *Equus* and fragments of an antler of *Cervus elaphus*. These were all in the same mineralized condition and of the same color as the human bones.

When the pieces of the cranium were put together, it was possible to estimate the cranial capacity, which Woodward gave as not less than 1,070 cubic centimeters. The bones are tough and hard, and the walls of the brain case exceedingly thick, the average thickness of the frontal and parietal being at least 1 centimeter. The face and the greater part of the forehead are missing. The length of the cranium from glabella to inion is about 190 millimeters (7.5 inches), while the greatest parietal width is 150 millimeters (5.9 inches). The forehead is steeper and the brow ridge feebler than in the later Neandertal type. The cranium is low and broad, with a marked flatness on top, and the mastoid processes are relatively small.

The lower jaw is more primitive than the cranium (Fig. 207). The horizontal ramus is rather slender, resembling that of a young chimpanzee, especially in the region of the symphysis. Only two teeth, the first and second molars, were found, and these were in their sockets. They were said to be human although relatively of large size and narrow, thus requiring more linear space for their setting in the jaw. Each has a fifth cusp. The crowns are worn flat by mastication, indicating that the individual was of adult age. The ascending ramus is broad, and the sigmoid notch at the top in front of the articular process is shallow. In these respects the Sussex lower jaw approaches that found near Heidelberg. The feeble brow ridges, the small area for the insertion of the temporal muscles, and the rather insignificant mastoid processes point to a member of the female sex. Woodward regarded the skull as belonging to a hitherto unknown species of *Homo,* for which he proposed the name *Eoanthropus dawsoni.*

A study of the cast of the cranial cavity was made by G. Elliot Smith, one of the highest authorities on the human brain, who found that while it bears a similarity to the brain cases of Gibraltar and La Quina, both Paleolithic and supposedly female, the Piltdown brain case is smaller and more primitive in form than these. The most striking feature is the "pronounced gorilla-like drooping of the temporal region, due to the extreme narrowing of its posterior part, which causes a deep excavation of its under surface." This feeble development of that portion of the brain which is known to control the power of articulate speech is most significant. To Elliot Smith the association of a simian jaw with a cranium more distinctly human is not surprising. The evolution of the human brain from the simian type involves a tripling of the superficial area of the cerebral cortex; and "this expansion was not like the mere growth of a muscle with exercise, but the gradual building-up of the most complex mechanism in existence. . . . The growth of the brain preceded the refinement of the features and the somatic characters in general."

After Elliot Smith's study of the cast of the Piltdown brain case had been made, Symington presented to the Anatomical Society of Great Britain and Ireland a series of casts proving that the endocranial cast does not give definite information regarding the features of the brain in detail. Although it may give fairly accurate information regarding the larger features of the brain, the possibility of recognizing in an endocranial cast the anterior branches of the fissure of Sylvius, the parieto-occipital fissure, and certain other sulci is open to serious question.

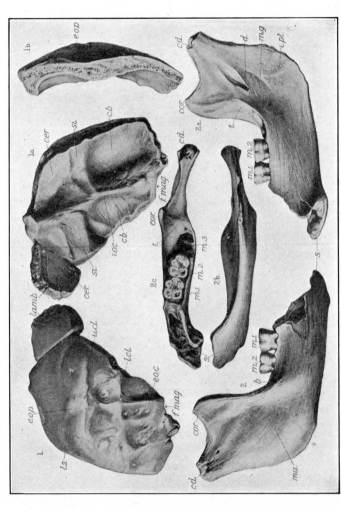

FIG. 207. PARTS OF THE PILTDOWN CRANIUM AND LOWER JAW. PRE-CHELLEAN EPOCH.

1, Imperfect occipital, outer view; 1a, inner view; 1b, broken vertical section, left side, showing extraordinary thickness; cb, cerebellar fossa; cer., cerebral fossa; e.o.c., external occipital crest; e.o.p., external occipital protuberance; f.mag., foramen magnum; i.o.c., internal occipital crest; lamb., portion of lambdoid suture; l.c.l., lower curved line; l.s., linea suprema; si, groove for lateral sinus; u.c.l., upper curved line. 2, right mandibular ramus, imperfect at the symphysis, outer view; 2a, inner view; 2b, inner view; 2c, upper view; b, ridge below origin of buccinator muscle; cd., neck of condyle; cor., coronoid process; d, inferior dental foramen; i.pt., area of insertion of internal pterygoid muscle; m.1, m.2, first and second molars; m.3, socket for third molar; m.g., mylohyoid groove; ma., area of insertion of masseter muscle; s, incurved bony flange of symphysis; t, area of insertion of temporal muscle. Scale, ca. ⅔. After Dawson and Smith Woodward.

The associated worked flints of Piltdown have been compared with the so-called eoliths from the North and South Downs. According to Lankester, "many of the flints of this Piltdown gravel have been worked by early man into rough implements. They are of flat shape, often triangular in area, and show a coarse but unmistakable flaking of human workmanship." He considers them ruder and earlier than any flint implements that can rightly be called Chellean.

Scientists have often remarked on the paucity of human remains that could with certainty be referred to a very early epoch, a condition which more than anything else has kept in check the science of prehistoric anthropology. After all, there is no evidence quite so incontrovertible as the presence of man's own skeletal remains. We may justly differ on the question of whether or not a given flint is an artifact. When, however, a human skull is found associated with rudely flaked flints, the nature of which might be questioned if occurring alone, the burden of proof is at once shifted from those who believe them to have been utilized by man to those who would call them the work of Nature. On the other hand, this does not by any means let down the bars to indiscriminate claims for the artifact nature of all primitive-looking flints.

The author has been for many years a believer in the prehistoric possibilities of southern England because of the occurrence of outcrops of flint-bearing Chalk which stretch from Dorset and Sussex on the south to Caddington and the Cromer Forest Bed on the north.

It is not generally known to Channel voyagers that the white cliffs at Beachy Head and again more than 80 kilometers (50 miles) farther east at Dover are the bases of a great anticlinal fold whose axis passes from Dungeness in a westerly direction through Hampshire. The crest of the fold, which once towered high over what is now the Weald, disappeared ages ago, leaving two slender tongues from the great Chalk plain of Dorset, Wiltshire, and Hampshire, the tip of one (the North Downs) being at Dover and that of the other (the South Downs) at Beachy Head. The scene of Dawson's epoch-making discovery is almost due north of Beachy Head, just beyond the South Downs plateau and hence near the southern limits of the Weald. The Ouse takes its rise in the Weald and flows southward, cutting through the South Downs and emptying into the Channel at New Haven.

The Piltdown gravels are 24.4 meters (80 feet) higher than, and 1.5 kilometers (nearly a mile) distant from, the present stream bed of the Ouse. The physiographic features of the valley have not changed since Roman times. The relation of the present bed to the bed that existed when the Piltdown gravels were formed, therefore, indicates for the latter a great antiquity. Although they may not be so old as the patches of red clay with rude flints on the Downs north of Ightham, as well as on the South Downs at Beachy Head and Eastbourne, some at least of the materials composing them may once have been a part of older deposits. The broken edges of all the bones, human as well as animal, show more or less wear. The remains of the Pliocene elephant (*Stegodon*) and especially *Mastodon* are most worn, and they are evidently derived from some older deposit, as they are typical Pliocene forms. The teeth of *Hippopotamus* might be either Upper Pliocene or Pleistocene, but the beaver teeth are probably Pleistocene. The gravel bed is probably to be regarded as Pleistocene.

The form of the Heidelberg cranium can only be conjectured. Comparing the parts in common, the Piltdown lower jaw is seen to be intermediate between the lower jaw of Heidelberg and that of a young chimpanzee. The height of the ascending ramus is somewhat greater, its breadth is less, and the sigmoid notch deeper in the Piltdown mandible than in the specimen from Mauer. The descending ramus is remarkable for the thickening of its antero-interior margin, this affording ample space for the insertion of the temporal muscle. The mylohyoid groove (*m. g.*) is behind rather than in line with the mandibular foramen. These and the complete absence of the mylohyoid ridge are characteristic of the apes rather than of man.

The transverse diameters of the first and second molars are less than in *Homo heidelbergensis*. While the antero-posterior diameters are identical in the two mandibles, the configuration of the horizontal ramus and the symphysis is such as to require a space of some 60 millimeters (2.3 inches) for the setting of the anterior teeth in the lower jaw of Piltdown, or 20 millimeters (0.7 inch) more than in the Heidelberg mandible.

A comparison of the cranium with other ancient human skulls throws new light on the anatomical make-up of the earliest races of man. The man of Piltdown has not the low, sloping forehead and prominent brow ridges of even so late a type as Mousterian man; and yet, according to Elliot Smith, its brain was the most primitive and most simian human brain thus far recorded. The Piltdown remains tend therefore to prove that in the Lower Quaternary the differentiation among the Hominidae had already progressed much farther than has been generally supposed, and that we shall have to go a long way back in the past to find the parting of the ways between the ancestor of man and his nearest of kin among the apes.

In the reconstruction of the Piltdown cranium, differences of opinion soon arose. Of the brain case, nine fragments, part of the frontal, parietal, occipital, and temporal, were found. From these Smith Woodward reconstructed a skull with a capacity of about 1,076 cubic centimeters. On the other hand, a reconstruction by Keith gave to the skull a brain capacity of 1,500 cubic centimeters, that is to say, that of a well developed modern European skull. After further study Smith Woodward acknowledged a small error. He found that the "longitudinal ridge along the outer face at the hinder end of the parietal region is not median, but one of a pair such as frequently occurs in the lower types of human crania." In the published reconstruction by Smith Woodward there should thus be a slight readjustment of the occipital and right parietal bones, "but the result does not alter essentially any of the conclusions already reached."

With this opinion Elliot Smith was in complete accord. From an examination of the original fragments he was able to determine the location of the median line of the skull. The persistence of slight traces of the sagittal suture in the regions of the bregma and lambda made this possible. The true median plane in this particular case, however, passes a little to the left of the junction of the coronal with the sagittal suture, owing to a slight deflection of the latter. Since this deflection is never more than a few millimeters (except where large bregmatic wormian bones are present, and they are not in this case), the bregma and lambda are good guides in locating the median plane. In line with the median plane as thus determined, the endocranial aspect of the frontal bone presents a

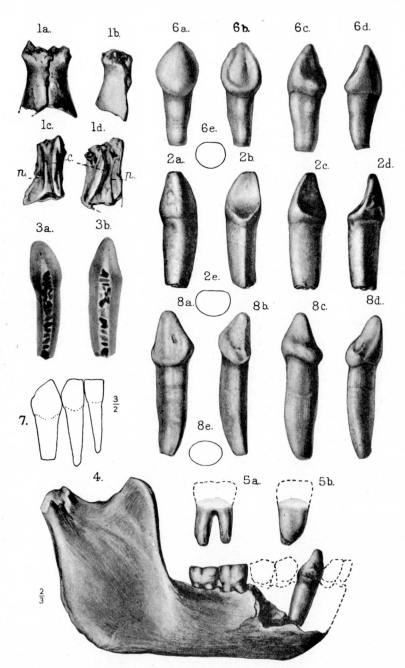

FIG. 208. THE NASAL BONES AND CANINE TOOTH FROM PILTDOWN.

1a, Nasal bones, front view; 1b, left side; 1c, left nasal from within; 1d, right nasal from within (scale, ⅓). 2a, Right lower canine, outer or labial view; 2b, inner or lingual view; 2c, anterior or median-interstitial view; 2d, posterior or lateral-interstitial view;

well defined longitudinal ridge, corresponding to the "place where the two halves of the frontal bone originally came together at the metopic suture." The cranial capacity of the Piltdown skull, then, is evidently not very much greater than the original estimate of 1,076 cubic centimeters.

In addition to exhaustive laboratory studies of the parts above mentioned, a painstaking and systematic search was made at the Piltdown site. The mandibular ramus had been found *in situ*. All the gravel *in situ* within a radius of 5 meters (16.4 feet) of this spot was "either washed with a sieve, or strewn on specially prepared ground for the rain to wash it; after which the layer thus spread was mapped out in squares, and minutely examined section by section." In this spread Father Teilhard de Chardin, assisting at the work for three days, in August, 1913, found a right canine tooth (Fig. 208). The human nasal bones and the turbinated bone were not recovered from this spread but from disturbed gravel within less than a meter from the spot where the mandible had been discovered.

The nasal bones are said to "resemble those of existing Melanesian and African races, rather than those of the Eurasian type." The nasal bones are by no means negroid in character. In thickness they correspond to the bones of the skull previously found. The canine tooth not only corresponds in size to the mandible, but belongs to the same half (right) as that recovered. It likewise agrees with the two molar teeth in the degree of wear due to mastication. The extreme apex is missing, but whether by wear or by accidental fracture is not determinable. The enamel on the inner face of the crown has been completely removed by wear against a single opposing tooth. The worn surface "extends to the basal edge of the crown, as indicated by the clear ending of the cement along its lower margin." This canine tooth is larger than

2e, transverse section of neck (scale, ¼). 3a, 3b, Radiographs of the canine tooth. 4, Right half of Piltdown lower jaw with canine tooth in position (scale, ⅔). 5a, 5b, impression of cavity for roots of lower third molar. 6a, Right lower milk canine of *Homo sapiens*, outer or labial view; 6b, inner or lingual view; 6c, anterior or median-interstitial view; 6d, posterior or lateral-interstitial view; 6e, transverse section of neck (scale, ⅔). 7, Lower milk canine and milk incisors of *Homo sapiens* (scale, ⅔). 8a, Right lower milk canine of orang (*Simia satyrus*), outer or labial view; 8b, inner or lingual view; 8c, anterior or median-interstitial view; 8d, posterior or lateral-interstitial view; 8e, transverse section of neck (scale, ⅔). After Dawson and Smith Woodward.

any human canine hitherto found, and interlocked with the opposing upper canine. It rose above the level of the other teeth and was
separated from the lower premolar by a diastema. On the other
hand, there is no facet due to wear against the outer upper incisor,
such as often occurs in the apes.

The various elements that make up the gravel bed at Piltdown
are better known to-day than when the first report was published;

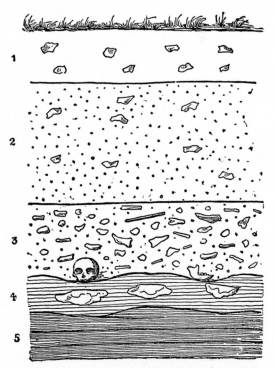

additional fossil ani-
mal remains have
also been recovered.
F o u r well defined
layers have been de-
termined (Fig. 209).
At the top is a de-
posit of surface soil
35 centimeters (13.6
inches) thick, con-
taining pottery and
flint implements of
various ages. The
second bed consists
of undisturbed gravel
varying from a few
centimeters to a me-
ter in thickness. The
prevailing color is
"pale y e l l o w with
occasional d a r k e r
patches." A r u d e
paleolith of the Chel-
lean type was found
in the middle of this
layer, which likewise

FIG. 209. SECTION OF THE GRAVEL PIT AT PILTDOWN.

The Piltdown remains were found at the base of the
third layer. The strata are as follows: 1, surface soil; 2,
pale-yellow gravel; 3, dark-colored gravel; 4, pale-yellow
clay and sand; 5, undisturbed strata of the Tunbridge
Wells Sand. After Dawson and Smith Woodward.

contained rolled, iron-stained, subangular flints. The third layer,
some 50 centimeters (19.5 inches) thick, is easily distinguished because of its dark ferruginous appearance. It contains rolled and subangular flints similar to those found in the layer above. All the
fossils (with the exception of the remains of the deer) were either

discovered in or have been traced to this third layer. So-called eoliths and at least one worked flint were likewise found here. The Piltdown remains came from it and near the uneven floor forming the upper limit of the fourth stratum. The latter has a thickness of about 25 centimeters (9.7 inches), is nonfossiliferous, and "contains flints of a much larger size than any of those in the overlying strata." Nothing that could be called an implement (eolith) has been reported from the fourth bed. Below (No. 5) are undisturbed strata of the Tunbridge Wells Sand (Cretaceous).

Our knowledge of the Piltdown fossil fauna has been supplemented by the finding of remains of one new form, a fragment of a tooth of *Rhinoceros,* in the same state of mineralization as the teeth of *Stegodon* and *Mastodon* previously described. While the specimen cannot be determined with absolute certainty, it belongs to either *Rhinoceros merckii* or *Rh. etruscus,* with the evidence rather favoring the latter. Additional remains of *Stegodon* (fragments of a molar) and *Castor* (fragments of a mandible) were likewise recovered. Judged from its fossil content, the third stratum at Piltdown would be classed as Pliocene were it not for the presence of man and the beaver. In view of the fact that the remains of these, although softer, are not so rolled and worn as the other fossil remains, the third bed, although composed in the main of Pliocene drift, was probably reconstructed in early Pleistocene time.

The prehistoric archeologist sometimes uncovers strange bedfellows; no other discovery is quite so remarkable in this respect as the assemblage from Piltdown. Nature has set many a trap for the scientist, but here at Piltdown she outdid herself in the concatenation of pitfalls left behind—parts of a human skull; half of an apelike lower jaw, a canine tooth, also apelike; flints of a Pre-Chellean type; fossil animal remains, some referable to the Pliocene, others evidently Pleistocene; all were at least as old as the gravel bed, and some of the elements apparently were derived from a still older deposit.

All the cranial fragments, including the nasal bones, are human and belong to the same individual. They were, however, so incomplete as to leave room for a difference of opinion especially in regard to the capacity of the brain case. Authorities are quite generally agreed that the cranium as well as the brain embodied certain primitive characters, even though the brow ridges were smaller and the forehead less retreating than in Mousterian man (*Homo neandertalensis*).

When it came to fitting the fragmentary lower jaw to the cranium, difficulties multiplied; it was the right half of the mandible, and the articular condyle was missing. Even had the condyle been present, there was no right glenoid fossa to receive it. A part of the left temporal bone, including its glenoid fossa, had been preserved, but it was typically human. The lack of the parts necessary to bring the mandible and cranial base into actual contact served to cloak the lack of harmony existing between the two. This lack of harmony was likewise further obscured by the incompleteness of the symphysial region.

The proximity of the brain case and lower jaw in the gravel bed; their apparent agreement in size and the nonduplication of parts present; the fact that they both bore the same marks of fossilization, showing "no more wear and tear than they might have received *in situ*"; the failure of previous discoveries to confirm the presence of higher apes among the European Pleistocene faunas; and, perhaps above all, the belief that a "generalized type" had been found, led inevitably to the association of cranium and mandible as parts of one individual or species. In dealing with the contents of a gravel bed, however, it is easy to overestimate the importance of proximity. Had Piltdown been a cave deposit or a camp site, the case for proximity might have been somewhat stronger; even in these, however, there is abundant opportunity for chance association. In any event, association can never be made to take the place of articulation.

From the start there were not lacking those who hesitated to accept the cranium and mandible as belonging to the same individual. This was the stand taken by Lankester on the occasion of the first report of the discovery before the Geological Society of London. On the same occasion Waterson was even more emphatic, saying it was very difficult to believe that the two specimens could have come from the same individual, since the mandible resembled that of a chimpanzee, while the skull was human in all its characters. In a later paper on the Piltdown mandible, he concludes that referring the mandible and cranium to the same individual would be equivalent to articulating a chimpanzee foot with the bones of a human thigh and leg.

Objections soon came also from France and Italy. Basing his opinion on the cranial characters, Anthony thought the specific name should have been *Homo dawsoni* instead of *Eoanthropus dawsoni*. About the same time a similar conclusion was reached by Giuffrida-Ruggeri. To Boule the Piltdown mandible is exactly like that of a chimpanzee; so that if this mandible had been found alone in the gravels of Piltdown associated with remains of Plio-cene animals, it would certainly have been called *Troglodytes daw-soni*. Without rejecting Smith Woodward's interpretation, which Boule considers to be within the realm of the possible, even of the probable, it would nevertheless seem to him prudent to leave the matter still open. He objects to the choice of the name *Eoanthropus*, and finally in his judgment Woodward's restoration does not ring true (*"elle sonne faux"*). It was this seemingly false note that impressed the author most of all on seeing the restoration for the first time. Dr. Gerrit S. Miller of the United States National Museum has compared the cast of the Piltdown mandible with casts of chimpanzee mandibles mutilated in the same manner, and finds not only similarity but "absolute" identity.

So much for the first two phases through which discussion of the Piltdown specimens has passed. In the first phase, the lower jaw, though admittedly apelike in character, was nevertheless re-garded as having belonged to the same individual as the skull. Then came a period when in the minds of many the mandible was referred to a genus entirely different from that of the skull, namely *Pan vetus*, a fossil chimpanzee. In the present or third phase, it is asserted by the British group that important and hitherto over-looked features of the lower jaw are positively human, and that the teeth are not only human but fundamentally unlike those of the great apes. Eight different characters are invoked to prove this: (1) The (long-crowned) Piltdown molars are more hypsodont than those of the chimpanzee. (2) The protoconid, metaconid, and hypoconulid (the three tubercles on the outer side of the molars) are larger than those in the largest chimpanzee tooth, and, further-more, the sulci dividing the cusps one from another are shorter and less conspiciously marked than in the chimpanzee. (3) In the

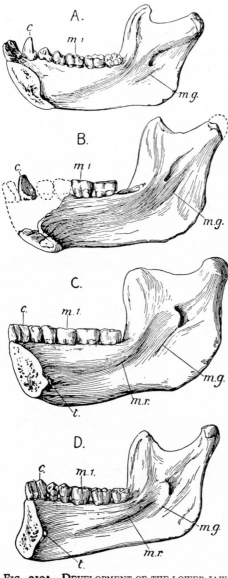

FIG. 210A. DEVELOPMENT OF THE LOWER JAW

Inner side of the lower jaw of Piltdown (*B*), compared with that of the chimpanzee (*A*), Heidelberg man (*C*), and modern man (*D*). *A* and *B* differ from *C* and *D* in the absence of genial tubercles (*t.*) and of the mylohyoid ridge (*m.r.*); also in the position of the mylohyoid groove (*m.g.*), which is back of the foramen instead of connected with it. Scale, ⅓. After Smith Woodward.

Piltdown teeth the crown passes almost insensibly into the root, and is not perceptibly wider or longer at its base than at its grinding surface, the reverse being the case with the molars of the chimpanzee. (4) Radiographs of the Piltdown jaw show that the molars are of the typical "taurodont" type, therein differing from the molars not only of the chimpanzee, but of all the great apes. (5) The enamel is thicker in the Piltdown teeth than in the chimpanzee. (6) The Piltdown jaw more nearly resembles that of the Kaffir than that of the chimpanzee. (7) The conformation of the inner surface of the body of the jaw forms in the chimpanzee two well marked types; between the extremes of these two types of chimpanzee lower jaw every gradation will be found, but in no case would there be any possibility of confusing the Piltdown fragment (or any similar fragment of a modern human jaw) with similar fragments of chimpanzee jaws. (8) In the chimpanzee the two lines, one drawn down the middle of the tooth row from the canine backwards, and the other drawn through the ascending ramus, converge in front of the canine, while in the Piltdown jaw they do not (Fig. 210 A and B).

Miller has attempted to demonstrate that not one of these eight characters brought forward is human in the diagnostic sense in which the Piltdown brain case and nasal bones are human. On the contrary, he believes them to be characters common to both man and the anthropoids. While it may be possible through additional material to prove that *Eoanthropus* is a valid genus, in his opinion neither the original fossils nor the circumstances of their chance association have yet been shown to demonstrate its existence.

In 1915 other human skeletal remains were found, but the circumstances of the find are not absolutely clear. The specimens are supposed to have come from a site similar to

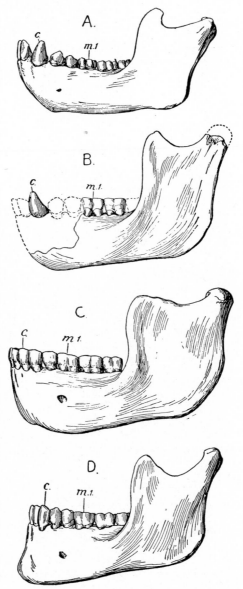

FIG. 210B. DEVELOPMENT OF THE LOWER JAW.

Outer side of the lower jaw of Piltdown man (*B*), compared with that of chimpanzee (*A*), Heidelberg man (*C*), and modern man (*D*) (see also Fig. 210A opposite). *c*, Canine tooth; *m.1*, first molar. Scale, ½. After Smith Woodward.

and about 3.2 kilometers (2 miles) distant from Piltdown. They include a small portion of the occipital, a portion of the right half of the frontal with a part of the orbital margin, and the first (or second) lower left molar. These pieces are in the same chemical condition as the human bones from Piltdown. They belong to the same physical type, even to the relatively great thickness of the cranial wall. In fact, the fragment of the frontal and the tooth might have belonged to the skull and lower jaw from Piltdown, although there is a very slight difference in tooth wear. But the occipital fragment must belong to another cranium, for the same part of the occipital is present in the Piltdown cranium. One can say that these last found fragments represent at least one additional individual.

If the Piltdown cranium and lower jaw belong to one individual, and if the tooth and cranial fragments from the new site all belong to a second individual, the evidence afforded by the new discovery, although based on small fragments, would confirm, in so far as it goes, Smith Woodward's original conclusion that the Piltdown cranium and lower jaw belong together; for it is scarcely within the range of probability that twice in succession an association of fossil human and chimpanzee bones should come to light, and that in each case the association should consist of identical parts, namely, the cranium of man and the lower jaw (or rather a tooth from the same) of chimpanzee, unless perchance the new lower left molar belongs to the lower jaw of Piltdown; if this be true, it would materially strengthen the view that the Piltdown cranium and lower jaw do not belong together. The cranium is everywhere thick and heavy for its size; the same is true of the additional occipital fragment. The lower jaw is of a wholly different type—graceful, shapely, relatively light in comparison with its size.

A radiographic study of the teeth of Piltdown, Heidelberg, Krapina, Ehringsdorf, and modern man, as compared with the teeth of the higher apes, supports the view that the teeth of the Piltdown lower jaw are human and not apelike. In man, both fossil and living, and in Piltdown the pulp cavities are relatively much larger than in the higher apes. The crown height of the Piltdown teeth is comparable with that of fossil man and considerably greater than that of the chimpanzee. Thus the Piltdown dentition is to be classed in this respect with that of *Homo heidel-*

bergensis, H. neandertalensis, and even with that of recent man, all of which are hypsodont, rather than with the low-crowned or chamaedont dentition of the chimpanzee. It is possible that the Piltdown jaw represents an older type of man than do the two Piltdown crania.

The latest reconstruction of the Piltdown skull by Elliot Smith and John Hunter has brought to light additional primitive cranial characters which are in keeping with the evidence furnished by the

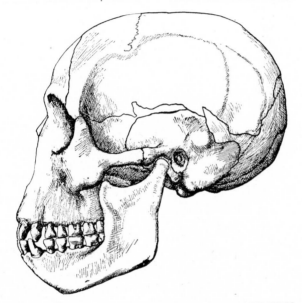

Fig. 211. Latest reconstruction of the Piltdown skull.

This new reconstruction reduces the supposed cranial dimensions and makes it seem more probable that the Piltdown cranium and the lower jaw belong together. Scale, *ca.* ⅓. After Elliot Smith and John Hunter.

discovery of the second individual and with the assumption that cranium and lower jaw belong together. They find that the occiput is much shorter than was originally supposed to be the case. This brings the cranium more in keeping with the lower jaw in so far as type is concerned, and gives a cranial capacity not exceeding 1,260 cubic centimeters (Fig. 211).

All the new evidence seems to point in one direction, the one that would lead to a correlation of cranium with lower jaw. The cranium is human, but belongs to a primitive type hitherto unknown, wholly different from the Neandertal type and even from the Heidelberg type. The lower jaw is apelike but not necessarily that of an ape; although more primitive than the cranium, the association of the two, if not based on positive data, is at least not an anatomic impossibility. There is also ground for referring the jaw to an earlier and more primitive type than the crania.

It was the author's good fortune to examine the original Piltdown specimens in 1922. In the exhibition room of the South Kensington Museum, Smith Woodward had placed on view a cast of the lower jaw (right half) fitted to the left half of the lower jaw of a chimpanzee. The author noted that the outer line of symphysial contact was longer on the Piltdown half than on the chimpanzee half, although the two rami were approximately identical in size. This means that the post-symphysial platform is even more pronounced in the Piltdown lower jaw than in the chimpanzee. In order to surmount such an anatomical obstacle, one must invoke a wider range of individual variation within the genus *Homo* (*Eoanthropus* included) than has hitherto been considered ample.

One can accept the association of cranium and lower jaw and still remain skeptical regarding the association of the canine tooth with the lower jaw. The second discovery did not include a canine tooth. The Piltdown canine was found later, but at a spot near where the lower jaw had been found. It is more apelike than the molars and is the only canine thus far discovered. Its right to a place in the dentition of the man of Piltdown comes partly through the exaggerated chimpanzee slope in the symphysial region of the lower jaw and partly through default of other claimants. This right would most certainly be questioned should another less apelike canine be found at Piltdown.

INTERGLACIAL MAN

Weimar.—Coincident with the outbreak of the World War appeared the announcement of an important discovery in a travertine quarry at Ehringsdorf near Weimar (Germany). The attention of prehistoric archeologists has long been turned toward the region of Weimar because of important discoveries made at Taubach and Ehringsdorf, both in the Ilm valley. Known since 1871,

the station of Taubach (back of the village of that name) was
systematically explored between 1876 and 1880. The deposits at
Taubach and Ehringsdorf are alike (Fig. 212). Their basis is a
layer of sand and gravel dating from the Third or Riss Glacial
Epoch (Obermaier). Above this is a lower travertine with remains
of the mammoth and woolly rhinoceros near the bottom, and those
of *Elephas antiquus* and *Rhinoceros merckii*, both witnesses of a
warm climate, toward the top. Next above at Ehringsdorf comes
the so-called "Pariser" (corruption from Pöröser) deposit, a sort
of loess. Higher still is a deposit of upper travertine with remains
of the stag and woolly rhinoceros; *Rhinoceros merckii* also occurs
at this level.

The human remains in question, consisting of a nearly complete

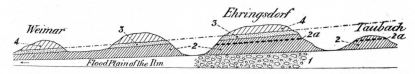

FIG. 212. GEOLOGICAL SECTION OF THE DEPOSITS AT EHRINGSDORF AND TAUBACH,
NEAR WEIMAR, GERMANY.

1. Gravels of the Riss Glacial Epoch; 2, lower travertine containing remains of a cold
fauna near the bottom and warm fauna toward the top; 2a, principal relic-bearing horizon
of the lower travertine in which have been found two human lower jaws, many flint
implements, and remains of a warm fauna; 3, "Pariser" loess, a deposit of less durable quality than the travertine; 4, upper travertine with remains of a cold fauna.

human lower jaw, formed the subject of a paper published in 1914
by Schwalbe of Strassburg. Like much of the archeological
material previously found at Taubach and Ehringsdorf, this lower
jaw is now the property of the Städtisches Museum für Vorgeschichte at Weimar. Because of its double association with that
city, Schwalbe proposed to call it the Weimar lower jaw (Fig.
213). For perhaps even better reasons, Virchow would call it the
Ehringsdorf lower jaw.

The lower jaw was found on May 8, 1914, at a depth of 11.9
meters (39.7 feet) below the surface in the lower travertine, 2.9
meters (9.5 feet) below the so-called Pariser loess, and 2.6 meters
(8.5 feet) above the gravel deposit left by the Riss glaciation. It
is from the Kämpfe quarry (Fig. 214) at Ehringsdorf and was
brought to light by means of a blast. In the circumstances it was

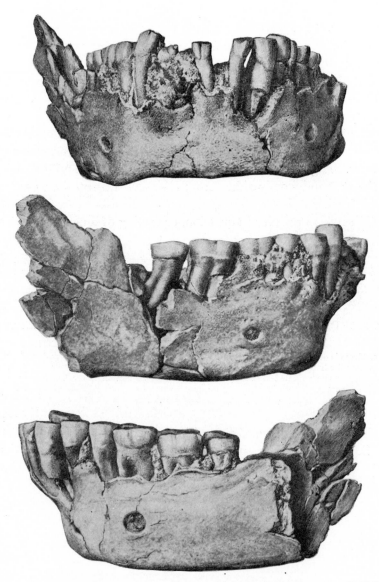

FIG. 213. FRONT AND TWO SIDE VIEWS OF THE ADULT LOWER JAW, PROBABLY
FEMALE, FOUND AT EHRINGSDORF, NEAR WEIMAR, GERMANY.

This important discovery was disclosed by a blast in the Kämpfe quarry in the interglacial deposits of the Lower (Warm) Mousterian or Acheulian Epoch. Original in the
Museum of Prehistory at Weimar. Natural size. After Schwalbe.

fortunate indeed that the lower jaw suffered no worse. All the teeth are intact and *in situ* save the two right incisors (in their place is a small mass of travertine containing a univalve shell). There is an abscess in the region of the left canine. Both halves of the body are practically complete. The right ascending ramus is in part present, although not enough remains to save the mandibular angle,

FIG. 214. THE KÄMPFE QUARRY AT EHRINGSDORF, GERMANY.

Two human lower jaws were found in the lower travertine just above the level of the men's heads in the picture. The thin soft layer of " Pariser " loess is visible about half way to the top of the section. The railroad track in the foreground rests on Riss gravels. Photograph by the author.

the coronoid and condyloid processes, and the mandibular or sigmoid notch. The left ascending ramus is completely gone.

A number of remarkable features are combined in the Weimar lower jaw. The absence of a chin is doubly emphasized because of the pronounced alveolar prognathism as shown in the figures, a condition not found in the lower jaws of Krapina and La Chapelle-aux-Saints, nor even in that of *Homo heidelbergensis.* Closely related to the alveolar prognathism is the sloping nature of the inner surface of the jaw in the region of the symphysis, the region called by Schwalbe *planum alveolare*. In all other lower jaws of the Neandertal type a median line in this field is much more nearly vertical. Below this planum alveolare is a spinous area but no distinct spine for the attachment of the genioglossal and geniohyoid muscles. Neither is there the customary ridge on the inner surface of each corpus for the attachment of the mylohyoid muscles. The absence of this mylohyoid ridge is even more marked than in the well-known mandibles of the Neandertal type.

The foramen mentale is unusually large. It is directly beneath the first molar (similar to the situation in *Homo neandertalensis*), whereas in recent man this foramen is located farther forward beneath the second premolar. In the Heidelberg lower jaw it is also large, but situated farther forward than in the specimen from Weimar.

Schwalbe lays special stress on the narrowness of the arch of the Weimar jaw. The breadth between the inner faces of the third molars is 48 millimeters (1.9 inches); the distance from the posterior surface of the third molar to the anterior margin of the median incisor is 69 millimeters (2.7 inches). The index derived from these two measures in the chimpanzee is 54.6. In the Weimar jaw this index is 69.5, while it is much larger in other known fossil human lower jaws: Heidelberg, 75.7, Krapina, 80, and La Chapelle, 100. Schwalbe admits, however, that the low index of the Weimar jaw might be due in part at least to post-mortem deformation. The jaw is also obviously of the female sex, which fact would help to account for the narrowness of the arch and for the low horizontal rami.

The teeth are much worn. Since the premolars are less worn than the canines, one is led to conclude that the points of the canines stood above the level of the premolars. There is no diastema between the canines and the first premolars. A notable feature is the relative smallness of the third molars. This unexpected condition proves that the tendency of the third molars to disappear is of much more ancient origin than other known jaws of the Neandertal and earlier types have led us to believe. The lower jaw bears distinct evidence of pyorrhea alveolaris.

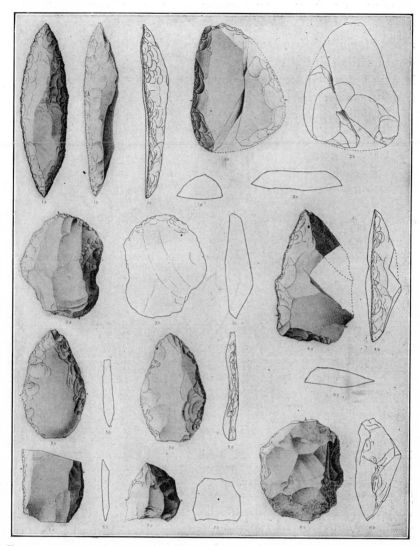

FIG. 215. FLINT IMPLEMENTS FROM THE LOWER TRAVERTINE (INTERGLACIAL) DEPOSIT AT EHRINGSDORF, GERMANY. ACHEULIAN OR LOWER MOUSTERIAN EPOCH.

1a–1d, Double point with outer face retouched along both margins; 2a–2c, sub-triangular scraper; 3a–3c, double scratcher; 4a–4c, combination scraper and point; 5a–5d, oval flat scraper with pronounced bulb; 6a–6b, fragment with retouched margins; 7a–7b, small nucleus; 8a–8b, disk-shaped scraper. Scale, ½. Originals in the Museum of Prehistory, Weimar. After R. R. Schmidt.

Without hesitation Schwalbe places the Weimar lower jaw in the Neandertal group, for which he proposed some years ago the name *Homo primigenius.* In the preliminary paper he does not describe the cultural remains found at the same level. He does, however, mention some of the numerous accompanying animals— *Rhinoceros merckii,* stag, horse, ox, and cave bear. There was

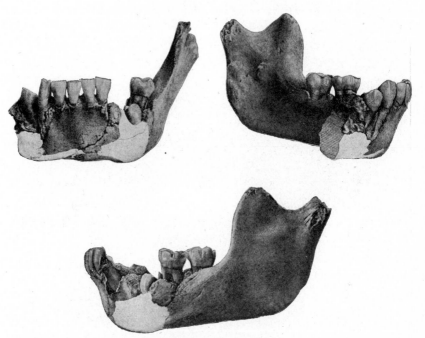

FIG. 216. FRONT AND TWO SIDE VIEWS OF THE LOWER JAW OF A CHILD SOME TEN YEARS OLD FOUND AT EHRINGSDORF.

This relic of interglacial man was found in the lower travertine at the Kämpfe quarry, the same horizon that yielded the adult lower jaw previously discovered (Fig. 213). These two lower jaws are with the exception of the Piltdown skull and the Mauer lower jaw, the most ancient human remains yet discovered. Scale, ⅔. Original in the Museum of Prehistory at Weimar. After Virchow.

also an abundance of charcoal and flint implements, the latter for the most part apparently retouched points and scrapers (Fig. 215).

Two human teeth (one of a child and one of an adult) had already been found in the lower travertine of Taubach. During the summer of 1908, Pfeiffer found human skull fragments in the same deposit at Ehringsdorf. In November, 1916, a second lower jaw was found, likewise after a blast and in the same quarry at Ehringsdorf, 25 meters (82 feet) north of the spot where the first lower jaw had been discovered and in the same horizon. It is that of a child of about 10 years (Fig. 216) ; with the lower jaw were brought to light two upper incisors, two upper milk molars, eleven ribs, two vertebrae, the epistropheus, the right clavicle, and the upper half of the right humerus. The fossil mammalian remains found near the skeleton of the child were, in the order of their abundance, *Elephas antiquus, Rhinoceros merckii, Ursus arctos, Equus germanicus, Cervus elaphus,* and *Alces latifrons.* Among shells, *Limnea peregra* and *Bithynia leachii* were the most abundant.

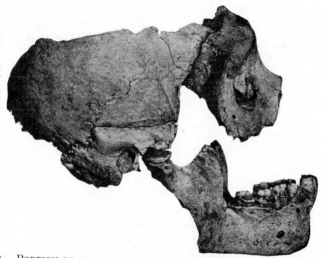

FIG. 217. PORTION OF ONE OF THE SKULLS FROM THE ROCK SHELTER OF KRAPINA, CROATIA. WARM MOUSTERIAN EPOCH.

Original in the musem at Agram. Photograph by Kramberger.

The two lower jaws from Ehringsdorf belong to the same epoch. Do they belong to the same race? The difficulties in the way of a categorical answer are two. Both are fragmentary; in addition, one is from a very small adult female while the other is that of a robust child. Nevertheless there are striking points in common, such as the absence of chin and the pronounced development of the planum alveolare. All the parts that are present in common speak for unity of race. Which race or species? Fortunately the lower jaw of the child furnishes certain data that are lacking in that of the adult. The ascending ramus is present, and it resembles the Piltdown more than it does that of the Heidelberg lower jaw. The dental crown relief admits of comparison with the same teeth from Krapina and Le Moustier, and here one finds similarity if not complete likeness.

During the summer of 1922 a human femur was found in the same quarry and at the same level as the two lower jaws.

Both Obermaier and Schmidt consider the lower travertine of Ehringsdorf (the deposit in which the lower jaw was recently

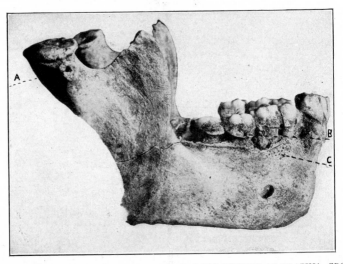

FIG. 218.   DISEASED LOWER JAW FROM THE ROCK SHELTER AT KRAPINA, CROATIA.

This jaw offers evidence of some of the diseases from which Paleolithic man suffered. It shows the effects of arthritis deformans (*A*), tartar (*B*), and caries (*C*).   Original in the museum at Agram.   Photograph by Kramberger.

found) and Taubach to be older than Mousterian. Although it contains no typical cleavers, on account of the character of the fauna as well as the industry Obermaier would call the artifacts from the lower part of the deposit at Taubach, Chellean. Schmidt, who has recently published examples of the industry, classes them as Acheulian.

In any case, all are evidently agreed that the deposit belongs to the Riss-Würm Interglacial Epoch. In that event, according to one school it might be Chellean, Acheulian, or early Mousterian; according to the school of Penck, it would have to be Mousterian. Whichever view is correct, on account of its anatomical characters, as well as the position of the deposit and the nature of the associated cultural and faunal remains, the anthropologist may justly claim for the Weimar lower jaws an antiquity surpassed perhaps only by the skull of Piltdown and the Mauer lower jaw (*Homo heidelbergensis*).

The human tooth found in the Klause cave at Neu-Essing, Bavaria, is thought by Obermaier to be of Acheulian age.

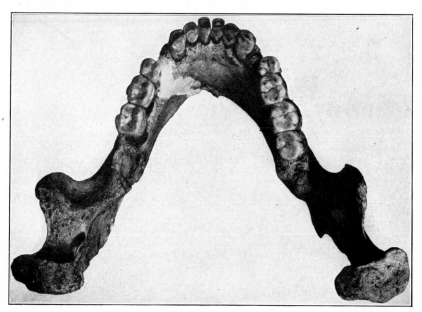

FIG. 219. VIEW OF THE DISEASED KRAPINA LOWER JAW FROM ABOVE.
See also Fig. 218. Photograph by Kramberger.

Krapina.—The rock shelter of Krapina in northern Croatia, known since 1895, was explored by Gorganovič-Kramberger of Agram from 1899 to 1905. It is situated on the right bank of the Krapinica just above the village of Krapina. The shelter was formed by the stream when the latter flowed at some 25 meters (82 feet) above its present level. The lowest deposit of the shelter consists of a pebbly layer similar to that in the present stream bed. Above the pebbly deposit is a series of eight ossiferous and culture-bearing zones, all of which probably belong to the Mousterian Epoch. These deposits, some 8 meters (26 feet) thick, were buried beneath a talus formed by the weathering of the overhanging soft

Miocene sandstone. Several thousand pieces of bone and about one thousand artifacts of flint, jasper, quartz, and chalcedony were found in this shelter. The topmost zone contained many complete bones of the cave bear; hearths, with partially burnt bones of the same animal, occurred in the lower zones. The bones of *Rhinoceros merckii* and *Bos primigenius* also were abundant. More than two thousand fragments of animal bones were found. The implements were chipped from pebbles similar to those in the bed of the Krapinica. Worked bones are likewise included among the industrial remains.

Of human bones some five hundred pieces, belonging to at least ten individuals, were found (Fig. 217). Many of these had been partially burned, and some had apparently been split open. The artifacts as well as the human skeletal remains represent a warm phase of the Mousterian Epoch. Kramberger calls attention to certain pathological phenomena in connection with the human remains from Krapina, including a distinct case of arthritis deformans in a lower jaw, caries, and tartar deposits (Figs. 218, 219). The distal end of a right ulna had sloughed off, due either to a wound or to disease.

MOUSTERIAN MAN

The Neandertal race is better known both through its skeletal and cultural remains than any race that preceded it. The type specimen, from the valley of the Neander, is only a fragmentary skeleton, but it is now reinforced by a score of other finds, each confirming and supplementing the others. The skeletal remains of this type are associated with the Mousterian industry and belong to the Mousterian Epoch. Hence Neandertal is the equivalent of Mousterian man. Schwalbe proposed the name *Homo primigenius* for this race, but *Homo neandertalensis,* proposed by King in 1864, is more widely used.

Gibraltar.—In 1848 a human skull was found at Forbes Quarry, on the north face of the rock of Gibraltar. It was taken to England in 1862 by George Busk, and by him was presented to the Royal College of Surgeons in 1868. The importance of this find, as representing a hitherto unknown fossil race of man, was not grasped until after the discovery of the skeletal remains in the Neander valley nine years later. Although by no means complete, the Gibraltar skull has shed new light on anatomical characters of the Neandertal race because of the well preserved base. The skull is probably that of a female, with a brain capacity estimated at from 1,200 to 1,296 cubic centimeters (Fig. 220).

In an attempt to throw new light on the age of this skull, Duckworth explored the north face of the rock of Gibraltar in 1910, giving especial attention to the Forbes Quarry site. In neighboring caves he found Mousterian flints, also a mixture of sand, limestone, and cement similar to that which fills the nose and orbits of the Gibraltar skull. The age of this skull, therefore, is no longer open to question.

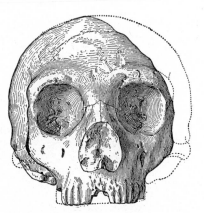

Neandertal.—Close to the railway station of the village of Hochdall, between Düsseldorf and Elberfeld, is the cave of Feldhofen, in Devonian limestone. It is situated some 30.5 meters (100 feet) below the level of the plateau and 18.3 meters (60 feet) above the water level of the Düssel River. This part of the valley is

FIG. 220. CRANIUM OF A FEMALE (PROBABLY) OF THE NEANDERTAL RACE, FROM FORBES QUARRY, GIBRALTAR. MOUSTERIAN EPOCH.

This was the first cranium of the Neandertal type to be discovered (1848). Scale, ⅓. Original in the Royal College of Surgeons. After Boule.

known as Neandertal, whence the name that has ever since been associated with the human remains found there. The cave, which is small, was filled to a depth of 1.5 meters (5 feet) with loess from the mantle on the plateau above, which had descended through a fissure (Fig. 221). Through this fissure the human bones might also have been washed to find lodgment in the cave below. As is usually the case, workmen uncovered the bones, which might have been lost to science had it not been for Dr. Fuhlrott of Elberfeld who recognized their importance and later turned them over to Professor Schaaffhausen of Bonn.

The skull was near the entrance; the other bones were farther in and lying in the same horizontal plane at a depth of about 50 centimeters (19.7 inches). The skeleton is thought to have been practically complete, but through ignorance of the workmen many of the bones were lost. The pieces rescued were the cranial cap, the two femora intact, the two humeri and ulnae almost entire, the right radius, the left half of the pelvis, a fragment of scapula, and remnants of five ribs. The discovery was announced in 1857. No artifacts or bones of animals were found with the human remains, but at a distance of less than 130 meters (426.9 feet), in a cave similarly situated, bones of the cave bear and *Rhinoceros merckii* were found, and the state of preservation of the animal and human

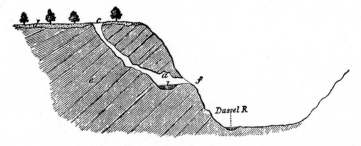

FIG. 221. SECTION OF THE CAVE OF FELDHOFEN IN THE NEANDERTAL, NEAR DÜSSELDORF, GERMANY.

The skeleton of Neandertal was found in the .oam at *b* in 1857. The referencesare as follows: *a*, cavern 18.3 meters (60 feet) above the *D*üssel and 30.5 meters (100 feet) below the ground level at *c*; *b*, loam covering the floor of the cave; *b–c*, vent connecting the cave with the upper surface; *d*, superficial sandy loam; *e*, Devonian limestone; *f*, terrace or ledge of rock. After Lyell.

bones is precisely the same. The Neandertal skeletal remains are preserved in the University Museum at Bonn.

In the discoveries at Gibraltar and Neandertal there were no associated remains that would serve to date the human remains. We do know, however, that the Gibraltar skull came from a matrix similar to that making up near-by Mousterian deposits; also that the loess deposit in which the Neandertal remains were found is the same as that of other Mousterian loess stations, namely, the so-called recent loess.

The relative flatness of the Neandertal type of skull is brought out clearly by Schwalbe's bregma and lambda angles plotted on the profile curve from the nasion to the inion (Fig. 222). From *c*, the highest point of the skull, line *ch* is drawn perpendicular to line *gi,* which connects the glabella with the inion. The ratio of *ch* to *gi* is much greater in the modern races than in the Neandertal, being 40.4 in the latter and 52 in the lowest types of recent man. Another striking difference is the retreating forehead of *Homo neandertalensis.* This may be determined by measuring the angle which the straight line drawn from bregma (*b*) to glabella makes with the base line *gi.* In the Neandertal skull the angle *bgi* is only 44°, while in *Homo sapiens* it never falls below 55°. The

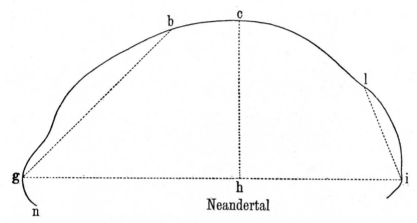

FIG. 222. SCHWALBE'S BREGMA AND LAMBDA ANGLES AS APPLIED TO THE NEANDERTAL CRANIUM.

The bregma angle is bounded by the lines *bgh;* the lambda angle by the lines *lih.* These angles are both much more acute in the Neandertal race than in all subsequent races. After Schwalbe.

lambda angle measures from 78° to 85° in recent man, while it is only 66° in the Neandertal specimen.

An increase in the size of the bregma and lambda angles would, of course, mean a marked increase in the length of the medial cranial curve *gbcli.* In respect to the relative length of this curve, the Neandertal skull resembles the ape skull more closely than it does that of recent man. In the latter the median curve is greater than any curve not in a median line and connecting the glabella with the inion. In the apes and the Neandertal race the median curve is shorter than the curve passing over the upper margin of the temporal bone. These two curves on the Neandertal skull are about of equal length.

Arcy.—There are half a dozen caves at Arcy-sur-Cure (Yonne). In one of these, the Grotte des Fées, part of a human lower jaw including the retreating chin was found by the Marquis de Vibraye about 1858. It was in the lowest of the relic-bearing deposits, associated with the remains of the cave bear, cave hyena, mammoth, and woolly rhinoceros. The Arcy lower jaw is in the Vibraye Hall at the Natural History Museum, Paris.

La Naulette.—In 1866, in the cave of La Naulette on the left bank of the Lesse in the township of Furfooz (Namur), Belgium, Dupont found an ulna, a canine tooth, and the well-known lower jaw, in association with the remains of the mammoth, woolly rhinoceros, and reindeer. They came from next to the lowest ossiferous level (third from the top), and are supposed to belong to a female. They are now in the Museum of Natural History, Brussels.

Šipka.—A specimen from the cave of Šipka (Moravia) is a portion of the lower jaw of a youth or a young adult, found in 1880 by Maška. It was at first supposed to belong to a child because the right permanent canine and the right premolars had not yet erupted; this was, however, simply a case of belated eruption, for the incisors bear distinct evidence of wear. Besides being relatively massive, the jaw shows other primitive features, such as absence of chin. It was, moreover, associated with an industry that has been classed as of Mousterian age. The Šipka lower jaw is in the Zemské Museum at Brünn.

Spy.—Perhaps no other discovery of fossil human remains has done more to crystallize our ideas as to the differences between recent man and Quaternary man than that of Spy. The cave of Spy is on the left bank of the Orneau some 14.5 meters (47.6 feet) above the river and near the railway station of Onoz-Spy (Namur). In the terrace in front of the cave, at a depth of about 4 meters (13 feet), Marcel de Puydt and Max Lohest found parts of two human skeletons in 1886.

An account of the discovery and an exhaustive study of the specimens were soon published by Fraipont and Lohest. Following this publication, there came a general recognition of the existence of the so-called Neandertal type of fossil man, since in Spy are united the racial characters of the Naulette absence of chin with cranial and limb-bone characters of the skeleton from the Neander valley.

The skeletons were found in the next to the lowest of the deposits, one 6 meters (19.7 feet) from the entrance and the other 8 meters (26.3 feet). They are known as Spy No. 1, and Spy No. 2. The latter was resting on its side, the hand in contact with the lower jaw. It could not be determined whether this attitude also applied to No. 1. Of No. 1 there were recovered the vault of the cranium, part of the upper jaw including nine teeth (five of them molars), the lower jaw almost complete, with sixteen teeth, the left clavicle, the right upper arm bone and the shaft of the left, the left radius, the heads of the two ulnae, an almost complete right femur, complete left tibia, and the right calcaneum. The parts belonging definitely to No. 2 include the cranial vault, two pieces of the upper jaw with teeth, two pieces of the lower jaw with teeth, fragments of the shoulder blades and left clavicle, incomplete humeri, the shaft of the right radius, parts of the ulnae, upper part of the left femur, the left calcaneum, and the left astragalus. In addition there are several vertebrae and small bones of the feet and hands which cannot be referred with certainty to either of the two skeletons, although the presumption is that they belong to one or both.

The two skulls represent individual variations of one type; No. 1 was probably a female, No. 2 a male. Both were adult, No. 1 having reached middle life and No. 2 being somewhat younger. The first is almost an exact replica of the Neandertal cranium (Fig. 223) except in point of size and rugged quality. The second represents a variation toward the more modern type of skull. In the norma verticalis both skulls are relatively long and ovoid, these characters being more marked in No. 1 with a cephalic index of 72.4. The cephalic index of No. 2 is approximately 77. While relatively thick in both, the cranial walls are thicker in No. 1 than in No. 2.

The femora are characterized by massive head, neck, and shaft. The shaft of the nearly complete femur has a pronounced curvature combined with an almost complete absence of the linea aspera (Fig. 224). Its condyles are adapted to the sharp backward inclination of the internal condyle of the tibia and point to habits of posture that were abandoned before the close of the Paleolithic Period. The lengths of femur and tibia correspond roughly to a stature of 1.59 meters, or slightly more than 5 feet 3 inches. The Spy skeletons escaped the ravages of the World War and are still at Liège in the private collection of Max Lohest. Most of the industrial remains from Spy are in the Marcel de Puydt collection at Liège (Musée Curtius).

The deposit bearing the skeletons contained also numerous remains of the woolly rhinoceros, horse, *Bos primigenius,* and the cave hyena. The remains of woolly elephant were common, but

FIG. 223. CRANIUM NO. I FROM THE CAVE OF SPY, NAMUR, BELGIUM. NEANDERTAL RACE. MOUSTERIAN EPOCH.

This cranium belongs to one of the two incomplete Mousterian skeletons found here. Scale, ⅖. Originals in the possession of Dr. Max Lohest, Liège. After Fraipont and Lohest.

those of the cave bear, reindeer, and stag were rare. Typical Mousterian flint implements were associated with the human bones.

Bañolas.—The history of the Gibraltar skull is almost paralleled by that of another discovery in Spain, province of Gerona, near the eastern end of the Pyrenean chain of mountains. Some 23 kilometers (14.3 miles) north-northwest of Gerona, the capital of the province of the same name, in the center of a depression, lies the lake of Bañolas, now only a remnant of what it once was. Immediately to the east of the southern end of the lake is the town of Bañolas, built on travertine beds left by the former greater lake. These beds rest on early Quaternary red clays which have been exploited extensively for building purposes. The quarry of Don Lorenzo Roura is near the northern limits of the town in what is called "Llano de la Formiga." Here in April, 1887, he encountered a human lower jaw embedded in the hard travertine at a depth of from 4 to 5 meters (13 to 16.4 feet). Fortunately Roura left the fragile jaw, almost complete, in its stone matrix and turned the block over to a Bañolas pharmacist, Don Pedro Alsius, who under-

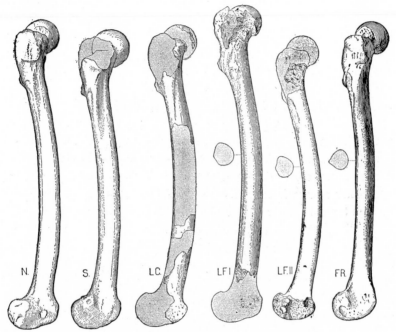

FIG. 224. MOUSTERIAN FEMORA COMPARED WITH THAT OF A MODERN FRENCHMAN.

Profile views. *N.*, Neandertal; *S.*, Spy; *LC.*, La Chapelle-aux-Saints; *LF*.I, L. Ferrassie I; *LF*.II, La Ferrassie II; *FR.*, modern Frenchman. Scale, *ca.* ⅛. After Boule.

took the preparation of the specimen by the careful removal of the matrix from the bone. The relic is still in the private collection of Alsius, or rather of his family, for he died early in 1915. Although he published nothing concerning the specimen, Alsius recognized its archaic character. The first printed notice seems to have been that in *Annuari del Institut d'Estudis Catalans,* Barcelona, 1909, by Professor Cazurro. Another note, by Professor Harlé, appeared in 1912 in the *Boletin del Instituto Geológico de*

España (Madrid). It was followed by an exhaustive study entitled *La Mandibula Neandertaloide de Bañolas* by Hernandez-Pacheco and Obermaier.

On account of its fragile character no attempt has been made to separate the lower jaw wholly from its matrix, and its inner surfaces are therefore not accessible. The outer surfaces, including a full set of sixteen teeth, are laid bare. The bone is of the same color as the matrix and highly fossilized. The right side is fairly well preserved. The condyloid process, however, is entirely gone. The anterior portion of the coronoid process is nearly complete, but its highest point cannot be definitely fixed. A small piece is missing from the angle at the junction of the horizontal with the ascending ramus, but its negative is so well preserved by the tufa that the gonion can be determined with accuracy.

The left half of the jaw was broken in seven pieces when discovered, but they have been successfully united. Owing to a very early break, the whole left half is shoved outward and backward to a slight degree, a defect which cannot be remedied. The left ascending ramus is not in so good a condition as the right. Although the coronoid and condyloid processes are missing, the transverse diameter of the latter can be measured because of the tufa negative. Nearly the whole of the condyle lies inside the plane of the outer surface of the ascending ramus if extended, as is the case with the lower jaw of La Chapelle-aux-Saints.

The neck of the condyle is short; the coronoid process, low and blunt, is seen in the nearly intact right ramus. The ascending branches are relatively low and broad. The body of the lower jaw is also low but robust. The chin is at least rudimentary if not wholly lacking; the angle of symphysis is 85°, placing Bañolas in the same class with La Ferrassie.

The lower jaw of Bañolas belonged to a male who had reached the age of about forty years. Morphologically it falls within the Neandertal group. Unfortunately it was associated neither with other skeletal remains nor with artifacts. The travertine and the lower jaw itself are undoubtedly Pleistocene. If not so archaic as the Gibraltar skull, it might well be as old as the remains from La Ferrassie, which were associated with a typical Mousterian industry. Bonarelli seems to be alone in believing the lower jaw of Bañolas to belong in the same class with that of Heidelberg.

Malarnaud.—In 1889 the discovery of a human lower jaw in the cavern of Malarnaud, some 40 meters (131.3 feet) above the bed of the Arize near Montseron (Ariège), was announced by Filhol. The lower jaw and associated fossil remains, including woolly rhinoceros, cave bear, cave lion, and cave hyena, came from a deposit of fine, red, ossiferous clay, which was capped by a thick layer of stalagmite. The absence of all but one tooth, and the filling of the sockets with the red clay, indicated in a convincing manner that the lower jaw had been transported with the clay to the spot. The sockets for the canines are quite large. The specimen is practically complete except for fragments lost from the condyles and coronoid processes. It belongs to the Museum of Natural History, Paris.

Gourdan.—There are a number of caves at Gourdan (Haute-Garonne), the chief one of which Piette named the Grotte Murée. In the lowest culture level of this cave, which was of Mousterian age, were found the cheek bones and maxillaries of a human female. The cheek bones resemble those of Gibraltar, and the orbits are apparently about as high as they are broad, a characteristic of the Neandertal race. A piece of lower jaw belonging to another individual was found at Gourdan; it is of the same type as the one from Arcy. At a higher level at Gourdan, Piette found human bones of Magdalenian age.

Le Moustier.—At a level some 10 to 12 meters (32.8 to 39.4 feet) lower than the classic station at Le Moustier is the station known as the lower rock shelter (see Figs. 54, 55 and 56), where Hauser began work in 1907. In March, 1908, he came upon human bones. Immediately thereafter a committee of French officials were called for purposes of verification. The exposed bones were then recovered to wait the arrival of a party of German anthropologists in August, 1908, when the skeleton was uncovered and removed under the direction of Klaatsch of Breslau. The bones, which came from a level 0.46 meter (1.5 feet) beneath the surface of the deposit, represent a burial. The right arm was flexed so that the right side of the skull rested on it. The left arm was extended; from a spot in close proximity to the left hand, Hauser had removed a beautiful cleaver 17 centimeters (6.7 inches) long and a typical Mousterian scraper before he was aware of the presence of the skeleton. Both the cranium and the lower jaw had been altered in shape by earth pressure. About the skeleton were seventy-four flint artifacts and various bones of the contemporary fauna, a large burnt bone of *Bos primigenius* lying directly on the skull.

The skeleton is that of a youth about sixteen years old. At this age, sex cannot be determined definitely from the bones alone, but Klaatsch believes the sex to have been male. The race characters are not so distinct as they would be at full maturity but point unmistakably to the type of Neandertal. The skeleton from Le Moustier is now in the Museum für Völkerkunde, Berlin.

La Ferrassie.—In the commune of Savignac-du-Bugue is the important Paleolithic site of La Ferrassie, consisting of a rock shelter and a small cave, known to Peyrony since about 1898. From 1909 to 1912 four skeletons of Mousterian age were found in the rock shelter. Two were those of adults and two of children. The Mousterians dug pits in the Acheulian deposit and there buried their dead. The remains of one child were less than a meter from those of an adult. In this Acheulian deposit near the two skeletons, the author found a cleaver in 1912. By the hand bones of the female adult, a Mousterian scraper was found. The limb bones of the skeleton were sharply flexed, resembling Neolithic burials. In 1920 the skeleton of a newborn infant was found in Mousterian deposits. The sixth skeleton, that of a child some seven or eight years old, was discovered likewise in Mousterian deposits in 1921. The position of the calotte was more than half a meter from the sepulture containing the rest of the skeleton, which was covered by a large stone with *cupules;* the pitted surface of the stone was underneath. The skeletons are in the Museum of Natural History, Paris. A study of them by Boule has not yet been completed.

The chief site at La Ferrassie is a vast rock shelter, the overhanging rock of which fell before the close of the Quaternary. The various horizons are easily determined by their color. There is a pure spring in front of the rock shelter.

La Chapelle-aux-Saints.—A discovery of Paleolithic human remains was made on August 3, 1908, by the Abbés J. and A. Bouyssonie and L. Bardon, assisted by Paul Bouyssonie, a younger

brother of the first two. It is in many respects one of the most
satisfactory, particularly on account of the pieces being so nearly
complete. The locality is the village of La Chapelle-aux-Saints, in
the valley of the Sourdoire, 22 kilometers (13.6 miles) south of
Brive, in the department of Corrèze, which forms a part of one of
France's celebrated cavern belts, including Dordogne and Charente
to the west. The fauna of the horizon to which the human skeletal
remains belong includes reindeer, horse, ox, *Bison, Rhinoceros
tichorhinus,* fox, marmot, birds, and wild goat.

The human bones were found near the entrance to a cave. They
seemed to lie in a rectangular pit sunk to a depth of 30 centimeters
(11.7 inches) in the floor of the cave. They were covered by a
deposit intact, 30 to 40
centimeters (11 to 15
inches) thick, consist-
ing of a magma of
bone, of stone imple-
ments, and of clay. The
artifacts are typical of
the Middle Mousterian.
Directly over the hu-
man skull w e r e the
foot bones of a large
bovine animal still in

FIG. 225. SECTION OF THE CAVE OF LA CHAPELLE-
AUX-SAINTS, CORRÈZE, FRANCE.

The skeleton was found in the rectangular burial pit
sunk in the floor of the cave. 1, Fossil-bearing layer; 2,
clay; 3, light sandy clay; 4, rock; 5, undisturbed soil
(limestone and greenish clay). After Bouyssonie and
Bardon.

contact—proof that this piece had been placed there with the flesh
on and that the deposit had not since been disturbed. Here again
we have to do with a burial, the same as at La Ferrassie (Fig. 225).

Beyond the loss of teeth, due principally to old age, the skull
is so nearly intact as to make possible the application of the usual
craniometric procedure, thus leading to a more exact comparative
study than had been possible hitherto. This is particularly true of
the basi-occipital region, the upper jaw, and the face bones. We are
thus enabled to supplement our knowledge of Mousterian crani-
ometry at several points and to correct it at others. This is the
first case, for example, in which the foramen magnum has been
preserved in human crania of the Mousterian type. It is found to
be elongated, and is situated farther back than in modern inferior
races. The character of the inion and its relation to the cranial

base is revealed for the first time. The same may be said of the palate, which is relatively long, the sides of the alveolar arch being nearly parallel; that is to say, the palate is hypsiloid—one of the two characteristic simian forms. Boule also notes the absence of the fossa canina. The nose, separated from the prominent glabella by a pronounced depression, is relatively short and broad. The lower jaw is remarkable for its size, for the antero-posterior extent of the condyles, the shallowness of the incisura mandibulae, and the absence of a chin (Figs. 226, 227).

FIG. 226.   THE MIDDLE MOUSTERIAN SKULL OF LA CHAPELLE-AUX-SAINTS.
NEANDERTAL RACE.
Scale, ⅓.   Original in the Museum of Natural History, Paris.   After Boule.

The human skeletal remains all belong to one individual and represent a burial. They comprise the cranium and lower jaw, twenty-one vertebrae or parts of vertebrae, portion of the sacrum, some twenty ribs or parts thereof, the left collar bone (incomplete), the right and left humerus, parts of both radii, the two ulnae in part, the third and fifth right metacarpals, two first phalanges, portions of the two iliac bones, right femur almost complete and part of the left femur, broken left tibia, part of right tibia, left astragalus and calcaneum, the five right metatarsals, two fragmentary left metatarsals, and a toe bone. These are preserved in the Natural History Museum, Paris. From the same deposit about one thousand Mousterian flint implements were taken.

La Quina.—In the valley of the Voultron, about 3 kilometers (1.8 miles) northeast of Villebois-Lavalette (Charente), is the extensive fallen rock shelter of La Quina. This important site belongs to Dr. Henri Martin, who has explored it intermittently since 1905. Recently he ceded the richest portion of the site to the French Government. Near this station and to the southwest is another rock shelter which is of Aurignacian age and should not be confounded with La Quina proper.

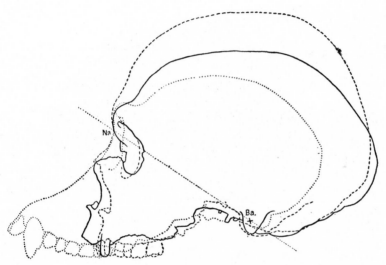

FIG. 227. STAGES IN CRANIAL DEVELOPMENT.

Profiles of the crania of a chimpanzee (dotted line), the Mousterian of La Chapelle-aux-Saints (solid line), and a modern Frenchman (dashed line). All are placed on a common basi-nasal line of equal length in each case (*Na.–Ba.*). They especially emphasize the differences in length of muzzle and in cranial height. After Boule.

La Quina represents Middle Mousterian culture. The deposits are several meters thick and more or less homogeneous in character. In the lower horizons Martin found more bones of the horse than of the reindeer. Bone fragments abound at all levels, those occurring most frequently being of *Bos,* reindeer, horse, bison, and fallow deer (*Cervus dama*). A remarkable feature is the large number that are flint-marked. Some of these are veritable chopping blocks (*compresseurs*), others are simply scarred at the points where tendons had been severed. The Martin collections from La Quina are preserved in his laboratory at Le Peyrat, near Villebois-

Lavalette, and at his house in the suburbs of Paris. The industrial remains of stone and of bone will eventually go to the Musée des Antiquités Nationales at Saint-Germain-en-Laye; the human remains will be deposited eventually in the Natural History Museum in Paris. Smaller but representative series of industrial (both stone and bone) and fossil remains have found their way into vari-

FIG. 228. ROCK SHELTER OF LA QUINA, CHARENTE, FRANCE.

The discovery here of tens of thousands of flint implements and utilized bones of Mousterian age, as well as the remains of parts of several human skeletons, makes this one of the richest of Mousterian stations. The skeleton of an adult female Mousterian (see Fig. 229) was found at the spot marked by the square piece of board. Photograph by the author.

ous museums of Europe and America, partly through exchange and partly through Dr. Martin's generosity.

In September, 1911, Martin discovered part of a Mousterian human skeleton near the base of the deposits at La Quina, in a greenish sandy clay silt which had been deposited in the bed of the Voultron when it flowed nearer to the base of the rock shelter than it does at present (Fig. 228). The skeleton is that of a woman about 25 years old. The teeth are well preserved and relatively

large. A deposit of tartar, although not abundant, is found principally about the necks of the molars. Traces of the habitual use of
a toothpick were found between the first and second molars. According to Anthony, who has made a study of the encephalic cast,
the brain of the woman from La Quina presents the same ensemble
of inferior characters as those of La Chapelle-aux-Saints. With
an antero-posterior diameter of 203 millimeters (7.9 inches) and
a maximum transverse diameter of 138 millimeters (5.4 inches),
the cephalic index of the skull from La Quina is 68.2 (Fig. 229).

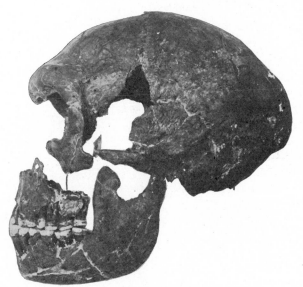

FIG. 229. SKULL OF AN ADULT MOUSTERIAN, FEMALE, FROM THE ROCK SHELTER OF
LA QUINA. NEANDERTAL RACE.
Original in the possession of Dr. Henri Martin, Paris. After Henri Martin.

From 1908 to 1921 discoveries of human skeletal remains at
La Quina have been reported as follows:

1. *1908*—Two astragali found in fairly close proximity, right and
left presumably belonging to the same individual (found in Martin's
couche [2] C3).

2. *1908*—Fragment of an occipital with the right half of the occipital protuberance (found in Martin's *couche* B2).

---

[2] All the horizons at La Quina are of Mousterian age; Martin numbers
from the top down, so that C3 means section C, third horizon from the top.

3. *1910.*—Dorsal vertebra (found in Martin's *couche* C2).

4. *1911*—Second lower right molar and third lower left molar (found about 1 meter apart in Martin's *couche* C2), both belonging presumably to one individual.

5. *1911*—The skeleton mentioned above (found in Martin's *couche* B3).

6. *1912*—Fragment of a parietal (found in Martin's *couche* B2).

7. *1912*—Part of the right half of a frontal with a considerable portion of the brow ridge (from Martin's *couche* C2).

8. *1912*—Part of the left half of a frontal with a portion of the brow ridge (from Martin's *couche* C2).

9. *1912*—Left ramus of a lower jaw with the two premolars and three molars in place (from Martin's *couche* C2).

10. *1913*—Left temporal (from B3).

11. *1913*—Fragment of the occipital (from M2, same as Martin's M gamma).

12. *1913*—Fragment of left frontal (from M2).

13. *1913*—Left parietal (M2).

14. *1908*—Left parietal (H1).

15. *1913*—Right parietal of a youth (B2).

16. *1913*—Median fragment of left temporal (C2).

17. *1913*—Lower left canine (C1).

18. *1915*—Cranium of a child about eight years old, discovered by Mme. Henri Martin (C2).

19. *1920*—Left patella (B3).

20. *1921*—A second lower left molar, a third lower left molar, a first upper right premolar with two roots; all apparently belong to the same individual. Found September 8 (C2). Incisor found September 9, in same deposit.

Of these the discovery of the skull of a child about eight years old is second in importance only to that of the adult skeleton in 1911. The cranium of the child is almost complete and not altered by postmortem pressure in the deposit. It is so nearly perfect as to throw new light on the ontogeny of Neandertal man. Practically all the distinguishing features of the race are present, so that the Neandertal child of eight years resembled the adult of his race more closely than the modern child resembles the adult of his race. There is present, however, one feature of the face that had not been anticipated. All Neandertal crania thus far discovered are more or less fragmentary; among the parts easily lost and usually missing are the nasal bones. Fortunately the nasal bones are present in this child and by their shape prove that Mousterian man had a fairly well developed nasal bridge (Fig. 230).

From a study of the bones and their mode of occurrence Martin concludes that the Neandertal type persisted at La Quina until the end of the Mousterian, that cannibalism was not practiced, nor did the Mousterians of La Quina bury their dead. The various human bones and teeth came from two absolutely different archeologic

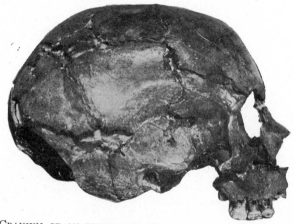

FIG. 230. CRANIUM OF AN EIGHT-YEAR-OLD UPPER MOUSTERIAN CHILD FROM LA QUINA. NEANDERTAL RACE.

This cranium exhibits practically all the distinguishing characteristics of the Neandertal race. The nasal bones which accompany this cranium were the first discovered on Neandertal crania; they indicate that Mousterian man had a fairly well developed nasal bridge. Original in the possession of Dr. Henri Martin, Paris. After Henri Martin.

deposits—the finds 1 and 5 from an ancient muddy bed of the Voultron, all the others from a deposit made up largely of débris fallen from the cliff above combined with kitchen refuse.

Le Pech de l'Azé.—Of minor importance was the discovery by Peyrony in a cave some 5 kilometers (3 miles) from Sarlat (Dordogne). At a depth of 10 centimeters (4 inches) in a deposit of Upper Mousterian age, surrounded by an abundance of bones broken intentionally and teeth of *Bos,* stag, horse, and reindeer, also flint points and scrapers, was found the skull of a child some five or six years old. In the deposit immediately below the skull Peyrony found flint cleavers belonging presumably to an earlier phase of the

Mousterian Epoch. Associated in the same deposit with the human skull were bones of *Cervus elaphus, Bos priscus, Rangifer tarandus, Equus caballus, Capra sp.*

Le Petit-Puymoyen.—Portions of an upper and a lower human jaw were found by Favraud in the station of Le Petit-Puymoyen (Charente). These were published by Dr. Siffre, a dentist, and are of Upper Mousterian age.

Isturitz.—A human lower jaw was found in 1895 at Isturitz (Basses-Pyrénées), associated with numerous bones of the cave bear and woolly rhinoceros. In type the lower jaw approaches that of Malarnaud.

Estelas, Aubert, Salle-les-Cabardes.—About this time fragmentary human remains presumably of Mousterian age were reported from the caves of Estelas and Aubert, both in Ariège, and from Salle-les-Cabardes (Aude).

Island of Jersey.—The discovery in 1910 of thirteen human teeth of Mousterian age was made by Nicolle and Sinel in the cliffs overlooking Saint Brelade's Bay, south coast of Jersey. Other faunal remains included the reindeer, woolly rhinoceros, and two varieties of horse. The human teeth are described by Keith as being the most primitive examples of the Neandertal type yet discovered. Obermaier, however, would refer them to the Upper Mousterian Epoch. The cave is known as La Cotte de St. Brelade.

Ochos.—The fragment of a lower jaw found about 1905 in the cave of Schwedentisch (Moravia) is undoubtedly of Mousterian age. Whether it antedates the Upper Mousterian Epoch is still an open question. According to Rzehak, it bears a close resemblance to Spy No. 1.

Ipswich No. 2.—In the Bolton and Laughlin brickfield at Ipswich, in what Reid Moir calls "floor C" at the top of the brick earth, workmen found the shaft of a right human femur, the shaft of a left humerus, and a cranial fragment, all belonging to one or more adults. Mousterian flint implements also occur at this level. The femur is neither so round in section nor so short and massive as the femur from Spy. It belongs to what might be called an attenuated Neandertal type, and is not to be confused with the skeleton from the same brickfield described near the end of this chapter.

Broken Hill.—Discoveries of fossil human remains are of rare occurrence. Since all the world is more or less interested in them, new discoveries of this kind should be made known promptly, and as widely as possible. In this respect British anthropologists have more than once set a good example by publishing preliminary reports with ample illustrations. Such a report followed the discovery in 1921 of some remarkable human skeletal remains in Bone Cave at Broken Hill, northern Rhodesia.

The cave is appropriately named because of the hundreds of tons of animal bones found therein. In so far as they have been examined, these bones belong to species scarcely distinguishable from those still living in Rhodesia. At the bottom of a great mass of loose débris containing animal bones and at a depth of 27.45 meters (90 feet) below the entrance to the cave, parts of at least two human skeletons were found in association with implements of stone and bone and broken bones of animals. Both human and animal bones were in a good state of preservation due to a coating of zinc silicate (hemimorphite).

The human remains include a nearly complete adult male cranium, part of the upper jaw of a slightly smaller cranium, the two ends of a large left thigh bone, the shaft of a smaller thigh bone, a large left shin bone (tibia), a sacrum, and part of an adult male left innominate. The cranium is complete except for the loss of a portion involving the right temporal and the right half of the occipital, including part of the margin of the foramen magnum. The most striking aspect of the cranium is the facial; seen either from the front or the side, it approaches the gorilliod type more nearly than is the case in any other known human cranium. This is especially true of the massive brow ridges and is apparent even in minor details. The depth of the maxillary measured from the anterior nasal spine to the median point on the alveolar margin is excessive and serves to accentuate the prognathism. The nasal bridge is flatter than in Neandertal man (Figs. 231 and 232).

The dentition is typically human; the teeth, originally sixteen in number, are set in an alveolar arch quite human in shape and outlining a handsomely domed palate. The teeth are worn in primitive fashion, the upper incisors having met the lowers squarely edge to edge. The third molars are somewhat smaller than the other molars, but this is also the case in the adult lower jaw from Ehringsdorf. Several of the teeth had suffered from caries, and there are maxillary abcesses especially in the region of the molars. The posterior nasal spine is as well developed as in recent man.

The cranium is low and the cranial capacity correspondingly small, not exceeding 1,280 cubic centimeters. The transverse occipital crest in the region of the inion is better developed than in any known example of *Homo neandertalensis*. The mastoid processes are relatively so small as almost to approach the anthropoid type. The generous allowance of space for the implantation

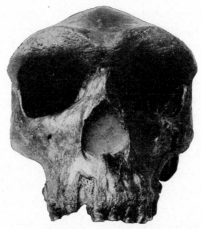

FIG. 231.   FRONT VIEW OF THE CRANIUM FROM BROKEN HILL, RHODESIA.

This aspect shows three remarkable characteristics: enormous brow ridges, much larger than in Neandertal crania from Europe; large orbits; and high upper jaw. Scale, *ca.* ⅓. Original in the Museum of Natural History, South Kensington, London. After Smith Woodward.

of the temporal and nuchal muscles is in keeping with the facial characters. Both connote a preponderance of physical over mental capacity. The foramen magnum is situated farther forward than in the cranium from La Chapelle-aux-Saints, indicating that the race of Broken Hill had succeeded in attaining more completely the erect posture than had the race of Neandertal.

The failure to recover the lower jaw is a misfortune. We know, however, that it must have been of enormous dimensions, for Smith Woodward finds that the largest known fossil human lower jaw, that of Heidelberg, is both too narrow and too short to fit the cranium from Broken Hill.

The tibia seems to be somewhat too large for the large femur and corresponds to a stature of approximately 1.83 meters (6 feet). The articular surfaces at the knee joint of both femur and tibia indicate by the direction of their planes that the posture was not erect.

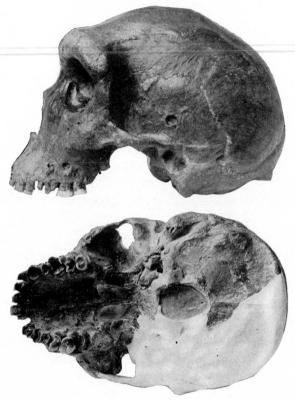

FIG. 232. SIDE AND BASAL VIEWS OF THE CRANIUM FROM BROKEN HILL, RHODESIA.

Although this cranium possesses Neandertal characteristics, even in exaggerated form, it is of uncertain date. The lower jaw has unfortunately not been discovered, but to fit the cranium, it must have been larger than any known fossil human lower jaw. After Smith Woodward.

Although the middle portion of the shaft of the femur is missing, enough is present to show that the shaft was comparatively round with only a slightly developed linea aspera. In other words, the femur comes nearer to the Neandertal type than does the tibia, which in its general aspect does not fall wholly within the range of any known tibial type.

There is at least one lesson to be drawn from the discovery at Broken Hill, and that is the danger of being misled by individual variation in a series so woefully incomplete as is our present list of fossil human skulls. With due allowance for such variation, it would seem reasonable to assume that the man from Broken Hill is a variant of the Neandertal type.

The industry from the Broken Hill cave is amorphous, with certain features suggestive of the Mousterian as seen in the spheroidal stones. The several flint flakes, three quartzite flakes, one flake of quartz crystal, and the bone point are all nondescript.

Shara Osso Goh.—Fathers Licent and Teilhard de Chardin have just announced an important discovery of human skeletal and cultural remains at a depth of 60 meters (197 feet) in a river deposit in northern Kansu, China, through which the River Shara Osso Goh has cut a deep gorge. They report the finding of parts of six skeletons including one well fossilized skull with retreating forehead, low vertex, and large orbits, in association with numerous small rude implements of quartzite and bones of the elephant, rhinoceros, bison, camel, deer, horse, and other mammals. One horse is said to be no larger than a collie dog. These human skeletons and the associated culture are probably of the same age as the artifacts recently found by Father Teilhard in the province of Shensi, China (see p. 152), and attributed by him to the Mousterian Epoch.

Summary of the Physical Characters of Mousterian Man.— An ensemble of anthropoid characters distinguishes the Mousterian skull from that of any primitive race now living. The facial region is long and prominent, the cheek bones flat and tapering. The diameter from front to back is long in comparison with the transverse diameter—a condition known as dolichocephaly. The height is low in proportion to the length; in other words, the skull is platycephalic. The brow ridges are enormously developed throughout their entire length. The nasal bridge is more prominent than might be expected in a type containing so many pithecoid characters. The distance from brow ridges to chin is great in proportion to the total length of the face. The lateral extension of the brow ridges and the prominent zygomatic arches add to the breadth of the face. The orbits are large and round. The dis-

tance from the anterior margin of the foramen magnum to a
median point on the alveolar margin is likewise relatively great,
giving to the Neandertal race a prognathism not equaled in living
races. The superior maxillaries are expansive but featureless. The
mastoid processes are not prominent. The position of the foramen
magnum is back of the center of the cranial base and, in keeping
with this position, its axis is inclined slightly forward. The glenoid
cavities are large and shallow. The superficies of the palate exceeds
that in modern man.

The face bones below the brow ridges are as a rule not so well
preserved as the remainder of the skull. The result is that we
have for purposes of study a greater number of lower jaws than
of upper jaws. It was the cumulative discoveries of lower jaws
consistently following a given type that led gradually to the determination of the Neandertal species. The most striking features
of the lower jaw are massiveness and absence of chin. The angle
of symphysis is always over 90° and often surpasses 100°. The
ascending branches are broader than in living races; the coronoid
processes are lower and the sigmoid notches shallower. The condyles are relatively flat in conformity with the shallow, expansive
glenoid fossae into which they fit.

The Mousterian dentition resembles more closely that of living
races than do the jaws or even the skulls. The alveolar arch of
the upper jaw approaches more nearly the hyperbolic or parabolic
than it does the hypsiloid form. The close approach to the *sapiens*
type is due to the reduction in the size of the canines which came
at a very early period in the evolution of the race, or may even
date from a prehuman stage.

In size and shape the lower dental arch follows more closely
that of the upper than is the case in *Homo sapiens*. When the
jaws are closed the teeth of the lower jaw hit squarely against
those of the upper jaw. A condition approaching this is to be
noted among living savage races.

While all the teeth are relatively large, there is plenty of room
in the dental arches to prevent crowding, and there is often room
to spare behind the third molars. There is naturally no diastema
since the canines had long before been reduced practically to the
level of the premolars and incisors.

Neandertal man was not a stranger to alveolar and dental maladies, although not so great a sufferer as are the civilized races to-day, or even living savage races. The man of La Chapelle-aux-Saints had lost many teeth prior to his decease. The losses might have been due wholly to pyorrhea alveolaris, but it would be rash to assume that because the only two teeth still *in situ* are sound, none of the other thirty had suffered from caries.

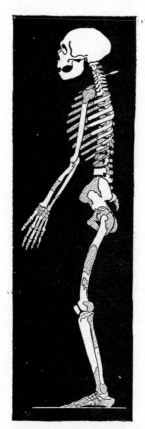

FIG. 233. RESTORATION OF THE MOUSTERIAN SKELETON OF LA CHAPELLE-AUX-SAINTS.

Note the posture, especially the slightly flexed legs and the low degree of curvature of the spinal column; the spinous processes of the neck vertebrae stand out almost horizontally. Scale, *ca.* $\frac{1}{15}$. After Boule.

Pithecoid characters are likewise prominent in the rest of the skeleton. The bones in general are short and robust, bearing witness to a powerful musculature. There are certain differences to be noted between the vertebral column of a recent European and that of Mousterian man. In the cervical region the spinous processes are bifid and slope downward, making an acute angle with the vertebral axis. On the other hand, those of Neandertal man stand out almost at right angles with the vertebral axis and are non-bifid as in the chimpanzee (Fig. 233). There are corresponding differences in the obliquity of the articular facets.

The lumbar curve is less pronounced than in recent man; so also is the curve of the sacrum, and the sacral vertebrae are narrower (Fig. 234). Pithecoid characters are to be noted in the conformation of the anterior border of the shoulder blade and in the long, slender, and arched collar bones.

The Mousterian humerus might well pass for one of recent date. This, however, is not true of the bones of the forearm. Radius and ulna are both markedly curved and in such a manner as to leave a wide

interosseous space, and the bicipital tuberosity is situated lower
than in recent man.   There is also a difference in the angle that
the axis of the humerus makes with the
axis of the forearm when the arm is fully
extended.   In Neandertal man the axis of
the forearm is approximately a prolonga-
tion of the humeral axis in a transverso-
vertical plane.   With us the axis of the
forearm veers laterally to the extent of
some 10°.   On the other hand, the elbow
of Neandertal man when fully extended is
more bent in respect to a vertico-longitu-
dinal plane than with us.   The forearm
is short in comparison with the length of
the humerus.   Boule notes that in the few
examples extant the right arm bones average larger than the left.   From this he
concludes that the tendency toward right-
handedness is a character acquired at an
early period in the history of mankind.   A
study of an important series of arm bones
from ancient graves of the Peruvian highlands led the author to a similar conclusion
respecting the Incas.

With the possible exception of the relative smallness of the carpals and shortness
of the thumb bones, the hand resembles
our own.

Our knowledge of the Neandertal pelvis is based on fragmentary material; yet
Boule has been able to note certain simian
characters, such as great length in proportion to breadth, flat ilia, reduced sciatic
spine and prominent ischial tuberosities.

The characters of the leg bones are
in accord with those of the vertebral
column and pelvis as well as the position of the foramen
magnum.   They point to a posture which can be described as not

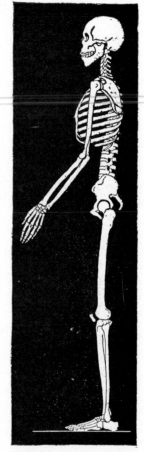

Fig. 234.   Skeleton of
a modern Australian.

The posture is more
nearly erect than in the
man of La Chapelle-aux-
Saints and the compound
curve of the spinal column
is more pronounced; the
stature is greater but the
bones are less massive.
Scale, ca. $\frac{1}{15}$. After Boule.

wholly erect. This is seen especially in the pronounced curvature and roundness in section of the femoral shaft, in the retroversion of the tibial apophysis at the knee, the relative robusticity of the fibula, the deviation of the neck of the astragalus, the digression of the axes of the great-toe bones, and the direction of the axis of the posterior surface of the heel bone (Figs. 235 and 236). The tibia is short in comparison with the length of the femur.

Elaborate studies have resulted in tables whereby one can determine approximately the stature by means of the long bones. These tables are, however, based on large quantities of recent material. They are therefore not applicable to the same degree when it comes to the small list of known Mousterian long bones. But

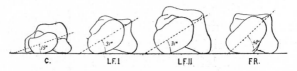

FIG. 235. SERIES SHOWING HOW THE ANGLE OF TORSION OF THE ANKLE BONE SHIFTED AS MAN ATTAINED ERECT POSTURE.

Angle of torsion of the astragalus (ankle bone) of: *C*, La Chapelle-aux-Saints; *L.F.*I, La Ferrassie I; *L.F.* II, La Ferrassie II; *FR.*, modern Frenchman. Scale, ⅓. After Boule.

the lack of material is compensated for in part at least by the greater homogeneity of the Mousterian species.

For the reconstruction of the stature of Neandertal man Boule has employed another method—the piecing together of natural-size drawings after cutting the paper at the axes of the various articular surfaces. He arrives at an average stature of 1.55 meters (5 feet) for the sexes combined, some 10 centimeters (4 inches) less than the average for the sexes combined in recent Europeans.

The cranial capacity loses its significance when not correlated with the size of the skeleton. Neandertal man was comparatively short of stature but of stocky build, hence the body mass was considerable. The brain case should be correspondingly ample. Is this ratio as great in Mousterian man as in primitive races to-day? Any answer to this question based on our present data must be considered as tentative. One is struck by the relatively great cranial capacity of La Chapelle-aux-Saints, but it would be unfair to compare this with modern averages based on large col-

lections of crania. With due consideration for the influence of
body mass and sex on cranial capacity and the paucity of Mousterian skeletal material, there would seem to be grounds for assuming that Mousterian cranial capacity was little, if any, inferior
to that of modern savages, that is to say, not far from 1,400 cubic
centimeters. The distinctive characters seen in the cranium of the
child from La Quina serve to indicate a remote origin for Neandertal cranial morphology.

Intracranial casts have been made from five Mousterian crania
—Gibraltar, Neandertal, Le Moustier, La Quina, and La Chapelle-

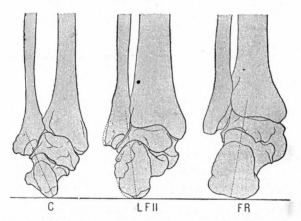

C   L F II   F R

FIG. 236. SERIES SHOWING THE DEVIATION OF THE AXIS OF THE POSTERIOR FACE OF
THE HEEL BONE AS MAN ATTAINED ERECT POSTURE.

The illustrations are of the left heel and lower leg bones (fibula and tibia): *C*, La Chapelle-aux-Saints; *LF* II, La Ferrassie II; *FR.*, modern Frenchman. In La Chapelle the
heel axis forms an acute outer angle; in La Ferrassie, the same angle is almost a right
angle; while in the modern Frenchman the angle is greater than a right angle. Scale, ⅓.
After Boule.

aux-Saints. It should be remembered that these represent the
encephalic volume rather than the size of the brain, for to the
latter must be added the cerebral membranes and cerebral fluids.
What conclusions may one be permitted to draw from a study of
these casts? All five are remarkably uniform as to general characters—very low in proportion to length and breadth, with traces
of circumvolutions much simpler than those in casts of recent
crania. The left hemisphere is slightly larger than the right, which
is in accord with the greater development of the right arm. The
superficies of the frontal lobe is only about 36 per cent of the

total surface of the corresponding cerebral hemisphere. In recent man this ratio is 46 per cent and in anthropoids 32 per cent. In respect to the development of the frontal lobe, therefore, *Homo neandertalensis* stands closer to the anthropoids than to *Homo sapiens.* Pithecoid characters are also noted in the conformation of the third frontal circumvolution, the occipital lobes (*sulcus lunatus*), and the cerebellum. As an organ of cerebration the brain

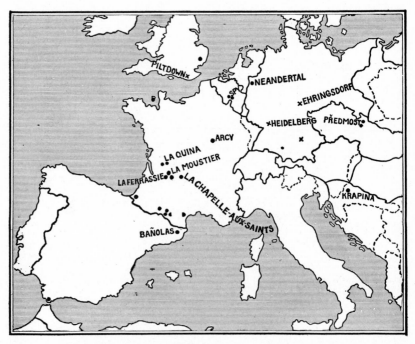

FIG. 237. MAP SHOWING THE GEOGRAPHIC DISTRIBUTION IN EUROPE OF LOWER PALEOLITHIC AND MOUSTERIAN HUMAN SKELETAL REMAINS.

Lower Paleolithic discoveries are indicated by crosses; Mousterian, by circles.

of *Homo neandertalensis* was evidently much inferior to the brain of any living group of *Homo sapiens,* however lowly.

Mousterian industrial remains force one to the same conclusion; these are confined almost wholly to a few simple types of chipped stone implements and the merest rudiments of an industry making use of bone. Mousterian man's most noteworthy possession was the use of fire; even this he probably inherited from a still older culture.

Attempted Reconstructions of Fossil Man.—The first genuine awakening concerning the physical characters of fossil man came with the discovery of human remains at Neandertal in 1857. Schaaffhausen even went so far as to publish a reconstruction.

A new impetus was given in this direction in 1895 after Dubois brought back from Java the remains of *Pithecanthropus.* More recently other anthropologists have attempted reconstructions of various early races of mankind. Although such reconstructions may serve a certain purpose in stimulating interest in and focusing attention on the matter in question, they are more apt to be misleading than would be reconstructions of other fossil animal forms because of our more intimate acquaintance with the human features and with portraiture. The reconstruction of the fossil man is at best but a mass of hypothetically coördinated details superficially stamped with all the precision of a genuine portrait. But it cannot in the nature of things be a portrait, which implies on the part of the artist an intimate personal knowledge of the sitter. Neither can it be a composite picture, which implies an intimate acquaintance with each of many individuals.

The size and shape of the mouth and lips, the soft parts of the nose, and the expression of the eyes are all of prime importance in the makeup of the human physiognomy. For these there are not sufficient data on which to base a reconstruction; the same is true of the ears.

Aurignacian Man

Mousterian man was superseded by a race of a type more nearly akin to modern man. The appearance of the new race was coincident with a marked change in the character of the cultural remains and especially with the origin of art. Physically the gap between the Mousterian type and the Aurignacian type has not yet been definitely bridged. Since Aurignacian times the succession of racial types forms a more or less unbroken series. On anatomical grounds alone it would be difficult to differentiate between the Aurignacian and the Neolithic population that buried its dead in natural or artificial caves and in dolmens. Emphasis is rightly placed on cultural and stratigraphic evidence in any attempt to distinguish among Aurignacian, Solutrean, Magdalenian, Azilian, and early Neolithic races.

Cro-Magnon.—For historic reasons we are justified in taking for our point of departure the discovery of Cro-Magnon.[3] In doing so we but follow the precedent set by basing our Mousterian race on the discovery made at Neandertal. The rock shelter of

FIG. 238.   ROCK SHELTER OF CRO-MAGNON, IN THE VILLAGE OF LES EYZIES, DORDOGNE, FRANCE.

The first discovery of skeletal remains of Aurignacian man was made here in 1868. The spot where the human skeletons were found is indicated by the arrow to the left of the building.   Photograph by the author.

Cro-Magnon is in the village of Les Eyzies near the point where the highway crosses the railroad tracks (Fig. 238). The shelter was completely buried beneath talus. When the railroad was built through Les Eyzies, this talus was removed in part and employed as filling for the roadbed. Later, toward the end of March, 1868, two local contractors, Berthoumeyrou and Delmarès, removed more of the talus for use on the adjacent highway, and in doing so they

---

[3] The patois for great hole (or pit).

uncovered the overhanging rock and came upon unexpected riches in the way of chipped flints, broken animal bones, and finally human crania (Fig. 239). Grasping the importance of the association, they immediately called Dr. Laganne from Bordeaux. A few days later, in the presence of witnesses from Périgueux, Laganne exhumed two crania and other skeletal fragments as well as worked bones of the reindeer and flint implements.

At this point Louis Lartet (son of Edouard) took charge of the investigations. He was able to determine a superposition of hearths, the lower ones alternating with sterile layers. The topmost relic-bearing layer included an unbroken series of hearths which actually touched the overhanging rock except for a short space at the very back. In this last relic-bearing deposit, and at the back, the human skeletons were f o u n d. The skull of the old man was the only one found in the space not wholly filled by débris (Fig. 240). The o t h e r skeletal remains, including parts of

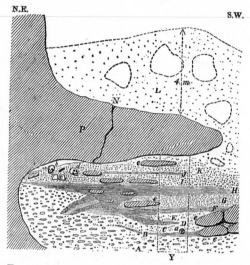

FIG. 239.   SECTION OF THE ROCK SHELTER OF CRO-MAGNON.

A, débris of soft limestone; B, first layer of ashes; C, calcareous débris; F, third layer of ashes; G, red earth, with bones; H, thickest layer of ashes and bones; I, yellowish earth, with bones, flints, etc.; J, hearth; K, calcareous débris; L, talus; N, crack in the limestone roof; P, projecting shelf of hard limestone; Y, place of pillar made to support the roof; a, tusk of an elephant; b, bones of an old man; c, block of gneiss; d, human bones; e, e, slabs of stone fallen from the roof at various times. After Lartet and Christy.

four other individuals, were found within the arc of a short radius from the old man.

Later researches seem to prove that these skeletons were contemporaneous with an Upper Aurignacian horizon, on the surface of which they were deposited. Owing to the circumstances of the find, there are still those, Rutot for example, who would class the Cro-Magnon skeletons as of Magdalenian age. There are others, including de Mortillet, who believe these burials to be intrusive and to belong to the Neolithic Period. In some respects the skele-

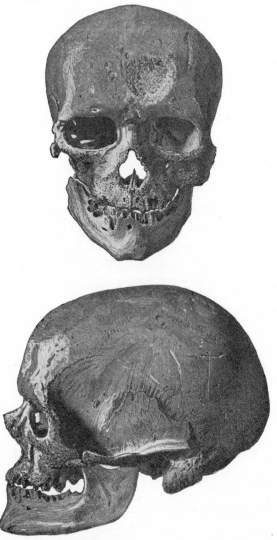

FIG. 240.   FRONT AND SIDE VIEWS OF THE SKULL OF THE OLD MAN OF CRO-MAGNON.

It is probable that this skull belongs to the late Aurignacian Epoch.   Scale, *ca.* ⅓.
Original in the Museum of Natural History, Paris.   After Lartet and Christy.

tons do resemble the early Neolithic races more closely than do certain Aurignacian skeletons of which the age is beyond question.

Broca points out several characters of superiority—volume of the brain, development of the frontal region, large facial angle. With these are united certain characters of inferiority—breadth of face, alveolar prognathism, enormous development of the branches of the lower jaw, evident strength of masticatory muscles. The femur of the old man had been injured and healed; the female had apparently been killed by a blow on the head. The stature was greater than that of modern Frenchmen, that of the old man being estimated at 1.80 meters (5 feet 11 inches); the cephalic index of the old man's cranium is 73.7, the cranial capacity well up to the average for *Homo sapiens*.

The skeletal remains from Cro-Magnon look as fresh as if they had been found in cave deposits of Neolithic age. The old man (cranium No. 1) had suffered from pyorrhea alveolaris as well as from bone cysts, especially in

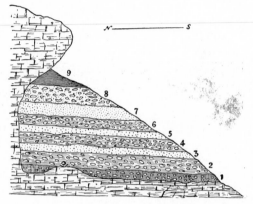

FIG. 241. SECTION OF THE ROCK SHELTER OF COMBE-CAPELLE, NEAR MONTFERRAND, DORDOGNE, FRANCE.

Near the base of the second horizon the well preserved skeleton of an adult Aurignacian was found (see Fig. 242). 1, Mousterian horizon (0.25 meter); 2, Lower Aurignacian with the human skeleton (0.3 meter); 3, sterile deposit (0.15 meter); 4, Middle Aurignacian (0.25 meter); 5, sterile deposit (0.15 meter); 6, Upper Aurignacian (0.3 meter); 7, sterile deposit (0.3 meter); 8, Solutrean, willow-leaf flint points with a single lateral notch at the base (0.6 meter); 9, humus (0.5 meter). Redrawn from Klaatsch and Hauser.

the lower jaw. The specimens are preserved in the Museum of Natural History, Paris.

Combe-Capelle.—The rock shelter of Combe-Capelle is near the top of an escarpment bordering on the valley of the Couze near Montferrand (Périgord) and some 40 meters (131.3 feet) above the Beaumont-Montferrand highway. This is one of the stations leased by Hauser. He began work here in February, 1909, and soon laid bare five culture levels (Fig. 241). The human skeleton was found near the base of the Lower Aurignacian horizon by workmen in August, 1909. On that day Hauser was away but was recalled by telegram, arriving early the following morning. The workmen had laid bare only enough of the skull to prove its nature, then covered it once more and awaited the arrival of their employer.

Photography was freely employed during the process of removal, after which all the parts were reassembled in a position approximating the original; the accompanying artifacts were added to complete the ensemble and the whole was photographed (Fig. 242). Many small shells, some of them perforated, were in close

FIG. 242. AURIGNACIAN SKELETON REASSEMBLED AS DISCOVERED IN THE ROCK SHELTER OF COMBE-CAPELLE.

Note the position of the shells (about the head), and flint implements (near the head, right arm, right leg, and left foot). Original now in the Museum für Völkerkunde, Berlin. After Hauser.

association with the skull. Typical Aurignacian flint implements and the metatarsal of *Sus scrofa* had likewise been buried with the body, which lay at a depth of 2.48 meters (8.14 feet). The removal of the skeleton was tedious because of the indurated character of the matrix. The bones, however, were wonderfully well preserved and the skeleton practically complete. Klaatsch of Breslau took part in the removal and later published an exhaustive report.

The skeleton is that of a well developed adult male with a stature estimated at 1.60 meters (5.25 feet). Klaatsch does not believe that the race represented by this skeleton could have been developed from the Neandertal race. They are so different that a zoölogist would class them as of separate species; and yet during their sojourn together in western Europe there was opportunity for a mixture of the two races to take place, and this probably did occur. The Chancelade skeleton, for example, may be the

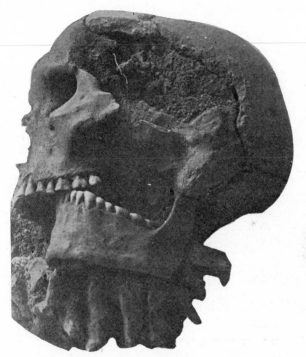

FIG. 243. THE AURIGNACIAN SKULL OF COMBE-CAPELLE BEFORE ITS REMOVAL FROM THE MATRIX.

The neck vertebrae are visible below the lower jaw; also a shell near the chin. After Hauser.

product of such a mixture. The human skeleton of Neandertal type found by Hauser at Le Moustier the previous year Klaatsch had named *Homo mousteriensis*. For the Combe-Capelle skeleton he proposed the name *Homo aurignacensis hauseri* (Figs. 243, 244). Both of these skeletons are in the Museum für Völkerkunde, Berlin.

Grimaldi.—Beginning nearest the French frontier, the caves of Grimaldi or Baoussé-Roussé are numbered from 1 to 9. The

first is the Grotte des Enfants, so called because of the discovery
there in 1874 and 1875 of the skeletons of two children. These
were buried so near together that Rivière believed them to have
been interred at the same time. On both, from the umbilicus to the
level of the hips, was found a belt consisting of more than one
thousand perforated shells of the species *Nassa neritea*. Near by
were a few flint chips. Rivière's statement that the two skeletons
came from a depth of 2.70 meters
(8.8 feet) has been questioned by
the Abbé de Villeneuve, who later
excavated the site and who found
a skeleton *in situ* at a depth of
1.90 meters (6.2 feet) on a bed
of tough clay. At this level and
at depths of 2.50, 4.20, 6.10, 7.05,
7.75, 8.90, and 9.80 meters, well
developed hearths were encountered (see Appendix I).

The skeleton found by de
Villeneuve at a depth of 1.90
meters (6.2 feet) is that of a
female advanced in years and of
small stature, estimated at 1.44
meters (4.73 feet). The cranium
is incomplete and fragile, the
lower jaw robust and with a well
developed chin. Taken as a
whole, the skeleton is not of the
Cro-Magnon type.

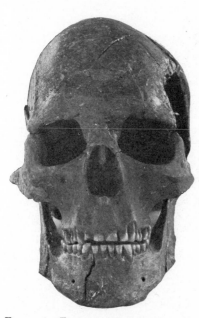

FIG. 244. FRONT VIEW OF THE AURIG-
NACIAN SKULL OF COMBE-CAPELLE.
See also Fig. 243. Scale, *ca.* ⅓. After
Hauser.

The skeleton of a man of tall stature was encountered at a
hearth level 7.05 meters (23.1 feet) from the surface (Figs. 245
and 246). Like the one above, this was a full-length burial. Both
the head and the feet had been protected, each by a stone. Above
the skeleton were found the bones of *Hyaena spelaea*. At the base
of the hearth on which the skeleton lay, there were encountered
rude implements of quartzite and limestone, each with one face
composed of a single plane. Elsewhere in the cave, and to a depth
of 8.90 meters (29.2 feet), the implements were of flint and Aurig-

nacian in type. Villeneuve noted the presence of hematite on the skull and Verneau found hematite underneath it. Peroxide of iron was likewise noted in the deposit immediately beneath the hearth in question. Near the skull were perforated deer teeth and *Nassa* shells also perforated. The stature, estimated from the long bones, is given as 1.92 meters (6.3 feet), and the cephalic index is approximately 76. Verneau was struck by the number of facial analogies existing between this subject and the skull of the old man of Cro-Magnon. They evidently belong to the same race.

The Grimaldi Race.—On the hearth 7.75 meters (25.4 feet) below the surface and separated from the hearth on which the tall man lay by a deposit of cave earth only 0.70 meter (2.3 feet) thick, there was discovered a double burial, the earliest sepulture in the cave. The skeletons are those of a young man and an old woman in close contact— the former face up and the latter originally face down. The legs were flexed, also the arms with the exception of the young man's left arm (Figs. 247 and 248). The skull of the youth rested in a shallow pit sunk through the hearth deposit, and both skulls

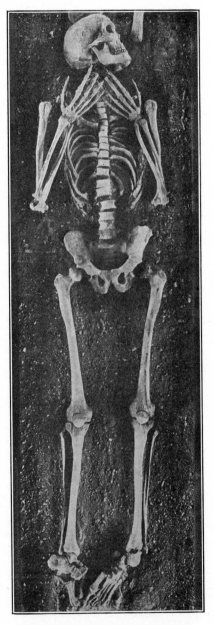

Fig. 245. Aurignacian male burial in the Grotte des Enfants.

were protected by a flagstone supported by two upright stones. The skeletons were contemporaneous with the hearth, for the overlying deposit had not been disturbed at the time of the burial.

Near the skeletons lay a considerable number of flint implements. In contact with and between the two crania were found two flint blades; a larger blade was found between the two pelves. There was likewise found on the right arm of the young man a flint blade, and near this blade two others and a scratcher. Two

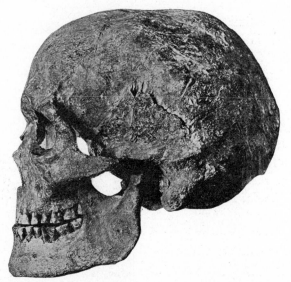

FIG. 246. SKULL OF THE LARGE AURIGNACIAN MALE FROM THE GROTTE DES ENFANTS.
See also Figure 245. Scale, *ca.* ⅓. Photograph by the Abbé de Villeneuve.

pebbles lay on the skull of the old woman, and above her pelvis a flint point with lateral notch at the base. This notch, however, was an accident, so that the piece is not referable to the Solutrean Epoch. About her left wrist were enough *Nassa* shells to form a bracelet. Traces of hematite were noted on the skeleton of the young man.

The young man had attained the age of from 15 to 17 years. An estimate places his stature at 1.54 meters (5 feet), and that of the female at 1.58 meters (5.2 feet). These individuals differ from those of Cro-Magnon not only in respect to stature but also in the slenderness of the skeletal parts and in cephalic characters.

FIG. 247. SKELETONS OF A NEGROID WOMAN AND YOUTH AS DISCOVERED IN THE GROTTE DES ENFANTS, GRIMALDI.

Villeneuve discovered this double burial at a depth of 7.75 meters (25.4 feet); 0.70 meter (2.3 feet) below the horizon of the burial of the tall male. The legs and arms were flexed, with the exception of the left arm of the youth. There were a few *Nassa* shells on the skull of the youth and two bracelets of *Nassa* shells on the arms of the female. Originals in the Museum of Anthropology at Monaco. After Verneau.

Morphologically the two crania are very much alike, with cephalic indices of 68 and 69 respectively. A nasal index of 51 places these crania in the mesorhinian class; the lower margins of the piriform aperture are rounded in such a manner as to suggest simian gutters. One is struck by the degree of prognathism, especially in the subnasal region. The chins are slightly receding. The dental arches are narrow, with almost parallel sides; the teeth are relatively large. In their ensemble of characters, these two individuals are negroid in type and differ from Pleistocene skeletons

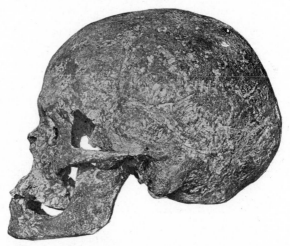

FIG. 248. SKULL OF THE NEGROID WOMAN FROM THE GROTTE DES ENFANTS.
AURIGNACIAN EPOCH.

See also Figure 247. Scale, *ca.* ⅓. Photograph by the Abbé de Villeneuve.

hitherto unearthed. Verneau therefore proposes for them a new name, *type de Grimaldi*. The Grimaldi race belonged to an early phase, the Cro-Magnons to a late phase, of the Aurignacian Epoch.

The first discovery by Rivière of Paleolithic human remains in a Mentone cavern was on March 26, 1872, in the fourth cave called Grotte du Cavillon. The skeleton lay at a depth of 6.55 meters (21.5 feet) from the surface. Powdered hematite overspread the skeleton and everything that had been buried with it, and a mass of the same coloring matter filled a furrow in front of the skull. The skeleton lay on the left side with its arms flexed so as to bring the left hand to the chin; the legs were only slightly flexed

(Fig. 249). Nearly every bone was recovered. The removal was difficult because of the indurated nature of the hearth materials in which the skeleton lay. The cranium was decorated with more than two hundred perforated shells of *Nassa neritea* and twenty-two canine teeth of the red deer (*Cervus elaphus*), also perforated for suspension. A bone implement 17 centimeters (6.6 inches) long, made from the radius of a deer, rested against the front of the skull; against the occipital two flint blades were found. On the tibia and fibula just below the left knee were forty-one shells of *Nassa neritea* perforated for purposes of attachment or suspension.

The Cavillon skeleton is that of an adult male with a stature approximating 1.85 meters (6.1 feet). The skull is dolichocephalic; its capacity is less than that of the old man of Cro-Magnon. The two most marked characters of the face are the rectangular orbits and the

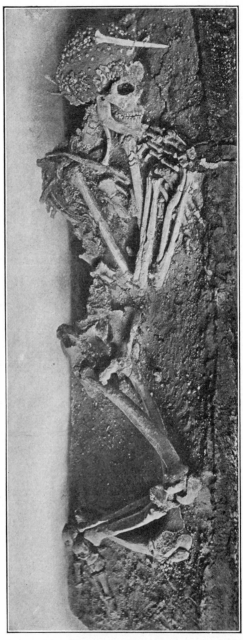

FIG. 249. AURIGNACIAN MALE BURIAL IN THE GROTTE DU CAVILLON.

complete absence of prognathism. The teeth were perfectly preserved, although much worn by use. This skeleton is preserved in the Museum of Natural History, Paris.

The only other human skeletal remains found in the cave of Cavillon consisted of a small quantity of bones and red ocher near the base of the deposits.

The Barma Grande (grand cave) is the fifth of the series. Human skeletal remains were found here at various times. Rivière reported a fragment of a lower jaw but did not give its stratigraphic position. The first skeleton was found by Julien and Bonfils in 1884. It lay on its back at a depth of 8.4 meters (27.6 feet); with it were three large flint flakes, one on top of the skull, the other two on the shoulders like epaulettes. The skull was covered with red ocher, and near it were a few flint nodules, flint chips, and teeth of the goat, red deer, and ox. The skull is preserved in the museum at Mentone.

In February, 1892, Abbo by chance uncovered part of a skeleton but left all but the skull in place until the arrival of Verneau some days later. Verneau found the burial to have been a triple one, near the center of the cave and at approximately the same depth, 8.4 meters (27.6 feet), as the skeleton previously discovered by Julien and Bonfils. The skeleton in front was that of a man of tall stature, the one in the middle that of a young woman, and the third that of a youth about 15 years old. The man rested on his back, full length, with face turned to the left. The other two rested on their left sides facing the man; their legs were extended, but in both the arms were sharply flexed so as to bring the hands near the chin.

About the neck of the man were fourteen canine teeth of the red deer, the crowns of which were ornamented with incised lines. Several fish vertebrae, the size of a trout's vertebrae, and some handsome bone pendants likewise decorated with incised lines were found at the same level. On his skull were several teeth of the red deer and fish vertebrae like those just mentioned and several perforated shells of *Nassa neritea*. Bone pendants and an object of deerhorn in shape like a double olive and ornamented with marginal striae were on the thorax. Somewhat lower on the body were perforated salmon vertebrae. Two large, perforated *Cypraea* shells, one on each side of the left tibia just below the knee, were removed by Verneau himself. The splendid flint blade, 23 centimeters (9 inches) long with a maximum breadth of 5 centimeters (2 inches) and with one end retouched, had been found by Abbo at the level of the man's left hand prior to Verneau's arrival.

The skull of the woman rested on the femur of an ox. She wore ornaments similar to those found on the male skeleton, but her ornaments were not so numerous. There was no necklace, for example, and no bone pendants on the head. In her left hand they found a flint blade 26 centimeters (10 inches) long, of greater size than the one in the left hand of the man but with fewer retouches.

The skull of the youth lay on a flint blade 17 centimeters (6.6 inches) long by 5 centimeters (2 inches) in breadth, chipped on one side only as in the two preceding cases. The more voluminous part of this blade is retouched in the form of a scratcher. On the forehead were several bone pendants, and the skull was covered with trout vertebræ and perforated *Nassa* shells. The handsome necklace was still in position; it is composed of a double row of fish vertebræ and a single row of perforated *Nassa* shells. This triple row was broken at intervals by the insertion of a canine tooth of the red deer (*Cervus elaphus*), each perforated[4] and incised. This attractive arrangement is maintained throughout and testifies to Aurignacian taste in such matters (see Fig. 109). After Verneau's departure there was found in the left hand of the youth a striated bone object in the shape of a double olive exactly like that found on the thorax of the man.

The skeletons of this triple sepulture lay in a bed deeply tinged by red ocher, and were covered by a coating of the same. Among the faunal remains immediately surrounding the sepulture those of *Cervus elaphus* and *Bos primigenius* were the most abundant. The height of the man is estimated at 1.94 meters (6.4 feet); that of the woman and of the youth, 1.64 meters (5.38 feet) and 1.65 meters (5.4 feet), respectively. All three of the crania are dolichocephalic. While not sufficiently complete to admit of exact measurements, one can say with certainty that they belong to the race known by the name of Cro-Magnon.

---

[4] Verneau's text mentions the perforations, but his illustration reproduced in Figure 109 does not show them.

Somewhat deeper than the triple sepulture, and under three flat stones, Abbo found another human skeleton. With it were a coiffure and a necklace similar to those worn by the other male. Finally, near the base of the cave deposit, on a hearth level, there was encountered a partially carbonized skeleton, accompanied by perforated *Nassa* shells.

Three burials have been reported from the cave of Baousso da Torre, the sixth cave of the series: the bones of a child near the base of the deposits at a depth of 4 meters (13.1 feet); the skeleton of an adult male, badly decayed, at a depth of 3.90 meters (12.8 feet); and the skeleton of a man at a depth of 2.75 meters (9 feet). In the last two burials mentioned the bodies had been covered with red ocher.

Paviland.—This station is a cave in the face of the precipitous limestone cliffs of Gower, overlooking the Bristol Channel in Glamorganshire. The skeletal remains of what Buckland supposed to be an ancient Briton of the period of the Roman occupation were found here in 1823. The bones were deeply stained by red ocher, and since they were originally supposed to be those of a female, they were dubbed the "Red Lady of Paviland." Apparently interred with the skeleton were numerous objects carved out of ivory and a mass of small sea shells. Ivory and shells were, like the human bones, stained with red ocher. These, as well as the flint implements from Paviland, are exactly comparable with Aurignacian remains in France. The fauna likewise points to the Aurignacian Epoch, since the horse, cave bear, and *Bos primigenius* predominate.

The skeleton was found by Buckland on the left side of the cave under 15 centimeters (5.8 inches) of earth. Some of the long bones were apparently exposed in part above the surface. The skull, vertebrae, and extremities of the right side had disappeared. On account of the magnitude of the head of the femur and of the tibia, Sollas believes the skeleton to be not that of a woman but of a man. Still better evidence might have been obtained from the left innominate bone, which Sollas reproduces but not in a position that would reveal the sex characters. The stature, estimated from the femur and tibia, was about 1.73 meters (5.75 feet), which is approximately the same as that for the Grimaldi and Cro-Magnon skeletons. The Paviland skeleton is in the museum at Oxford.

Halling.—Kent, and the valley of the Medway which traverses it, forms a fruitful hunting ground for the prehistorian. In 1912 a human skeleton was discovered at Halling on the left bank of the Medway in a deposit of brick earth. Between Halling and the Medway marshland is a natural terrace which rises 2.5 meters (8.2 feet) above the level of the marsh. A section of the terrace at the site in question discloses the following beds: (8) humus; (7) dark-red loam; (6) buff-colored brick earth; (5) buff-colored sand; (4) buff brick earth, at the top of which is an old land surface with hearths; (3) buff brick earth; (2) brown sand; (1) buff brick earth. The skeleton was found in bed Number 4 and at a depth of 1.8 meters (5.9 feet) from the present land surface. Competent authorities insist that the overlying beds had not been disturbed.

Dr. Edwards observed that the body had been laid on its back but was turned slightly so as to rest on its left arm. The limbs had been flexed on the trunk and the head had been bent on the breast, so that the extreme parts of the skeleton were less than 1 meter (3.3 feet) apart. The position was, therefore, not unlike that common to Neolithic burials. A logical conclusion would be that the Halling man had lived on the old land surface and when dead had been placed in a superficial grave on the spot.

The age of the old land surface can be determined largely from the nature of the artifacts and animal remains found on it and immediately beneath it. These point to the Aurignacian Epoch. Abundant remains of Pleistocene animals have been found in the corresponding terrace on the opposite bank of the Medway. The physical characters of the Halling man are such as one would expect to meet with in the Upper Paleolithic races. The bones of the cranium and lower jaw were badly broken but were replaced in part. In such reconstructions one must allow for a margin of error. The capacity is estimated at 1,500 cubic centimeters, the cephalic index at 75. Neither in the cranium nor the lower jaw are apelike characters discernible. The bones of the upper jaw and nose are lacking. The skeleton was that of a male not over 40 years old. For one of this age the teeth were much worn;

five of the molars had been lost from disease (abcesses), and one of the premolars had been likewise lost before death. All trace of animal matter was gone from the bones. On drying they became hard, turning first brown and then a light stone-gray color. They are referable with a fair degree of certainty to the Paleolithic Period, and probably to the Aurignacian Epoch.

Brünn.—Important loess finds were made in Brünn, Moravia, during the year 1891. A section of the loess reveals a superposition as follows, beginning at the top: (3) humus-loess, 1 meter (3.3 feet); (2) yellow loess, 3.5 meters (11.5 feet); (1) red loess. In this red loess was found the tusk of a mammoth and, immediately beneath it, a complete shoulder blade. By the side of this shoulder blade lay a human skull and parts of the skeleton belonging therewith. Near these and in the same red loess lay scattered more than six hundred *Dentalium* shell beads, a large, crude human figurine of ivory, bone and ivory disks, several rhinoceros ribs, and an angular block of syenite as large as one's head. In addition to the woolly elephant and woolly rhinoceros, the red loess contained remains of the horse and a fragment of antler, probably referable to the reindeer.

The skull is that of an adult male with an estimated capacity of 1,350 cubic centimeters and a cephalic index of about 66. It resembles very closely the one found in 1885 at a depth of 6 meters (19.7 feet) in the loess of Rother Berg near Brünn. In 1889 flint and bone artifacts were found at Rother Berg at a depth of 7 meters (22.9 feet). The Brünn skeleton was stained with ocher through intentional association of coloring matter with the body. It can be referred with certainty to the Upper Paleolithic Period, and with some degree of probability to the Aurignacian Epoch.[5] It will be recalled that bones of the Aurignacian skeleton from Paviland were stained in like manner.

Předmost.—Some 3 kilometers (1.9 miles) east of Prerau junction is the village of Předmost (Moravia). Near Předmost is an isolated calcareous rock known as Hradisko, the base of which is surrounded by fluviatile sands and gravels and which is capped by a deposit of loess 20 meters (65.7 feet) thick. In this

---

[5] Obermaier would class the Brünn skeleton as of the Solutrean Epoch.

loess at Hradisko a remarkable series of human, cultural, and animal remains have been discovered. The principal fossil is the mammoth, of which parts of about nine hundred individuals have been found. Among the faunal remains are included at least fifteen other species. The remains of the mammoth occur all the way from near the surface to a depth of 8 meters (26.3 feet).

The human skeletal and cultural remains of Paleolithic age, at first regarded as Solutrean but now definitely referred to the Aurignacian by such an eminent authority as Breuil, were encountered at a depth of from 2 to 3 meters (6.6 to 9.8 feet) in virgin loess, the archeological bed being from 50 to 80 centimeters (19.7 to 31.5 inches) thick. From this bed twenty-five thousand implements of flint, jasper, and other stone have been gathered. Characteristic for Předmost is a sort of flat club, in some examples as much as 50 centimeters (19.5 inches) long, made from the tibia or ulna of the mammoth. In the Maška collection are seven crude human figurines in sitting posture, carved out of the metacarpals of the mammoth and averaging 13 centimeters (5.1 inches) in length. Among other art objects there should be mentioned an ivory plummet decorated with an incised pattern and a stylistic female figure engraved on ivory.

The chief interest attaches to the human skeletal remains. Wankel found the fragment of a lower jaw. The series of remains discovered by Kříž include the cranium of a child about 12 years old, two fragments of youthful lower jaws, and fragments of bones belonging to at least six other individuals. Teeth and bones of the blue fox were found in contact (apparently fortuitous) with the forehead of a child. Maška found a communal sepulture completely bounded by stones and containing fourteen complete skeletons besides portions of six others. Ten dolichocephalic crania, six of them adult and four youthful, are completely reconstructed. The best preserved skulls are Nos. 3, 4, 9, and 10. Skull No. 3 is that of an adult male about 30 and is complete except in the region of the foramen magnum. Although the most primitive of the series, with relatively large brow ridges, it is not of the Neandertal type. The mastoid processes are large and the chin is fairly well developed. The bite is square. The lower left molar was lost before death. The teeth are all worn except the upper molar over the missing lower molar. The sutures are simple and wide open. Skull No. 9 is also that of a male; Nos. 4 and 10 are those of females. The Předmost skeletal remains are in the Zemské Museum at Brünn.

The only object deposited with the dead was a necklace consisting of fourteen small oval ivory beads on the neck of a child. The archeological layer covering the communal sepulture had not been disturbed since its original deposition.

La Combe and La Rochette.—During the summer of 1912 it was the author's good fortune to excavate the cave of La Combe near Les Eyzies (Dordogne). In the Aurignacian deposit we found a large lower left first or second human molar with an artificial conical perforation through the anterior root (see Fig. 72). The tooth is remarkable for its large size, roots as well as crown, and because of its being the first reported case of the use of human teeth as ornaments or charms during the Paleolithic Period. From the cave of La Rochette (Dordogne) have been taken isolated fragments of a skeleton and human teeth belonging to three individuals. These came from Aurignacian deposits.

Solutré.—The type station of the Solutrean Epoch also contains an Aurignacian horizon; in this horizon at the classic station of Crot-du-Charnier, a discovery of capital importance was made in 1923 by Drs. Mayet, Depéret, and Arcelin, who uncovered three Aurignacian burials *in situ* beneath the horse magma. The bodies had been left in extended position with the head in each case to the west. Near the first burial, that of a young adult female, there were found the fragmentary skeletons of two very young children. Burial Number 2 was that of an adult male as was likewise that of Number 3. The three formed a linear series running east and west; in each case two stones had been placed at the head (one on either side).

A detailed study of the three adult skeletons is being made by Dr. Lucien Mayet of the University of Lyons. The two males were tall—1.83 and 1.75 meters (6 feet and 5 feet 10.5 inches) respectively; their skulls have a cephalic index almost too large to bring them within the dolichocephalic class. They are said to be more like the Man of Cro-Magnon than that of Combe-Capelle. The stature of the female is estimated at 1.55 meters

(5 feet 1 inch). Some of the sepultures previously found in the Crot-du-Charnier were no doubt of Aurignacian age also, as Obermaier had already stated.

Camargo and Castillo.—The cave of Camargo near Revilla-Camargo (Santander) has three culture-bearing levels. In the lowest of these, which is of Aurignacian age, Sierra found a cranial calotte including the greater part of the frontal and the two parietals. The only human skeletal remains from the Aurignacian deposits in the great cave of Castillo near Puente Viesgo (Santander) are the lower jaw of a child and the molar of an adult.

Ojcow.—A human cranial calotte was taken by Czarnowski from an Aurignacian deposit in Oborzysko wielkie (black cave) near Ojcow (Poland).

Giesslingtal.—In the small valley of the Giessling near Spitz (Lower Austria), a loess station with two Aurignacian levels was recently brought to the attention of Dr. Josef Bayer. The finds included an ocher burial. According to the story of the peasant who found the bones, the skull, colored red by ocher, was taken to his house. His wife was so superstitious, however, that he broke the skull to pieces and threw them in the brook before the arrival of Bayer.

Solutrean Man

Anatomically the Solutrean would be difficult to distinguish from the Aurignacian races on the one hand and the Magdalenian on the other. Human skeletal remains that can be definitely dated as Solutrean are rare.

Laugerie-Haute and Badegoule.—The skeleton from the proto-Solutrean horizon in the rock shelter of Laugerie-Haute and the fragments of a child's skeleton from the Solutrean horizon of the rock shelter of Badegoule, both in the department of Dordogne, have little value from the viewpoint of ethnic characters.

Klause.—In the second cave of Klause near Neu Essing (Bavaria) Obermaier found a Solutrean sepulture. For the inhumation a slight pit had been sunk in the Mousterian deposit of this cave, the second from the top of the escarpment. The body was bent toward the left and the legs extended. Both above and beneath the skull was a mass of breccia composed of fragments of mammoth tusks. The body was completely surrounded by a mass of powdered ocher. The overlying deposit is of Magdalenian age. The burial had taken place after the close of the Mousterian Epoch and before the beginning of the Magdalenian. Aurignacian cultural remains are not represented in this cave, but those of the Solutrean Epoch are in evidence; hence the skeleton, that of a man about 30 years old, can be referred with a fair degree of assurance to the Solutrean.

Magdalenian Man

By Magdalenian man is meant the race that left its skeletal and cultural remains in Magdalenian horizons. Anatomically the Magdalenians were not very different from the Aurignacians and Solutreans; nor were they unlike the races that immediately followed. A number of sites have furnished skeletal remains that can be definitely determined as of Magdalenian age.

Laugerie-Basse.—On account of the remarkable series of portable art objects found there, Laugerie-Basse is of the first rank among Paleolithic stations. It is a great rock shelter on the right bank of the Vézère near Les Eyzies (Dordogne) which has been known since the first explorations of Lartet and Christy in that district. Here, in 1872, Massenat found a human skeleton *in situ*. It was removed in the presence of Cartailhac and Lalande from a depth of about 3 meters (9.8 feet) beneath the surface of the Paleolithic hearth deposits. The body rested on the left side; the vertebral column, bent forward, had been pinned down and crushed by a large fallen stone; the arms were flexed, the left hand under the head and the right on the neck; the legs were flexed so as to bring the knees near the elbows and the heels near the pelvis. Several perforated shells belonging to two Mediterranean species, *Cypraea pyrum* (or *rufa*) and *C. lurida,* were distributed over the body; four were on the forehead, two at each elbow, two above each knee, two on each foot.

The bones are those of an adult male. The skull was restored by Verneau (Fig. 250). It resembles those from Cro-Magnon, except that it is somewhat higher; it resembles more closely the skull from Chancelade found later. Massenat also found the skull of a child at Laugerie-Basse. In addition, isolated finds of human teeth, lower jaws, and broken long bones have been reported from various parts of the Paleolithic deposits of this site.

Raymonden.—An almost complete skeleton was found in the rock shelter of Raymonden, commune of Chancelade (Dordogne), in 1888. According to Hardy and Féaux, who found it, the skeleton lay at the base of a series of Magdalenian deposits and at a depth of 1.64 meters (5.3 feet) from the surface. Its position was only 1.65 meters (5.4 feet) above the level of the near-by stream, the Beauronne. The skull contained fine clay, marking a period of inundation.

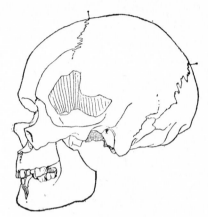

FIG. 250. RESTORATION OF THE MAGDA-LENIAN SKULL FROM THE ROCK SHELTER OF LAUGERIE-BASSE, DOR-DOGNE, FRANCE.

This skull resembles that of Cro-Magnon save that it is somewhat higher. Scale, ¼. After Hamy.

Evidence of periodic inundations was noticed at various levels in the deposits. De Mortillet believed the skeleton to be that of one who had met death by drowning. However, the posture of the body was more like that of a bundle burial. The head was bent forward and downward, the spinal column arched, the arms sharply flexed, bringing the hands to the head; the legs were flexed until the knees touched the dental arches, with heels against the ischial bones. This is similar to the forced attitude of a Peruvian mummy.

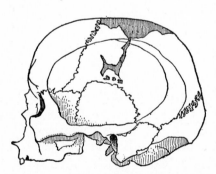

FIG. 251. MAGDALENIAN SKULL FROM THE ROCK SHELTER OF RAYMONDEN AT CHANCELADE, DORDOGNE, FRANCE.

Scale, ¼. Original in the museum at Périgueux. After de Mortillet.

The skull, which is fairly well preserved, is dolichocephalic, with an index of 72. The capacity was measured by means of shot (Broca's method), and the figure obtained by Testut was 1,710 cubic centimeters. The results obtained by this method are too large. Schmidt's reduction table for a capacity of this size calls for a reduction of 100 cubic centimeters; this would bring the capacity down to 1,610 cubic centimeters (Fig. 251).

The Chancelade skeleton belonged to a male about 55 years old, with an unusually large head and short stature (estimated at 1.50 to 1.59 meters, 4.9 to 5.1 feet). Among the characters of inferiority, noted by Testut, there should be mentioned the well developed lower jaw, the increase in size of the molars from the first to the third, the retroversion of the tibia, curved femora, and the angle which the great toe makes with the axis of the foot. The Chancelade skeleton is preserved in the museum at Périgueux.

Les Hoteaux.—The Abbé Tournier and Ch. Guillon found an early Magdalenian burial at the base of terrace deposits in front of a small cave near Rossillon (Ain). The body had been left full length on its back, enveloped in powdered ocher and accompanied by flint implements, a perforated tooth of the red deer, and a baton of reindeer horn. The burial was at the depth of 2 meters (6.6 feet) from the present surface, and in the lowest of six superposed hearths, and antedated all the hearths above it.

Sordes.—The skeletal remains of Magdalenian man were found at the base of the deposits in the cave of Duruthy at Sordes (Landes). The broken skull was surmounted by blocks of limestone. Near-by was a femur belonging evidently with the skull. Scattered about were other long bones. Some fifty canine teeth, three of *Felis* and the remainder of *Ursus ferox,* nearly all of them perforated and decorated with incised designs, were found in association with the skull. The engravings on four of the teeth represented a seal, a pair of glovelike figures, and two species of fish. The superficial deposit at Duruthy yielded thirty-three human skeletons of a Neolithic population closely resembling the Cro-Magnon race.

Placard.—Among the human skull fragments found by de Maret in the cave of Placard (Fig. 252), several were artificially shaped so as to be used as vessels (Fig. 253). These cranial vessels all came from Magdalenian horizons. The four found in 1883 were near the left wall of the cave and were in a row side by side, the concavities all turned upward, suggestive of their use as vessels. Of the two crania found in the Upper Solutrean horizon, one is pathologic and has a distinct metopic suture.

La Madeleine and Cap-Blanc.—In the type station of La Madeleine (Dordogne) there was found an incomplete skeleton including the greater part of the frontal bone, half of the lower jaw, and bones of the trunk and extremities. The rock shelter of Cap-Blanc (Dordogne) has also yielded part of a human skeleton of Magdalenian age. Cap-Blanc is now controlled by the French Government. In the stone house which protects the six mural figures of the horse is a case containing casts of the human skeletal remains found there.

FIG. 252. CAVE OF LE PLACARD, CHARENTE, FRANCE.

The interior of this cave was found to be especially rich in Solutrean and Magdalenian cultural remains (see Appendix I). The principal collections from Le Placard are now in the National Museum at Saint-Germain and at Toronto, Canada. Photograph by the author.

Obercassel.—A sepulture containing two well preserved skeletons was found in 1914 by workmen in diluvial deposits of Obercassel near Bonn (Rhine). The skeletons lay at a depth of 7 or 8 meters (23 to 26.3 feet) under a basalt rubble and in a bed of loam mixed with pulverized ocher. They had been protected by large flagstones of basalt. The two bodies were scarcely a meter apart, one that of a man, the other that of a woman. Under the head of one they found a bone polishing implement, 20 centimeters (7.9 inches) long, with the head carved to

resemble that of some rodent, probably the marten, and the flat shaft
ornamented with an incised pattern along the back.   In the sepulture
there was another art object, a tabular piece of bone engraved on
both sides to represent the head of a horse, in the style of figures *à
contour découpé* found in the French Pyrenees and characteristic of
the Magdalenian Epoch.   Two additional bone objects were found
in the sepulture, but there were no stone implements.

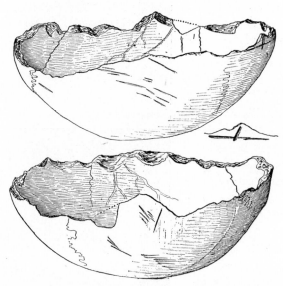

FIG. 253.   DRINKING CUP FORMED FROM A HUMAN CRANIUM FROM THE CAVE OF
LE PLACARD.

This was one of several crania altered by chipping to serve as vessels, which were left
by the Lower Magdalenians in a row at the side of the cave.   The upper view shows the
left side, the lower, the right.   Scale ⅓.   After Breuil and Obermaier.

The female skeleton is that of a young woman with a stature
of 1.55 meters (5 feet) and not more than 20 years old.   In life
the eruption of the teeth had been complete except for the one
upper and two lower third molars.   The dental arch is parabolic.
The horizontal circumference of the cranium measures 512 millimeters (20 inches) and the cephalic index is 70.   A metopic suture
is visible in the half-tone illustration reproduced by Bonnet.   The
orbits are rectangular.

The other skeleton is of a sturdy male some 50 years old.
The horizontal circumference of the skull is 538 millimeters (21

inches) and the capacity is estimated at 1,500 cubic centimeters. All but three teeth of the upper jaw had been lost during life. In the lower jaw only two teeth were lost during life and two after death. All remaining teeth were worn until but little of the enamel remains. The exposed dentine is black as ebony. In both skulls the face is broad and the lower jaws well developed.

Freudental.—The cave of Freudental in the canton of Schaffhausen, Switzerland, has yielded cultural remains of the Magdalenian Epoch. In 1874 human skeletal remains were found there, including eight skull fragments, fragments of lower jaws belonging to children, teeth, three vertebrae, a nearly complete sacrum, and a fragment of the pelvis. These have been the subject of a recent study by Schlaginhaufen, who sees no reason why they should not be classed as of Magdalenian age.

Miskolcz.—In the Balla cave near Miskolcz (Hungary) Hillebrand found the complete cranium of a child (sepulture) at a depth of 1.3 meters (4.3 feet) from the surface in Magdalenian deposits.

Castillo.—Two cranial fragments were found in Magdalenian deposits at the mouth of the cavern of Castillo near Puente-Viesgo (Santander), Spain.

Other Localities.—Skeletal fragments of slight importance have been reported from Magdalenian deposits at Aurensan and Espélugues (Hautes-Pyrénées), Grotte des Fées (Gironde), Brassempouy (Landes), La Mouthe, Les Eyzies and Limeuil (Dordogne), Gourdan and Monconfort (Haute-Garonne), and Lussac-le-Château (Vienne). A cranium from Mas d'Azil and three crania from the Grotte des Hommes at Arcy-sur-Cure are probably also of Magdalenian age.

Summary of the Physical Characters of Upper Paleolithic Man.—The Aurignacian population of Europe was not so homogeneous as the Mousterian. During the early Aurignacian Epoch two rather distinct types existed: (1) a negroid type represented by two skeletons found near the base of the deposits in the Grotte des Enfants near Mentone and known as the Grimaldi race; (2) the man of Combe-Capelle (Dordogne), which retains a slight suggestion of kinship with *Homo neandertalensis,* but which on the whole has the right to the title of *Homo sapiens;* (3) the skeletons from Předmost, especially skull No. 3 which is suggestive of a transition form between Mousterian and Aurignacian man. The so-called Cro-Magnon race is a late product of the Aurignacian Epoch and does not differ to any marked degree from the long-headed Neolithic (and later) races of Europe. Judging from the recent discovery at Solutré, some of the Aurignacians came very near to being brachycephalic.

The skeletal remains, which can with any degree of certainty be ascribed to the Solutrean Epoch, are comparatively few in

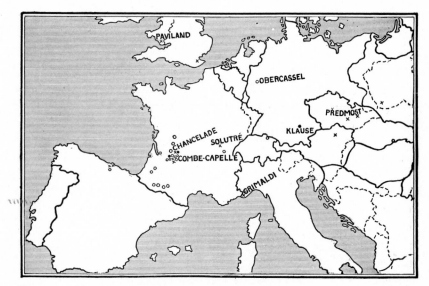

Fig. 254. Map showing the geographic distribution in Europe of upper paleolithic human skeletal remains.

Aurignacian discoveries are indicated by crosses; Solutrean, by black circles; and Magdalenian, by white circles.

number. Somatologically they differ in no marked degree from the human skeletal remains belonging to the epoch which preceded and that which followed.

There is little distinctive somatologic evidence on which to base a Magdalenian race as differing appreciably from the Solutrean and Aurignacian. If a Magdalenian type exists, it is probably best represented by the skeleton from Raymonden at Chancelade (Dordogne). One must not lose sight of the fact that the osteologic record of fossil man is even yet so fragmentary that there is grave danger of mistaking individual characters for those on which varieties or species should be based.

With due regard for the incompleteness of the record and for individual variation, there are elements of homogeneity sufficient to bind the Upper Paleolithic races into a more or less consistent whole; for this view the Upper Paleolithic culture also speaks with no uncertain voice. Both lines of evidence point to an African strain in both Aurignacian blood and culture (art as well as industry). A new cultural contribution, probably oriental, came in with the Solutreans without leaving any appreciable effect on the general physical make-up of the population; even the cultural influence gradually subsided, giving place to a sort of Aurignacian renaissance, better known as the Magdalenian Epoch.

Taking the Upper Paleolithic races as a whole, we find a curious blending of negroid and eskimoid characters, the latter being easily dominant, especially toward the end of the period. Paleolithic artists dealing with the human form made no attempt to mask the negroid characters of their models, particularly during the greater part of the Aurignacian Epoch; these are seen in the figurines from Grimaldi, Brassempouy, Lespugue, and Willendorf, and the bas-reliefs from Laussel. Unfortunately, the artist paid little attention to the features, and yet the head from Grimaldi has an unmistakable negroid aspect. Unmistakable also are such characters in the female as the large pendant breasts, adipose hips with a tendency toward steatopygy, and the excessive development of the nymphs (*labia minora*). The negroid strain is also seen in some of the representations of the male. On the other hand, in human representations by Magdalenian artists, the negroid features are practically nonexistent. Testut pointed out the resemblance of the Chancelade skeleton to that of the Eskimo of Greenland and Labrador. Hamy was one of the first to call attention to the physical and cultural kinship between the eastern Eskimo and the Magdalenians of western Europe. A like kinship would seem to exist between the Bushman of southern Africa and the Aurignacians; if one is not descended from the other, a common ancestry may at least be ascribed to both. The Aurignacians certainly rode on the crest of the culture wave of their time, no matter what their color may have been.

Human Skeletal Remains Presumably Paleolithic

An increasing knowledge of culture sequence and of the somatology of prehistoric races has helped immensely to date fossil human skeletal remains. This is especially true of current discoveries and will be even more so in the future. It does not, however, suffice to date accurately discoveries that were made in years gone by and in absence of proper control. To this class belong an impressive list of cases which, if made under circumstances of rigid scientific control, might have added much to our knowledge of human paleontology. Many of these are of sufficient importance historically and for other reasons to find a place in any comprehensive account of fossil man. Of the examples that follow, some may be recent. By far the greater number are of Quaternary age, although it is not possible to say to which culture epoch they belong.

Canstadt.—The earliest discovery of presumably fossil man was at Canstadt, a suburb of Stuttgart, in Württemberg. Here, in 1700, by order of Duke Eberhard Ludwig, excavations were made on the site of a Roman *oppidum*. From the underlying loess a quantity of fossil animal remains were taken, including those of the mammoth, woolly rhinoceros, cave bear, etc.

In 1835 Jaeger published a work on the paleontology of Württemberg, including therein the fossil animal remains from Canstadt and the figure of a cranial calotte. Whether this cranial fragment was originally found *in situ* in the loess of Canstadt or was from a Neolithic, Roman, or Merovingian burial will probably never be known. Hamy regarded it as representing an attenuated Neandertal type and probably of Pleistocene age. He even went so far as to make it the type specimen for a race called "Canstadt." In 1892 Hervé marshaled evidence sufficient to prove conclusively that the proper place for the Canstadt cranium is in the nonauthenticated class. The interest attaching to it will no doubt remain chiefly historic. It is preserved in the museum at Stuttgart.

Lahr.—As early as 1823 Ami Boué reported the discovery of fragments of a human skeleton found at Lahr in the Rhine valley near Strassbourg. These were sent to Cuvier, who identified them as human and stored them away so carefully that they were not rediscovered until many years later. The fragments included certain vertebrae, part of the left ilium, humerus, radius, parts of two tibiae, bones of the hands and feet, and nearly the entire left femur. The pieces, it will be noted, are practically all from the left side and do not include any part of the skull. They were extracted from indurated loess at a height of some 12 meters (39.4 feet) above flood waters of the Schutter. They have never been definitely dated. On anatomical grounds, Hamy referred them to his so-called "Canstadt" race. Since this race remains a part of prehistory's suspense account, the bones from Lahr must be disposed of in like manner.

Bize.—Even before the rediscovery of the Canstadt skull (1828), Tournal announced the discovery of human bones and pottery in one of the caves of Bize (Aude). These were apparently associated with remains of a fauna in part now extinct. According to Marcel de Serres, the human bones were in the same chemical state as the fossil animal remains. There are two caves near Bize. They were found to contain flint and bone implements and engravings, all apparently of Magdalenian age, and pottery of a later epoch. One of the caves yielded a fine Solutrean laurel-leaf point now in the Narbonne Museum, and one yielded the human bones in question. Whether these human bones belong to the Paleolithic or to a later period has not been determined.

In 1829 Christol published a notice on human bones found in a cavern at Pondres, near Nîmes (Gard). The cavern was completely filled with mud and gravel. This deposit contained a *mélange* of bones of extinct species of hyena and rhinoceros, fragments of pottery, and human bones. The age of these human bones, like those from Bize, remains a matter of uncertainty.

Engis.—In 1833 and 1834 Schmerling published his researches on the fossil bones found in the caverns of the province of Liège. The caverns explored included the one at Engis on the left bank of the Meuse some 13 kilometers (8 miles) southwest of Liège. The remains of at least three individuals were disinterred. Of these, the skull of a youth was embedded by the side of the tooth of a mammoth and, although entire, was so fragile that it could not be saved. The only skull in a state to be preserved was that of an adult, buried 1.5 meters (4.9 feet) deep in a breccia containing the

tooth of a rhinoceros and several bones of a horse and of the reindeer. This skull, still preserved in the University Museum at Liège,
is known as the Engis skull. A worked bone and several flint
implements were found in the Engis cave near the human bones.
The Engis skull is probably of late Paleolithic age. The cave in
which it was found was destroyed prior to 1860.

Denise.—In 1844 Aymard, curator of the Puy Museum, reported the discovery of human remains taken from a tufa deposit
of the extinct volcano of Denise near Le Puy (Haute-Loire).
These remains are authentic, but the geologic age of the deposit has
not yet been definitely established. Boule considers the deposit to
be of pre-Mousterian age and the man of Denise to be contemporary
with *Rhinoceros merckii.*

After investigating the site of the discovery, Pictet was satisfied
that the fossil bones belonged to the epoch of the last volcanic
eruption. At a scientific congress held at Le Puy in 1856, the
question of the age of the Denise fossil remains received exhaustive
consideration. In 1859 Hébert, Lartet, and Lyell all visited the
spot but were able to gather but little additional information. The
human remains consist of a "frontal and some other parts of the
skull, including the maxilla with teeth, both of an adult and young
individual; also a radius, some lumbar vertebrae, and some metatarsal bones," all embedded in a light, porous tufa. They are in the
museum at Le Puy. Félix Robert has stated that blocks of tufa
like the one containing the human bones frequently occur on certain
parts of the mountain, and that this is the deposit which contained
the remains of the cave hyena and *Hippopotamus major.*

Engihoul.—Shortly after Lyell's visit to Liège in 1860, Malaise
made additional discoveries in the second cave of Engihoul on the
opposite bank (right) of the Meuse from the Engis cave. In a
clay deposit 60 centimeters (2 feet) beneath the stalagmite, he
found parts of two human skulls. These were associated with
fossil bones of animals now extinct and in the same chemical state,
so that Malaise felt justified in referring all to the same period.
The human bones were examined by Hamy, who declared one of
the lower jaws to be strikingly like that of the old man of Cro-
Magnon. This mandible, known as Number 1, belonged to an adult
male. The second lower jaw, which belonged to an aged male,

presented the same characters, only somewhat exaggerated. The three cranial fragments were not large enough to throw much light on the head shape. The muscular imprints were mediocre, and the crania were evidently dolichocephalic. If the skull of Cro-Magnon is Paleolithic, those of Engihoul may justly claim a like distinction in the absence of positive evidence to the contrary.

Castenedolo.—The vault of a human skull was found in 1860 by Professor Ragazzoni at Castenedolo about 9.5 kilometers (5.9 miles) southeast of Brescia, Italy. The piece appeared to be *in situ* in Pliocene deposits between blue clay and a subjacent coralline deposit and at a depth of about 2 meters (6.6 feet). In 1880 scattered fragments of the skeletons of two children were found at the same level. A final discovery of a woman's skeleton in contracted posture was made in the stratum of blue clay at a depth of 1.06 meters (3.5 feet). The skull of this woman is the only one complete enough to admit of reconstruction. The skull is dolichocephalic with an index of 71.4. The capacity is estimated at 1,340 cubic centimeters. Both cranium and lower jaw are modern in type. The skeletal remains from Castenedolo are preserved in the Department of Anthropology, University of Rome.

Moulin-Quignon.—Boucher de Perthes explored the sand and gravel pits about Abbeville for nearly thirty years before finding human skeletal remains. He is said to have finally offered a reward of 200 francs for such. Eight days later, March 28, 1863, a workman in the Moulin-Quignon pit called him to remove with his own hands a human lower jaw which lay in a stratum at a depth of about 5 meters (16.4 feet) from the surface. The find immediately attracted international attention. The jaw was taken to the museum of the Royal College of Surgeons and there severed by a saw in order that the state of fossilization might be observed. The bone was found to contain 8 per cent of animal matter; but this should not be surprising in view of the fact that bones of animals from Pleistocene deposits have been found which contain as much as 30 per cent of animal matter. More serious was the knowledge that the workmen in the Moulin-Quignon pit had been forging flint implements and planting them *in situ* where they could be conveniently "found" in due time by prospective and unsuspecting visitors.

In view of its possible importance, an international inquest was held at Paris over this jaw some six weeks after its discovery. The five French members voted in favor of its authenticity; the four English participants believed it to be apocryphal. The English viewpoint has since prevailed. To-day Keith stands alone in championing the claims of the Moulin-Quignon lower jaw.

Olmo.—The specimen known as the Olmo cranium came from a railroad cut in the valley of the Arno, Val Chiana, above Florence and a little beyond Arezzo. It is said to have been found 15 meters (49.2 feet) beneath the surface in a compact blue lacustrine deposit. Above this layer there came, in succession, coarse ferruginous gravels, finer gravels, ancient alluvial and recent alluvial deposits. Near it were found fragments of charcoal and a flint implement, the latter being of the Mousterian type. Later, at the same level and at a distance of 2 or 3 meters (6.6 to 9.8 feet), the lower jaw of a horse and the tip of an elephant's tusk were found. All these specimens were deposited in the museum at Florence.

Professor Ignio Cocchi, who made the discovery in 1863 and who carefully studied the pieces (as well as the site), referred all to the Lower Quaternary. Forsyth Major was led to confuse them with the Pliocene remains that are found in the same locality. In 1897 Cocchi revised his opinion in regard to the Olmo cranium, referring it to the closing phases of the Quaternary, a view which is no doubt more nearly in keeping with the facts. The Olmo cranium is certainly not of the Neandertal type. According to Hamy, it belonged to a female. The forehead is well developed and the brow ridges are small even for modern crania.

Furfooz.—The cave of Frontal at Furfooz, province of Namur (Belgium), was explored in 1864 by Ed. Dupont, who found that the entrance had been blocked by an upright flagstone in front of which was a disturbed deposit of Magdalenian age. Within the cave were found Neolithic potsherds (Robenhausian) and a human sepulture, the age of which has not been determined with certainty.

Goyet.—In the third cavern of Goyet near Namur, Dupont distinguished five ossiferous levels. The lower two are apparently of Aurignacian or Solutrean age. The upper three have been referred to the Magdalenian Epoch. A few human bones were found in all three. The most important pieces, two incomplete lower jaws, came from the fourth level, that is, the second from the top. These two pieces differ in that the angle of symphysis of one is acute and that of the other is obtuse; in other words, the latter, with its absence of chin, is comparable with the lower jaw of Naulette, or the Neandertal type. The other agrees with the type that one would expect to find in a Magdalenian deposit.

Eguisheim.—Part of a human cranium was discovered by workmen in 1865 while digging a beer cellar at Eguisheim, in the upper Rhine valley some 70 kilometers (43.75 miles) south of Strassbourg. The fragments include the frontal and right parietal belonging to one individual. They were found in what appeared to be virgin loess, from which a metacarpal of a small horse, a frontal of the great stag, and a molar of the woolly elephant were likewise taken. These exhibited the same fossil and chemical state as the human cranium. The Eguisheim specimen, now preserved in the museum at Colmar, has been referred by some authors to the Neandertal race, and by others to the so-called Cro-Magnon race. It probably represents an intermediate stage.

Clichy.—The section of the pit at Clichy (Paris), is made up of four beds: (4) vegetal earth; (3) red diluvium (probably brick earth); (2) yellow clay-sands (middle ancient loess); (1) gray diluvium. At a depth of 5.45 meters (17.9 feet) and in a band of reddish sand at the base of the gray diluvium, E. Bertrand found part of a human skeleton in 1868. At the same level in this pit there have been found fossil remains of *Elephas, Bos, Equus, Rhinoceros,* and *Cervus.* The nearly complete vault of the cranium has a female aspect. It is low, long, and narrow, with a cephalic index of about 68.

Later, Reboux found in the sands of deposit No. 2 in the same pit, at a depth of 4.20 meters (13.8 feet) (1.25 meters, 4.1 feet, above the skeleton found by Bertrand), parts of two human skeletons—the lower jaw, part of the femur, and the twelfth thoracic vertebra of a child of seven or eight years, also a lower jaw and cranial fragments of an adult. According to Hamy, these remains differ in type from the skeleton found at the lower level, since the latter represents a race of low stature, probably brachycephalic. The discoveries at Clichy should not be lost sight of even if their significance is not yet clear.

Brüx.—Mention should be made of the fragmentary skeleton found in 1872 at Brüx in Bohemia. The skull was encountered at a depth of 1.42 meters (4.66 feet) in diluvial sands, and immediately below, additional parts of the skeleton including a tibia. Above the skeleton but still beneath the vegetal earth a fine Neolithic stone ax was found. According to Professor Rokitansky, the skull is very low, with pronounced brow ridges; nevertheless the Brüx skeleton has not been fully recognized as belonging to the Paleolithic Period. The skeleton, which is in a bad state of preservation, is in the collections of the Vienna Anthropological Society.

Grenelle.—Prior to 1874, Émile Martin found a human cranial cap in the Helie sand pit, avenue Saint-Charles, at Grenelle. It came from the gravels at a depth of 7 meters (23 feet) from the surface, 5.6 meters (18.4 feet) deeper than the other human bones from the same pit. These last were referred by Belgrand to the period of transition between the Paleolithic and the Neolithic. On the contrary, the calotte from the basal portion of the superior gravels belongs to the level from which remains of *Elephas antiquus* and the hippopotamus were taken, and is evidently of Quaternary age. According to Hamy, the Grenelle cranium is comparable with those from Canstadt and Eguisheim.

Cheddar.—There are several caves at Cheddar in Somersetshire. The one known as Gough's cave was discovered in 1877. At intervals since that date it has been excavated. The human skeleton was discovered by a workman in 1903, and the find was published the following year by H. N. Davis. The recent accumulations on the floor of the cave are about 15 centimeters (5.9 inches) thick; beneath this is a layer of stalagmite from 15 to 30 centimeters (5.9 to 11.8 inches) thick; and then comes a bed of cave earth 1 meter (3.3 feet) in thickness. This relic-bearing layer has yielded stone and bone implements, remains of *Bos, Equus,* and *Ursus* (perhaps *spelaeus*), and part of a human skeleton including the cranium and lower jaw.

The industrial remains resemble closely those of the Magdalenian Epoch in France. The anatomical characters of the human skeleton are such as one would expect to find in Magdalenian man. The baton of reindeer horn is said to have been found in association with the skeleton. Keith estimates the stature of the Cheddar "man" to be 1.62 meters (5.3 feet), which leads one to wonder whether the sex might not have been female.

Bury St. Edmunds.—In 1882 a cranial fragment including parts of the frontal and of the parietals was found at a depth of 2.3 meters (7.5 feet) in a pit opened for brick earth by the roadside between Bury St. Edmunds and Saxham, parish of Westley (Suffolk). The site is on an elevation between the valleys of the Lark and the Linnet. In section there is some 40 centimeters (15.7 inches) of surface soil, then brick earth to a depth of 2.4 meters (7.88 feet), lastly a bed of compact brown loam which rests on the Chalk. The cranial fragment came from near the base of the brick earth in one of the numerous pockets eroded in the old Chalk surface. A section of any pocket is like that of the others. These were filled when flood waters of the region were at a much higher level than at present. From the brick earth of an adjoining pocket, teeth of the mammoth were taken. From the same deposit in two other near-by pockets, flint cleavers of the Acheulian or Mousterian type were obtained. Henry Prigg, the first to be called to the site after the discovery of the calotte fragment, testified that the piece must have been *in situ* before its removal by the workman. The fragment is thought to be from a female skull.

Podbaba.—An incomplete human cranium was discovered by a workman in a loess deposit at Podbaba near Prague, Bohemia, in 1883. Previously, fossil animal remains had been found in the same layer at this site—the tusk of a mammoth, two skulls of the woolly rhinoceros, and bones of the reindeer and horse. The cranium is low and long, with pronounced brow ridges. However, de Mortillet is not inclined to accept it as of Quaternary age because of the presence in that vicinity of post-Paleolithic graves and because the cranium in question was found at a depth of only 2 meters (6.56 feet). It is now generally conceded as belonging in the doubtful list.

Tilbury.—In the course of extensive excavations at the docks of Tilbury on the left bank of the Thames below London, portions of a human skeleton were found in 1883. These formed the subject of a study by Sir Richard Owen which was published in 1884. The pieces included portions of a cranium, part of the lower jaw with teeth, portions of the leg and arm bones, a fragment of the pelvis, and a few small bones. They represent an adult male, which, according to Owen, as well as to de Mortillet, is referable to the Neandertal race, although the brow ridges were only of moderate size.

The skeleton of Tilbury lay at a depth of 10.4 meters (34.1 feet) below the surface and about 0.7 meter (27.6 inches) beneath an old land surface belonging to the submerged-forest age. Keith believes that the man of Tilbury belongs to an earlier phase of the Neolithic Period and links him with Neolithic human remains from cists on the island of La Motte off the south coast of Jersey. The Tilbury skeleton is preserved in the British Museum, South Kensington.

Marcilly-sur-Eure.—In the Eure valley, commune of Marcilly-sur-Eure (Eure), in 1883 workmen found a cranium under 7 meters (2.3 feet) of brick earth. Its position was near the base of the brick-earth deposit. Unfortunately the cranium was broken by the workmen and only a portion of the cranial cap was saved. Judging from the nature of the fragment, it belongs to the attenuated Neandertal type. No flint implements were found in the pit at Marcilly, but the deposit elsewhere is known to contain both fossil and industrial remains. The skull fragment is in the Doré-Delente collection at Dreux.

Talgai.—The fragmentary skull of a youth some 16 years of age was found in 1884 on a tributary of the Condamine River, Darling Downs, some 40 to 48 kilometers (25 to 33 miles) southwest of Brisbane, Queensland, Australia. The discovery was made by a workman some 0.9 meter (3 feet) from the bottom of a freshly washed-out channel in the valley of Dalrymple Creek. At the site two strata are present. The surface stratum is a fine black soil 1.8 or 2.1 meters (6 to 7 feet) thick covering one of reddish-brown clay. It was apparently on the surface of this latter formation that the skull was found.

The lower jaw is missing. The cranium is highly fossilized and brittle. With the possible exception of the frontal bone, its reconstruction is largely conjectural. The facial prognathism is obvious. The sockets of the canines are distinguishable, and between them the jaw is extremely broad and flat. The incisor teeth are arranged in an almost straight transverse line, and the molar-premolar series lie almost parallel. The teeth present include the two canines, the two left premolars, and the first and second molars on both sides. Stewart Arthur Smith points out that in the Talgai skull the palate is elongated whereas in the Neandertal man and the remains from Wadjak it is wide and well arched. There are also differences in the teeth of the two types. He believes the Talgai skull to be undoubtedly of the Pleistocene age, as is likewise the human molar found in a Wellington cave, New South Wales.

Bréchamps.—In 1888 a cranial cap and other skeletal fragments were found in the gravel pit of Beaudeval, commune of Bréchamps, near Dreux (Eure-et-Loir). On the same day all but the cranial cap were destroyed by an act of vandalism. In the same pit characteristic Mousterian flint implements had been found. The deposit is undoubtedly Pleistocene. The Bréchamps cranium, now belonging to the Société d'Anthropologie de Paris, was subjected to a careful study by Manouvrier, who found it to be that of a male of ripe age, thick-walled and dolichocephalic, and not to have suffered any posthumous deformation. It is an example of the attenuated Neandertal type and probably dates from the latest phase of the Mousterian Epoch. It is in every way comparable with the cranium from Marcilly-sur-Eure. The cranial capacity is estimated at 1,410 cubic centimeters and the cephalic index at 75.5. The vault of the cranium is low, with a vertical index of 67.

Galley Hill.—A much discussed, nearly complete skeleton was found in 1888 at Galley Hill near Northfleet (Kent). The bones were 2.5 meters (8.2 feet) from the surface and lay in close contact; in fact, the entire skeleton might have been found by painstaking search. The gravel deposit from which they came corresponds to what is called the "100-foot terrace" of the Thames valley. The gravel at this site yielded no other fossil bones, but remains from a neighboring pit are said to belong to a very early stage of the Pleistocene. The discovery of the skeleton was made by workmen, who notified Heys, master of the local school. It seems that while Heys was gone for his camera, Robert Elliott appeared on the spot, saw some of the bones before they were removed, and bought the entire lot. In 1894 the skeleton was shown by E. T. Newton at a meeting of the Geological Society of London. The ownership of the skeleton passed from Elliott to Frank Corner of London.

Newton believed the bones to have been found *in situ*. Keith and Rutot are inclined to the same view. On the other hand, Boyd Dawkins and Duckworth have not been willing to consider the skeleton to be as old as the gravel deposit in which it was found. That it exhibits no essential differences from those of the modern Englishman can hardly be contested. Duckworth makes a strong point of the distortion of the Galley Hill skull. Similar distortions are of common occurrence in case of interments, but are practically unknown in cases where skulls were buried through natural processes. Under the circumstances it is impossible to determine positively the age to which the Galley Hill specimen belongs, and hence it should not be taken as a type specimen for any particular period or race.

Wadjak.—In 1889 a human female skull was found in the district of Wadjak some 96 kilometers (60 miles) southeast of the famous Trinil site. Hearing of this, Dr. Eugene Dubois visited the locality and was so fortunate as to discover portions of the jaws and cranium of an additional individual. His account of these was published in 1921. The skull of the female is the more complete and its features are characteristic for Australoid races, except that the dimensions are much above the average. Allowing for the excessive thickness of the cranial wall, the cranial capacity is estimated at 1,550 cubic centimeters. The cranial length is 200 millimeters (7.8 inches) and the breadth is 145 millimeters (5.6 inches), hence dolichocephalic with a cephalic index of 72.5. The forehead and orbits are low, the brow ridges prominent. The palate of the male is much larger than that of the female. The palatal area enclosed within the outer border of the dental arch is about 41.4 square centimeters, much greater than the average in the modern European. In both, the teeth are set in the form of a horseshoe, differing in this respect from the dental arch of the cranium from Talgai, Australia. Nevertheless the Wadjak and Talgai skulls have many points in common; they probably represent a very primitive proto-Australoid type of *Homo sapiens*. The remains from Wadjak are in the private collection of Dr. Dubois at Haarlem.

Dartford.—A human cranium, practically complete except for the face bones, was found in 1902 in a gravel pit at Dartford

(Kent) on the west bank of the Darent, where that stream breaks through the terrace to enter the marshland on the south bank of the Thames. This pit, which had proved to be unprofitable commercially, had just been taken over by a local collector, W. M. Newton. The cranium was found by two of his workmen and had been removed by them. They had observed a "pothole" 2.5 meters (8.2 feet) from the surface just prior to the find. Then came a fall of gravel, with which the pothole disappeared, and the skull was found in the débris. At about the same level as the pothole, there is a black band in the gravel along which Newton later found flint implements of the Acheulian type. No fossil animal remains were observed at the Dartford pit. The circumstances of the discovery are not such as to admit of a definite conclusion concerning the antiquity of the Dartford cranium, which is that of a male, dolichocephalic, and of modern type. The capacity of 1,740 cubic centimeters is 250 cubic centimeters above the modern average. For every reason, therefore, it would not be safe to conclude that the man of Dartford belongs to the Acheulian Epoch. The specimen is now in the museum of the Royal College of Surgeons.

Langwith.—The Langwith cave is 4.8 kilometers (3 miles) south of Creswell Crags in Derbyshire. Its discovery dates from 1903. Its location is just behind the rectory of Langwith Bassett, by a brook known as the Poulter. The rector, Rev. E. H. Mullins, aided by his household, made a systematic exploration of the cave, which had been filled almost to the roof with floor deposits to the thickness of 3.6 meters (11.8 feet). The original floor of the cave was found to be about 2.5 meters (8.2 feet) above the level of the Poulter.

Mullins was able to distinguish three horizons. The topmost, which consisted of loam, yielded remains of recent fauna and a bone awl. From the middle layer were taken numerous remains of extinct animals, including the cave bear and the woolly rhinoceros. In the bottom or oldest deposit, Mullins found abundant evidence of man's occupation—ancient hearths at all levels, calcined stones, worked flints, a bone pin, and the human skull in question. The associated animal remains included the woolly rhinoceros, reindeer, cave bear, brown bear, lemming, Arctic hare, *Bos primigenius,* etc. The cranium lay close to the floor at the side of and near the entrance to the cave. No other bones were near it, and it was covered by what seemed to be a "natural arch" of fallen stones, although there was no indication of such a fall from the ceiling. A skull fragment of a young child was also found in the bottom layer.

The adult cranium was that of a small man with a cranial capacity of about 1,250 cubic centimeters. It has a cephalic index of 70 and a projecting occiput. The vertical diameter is relatively low and the brow ridges only moderately developed. The face bones are gone. It has been referred, and no doubt rightly, to the Aurignacian Epoch.

Le Moustier.—In 1905, prior to Hauser's discovery of a Mousterian skeleton in one of the rock shelters at Le Moustier, Rivière announced the discovery of a female skeleton almost entire in the rock shelter of Bourgès. These two rock shelters are contiguous. Rivière insists that the Bourgès deposits had not been disturbed prior to his own exploration and that the skeleton was found *in situ* and hence was contemporaneous with the fauna and industry found in the same deposit. According to him, this deposit belongs wholly to the Mousterian Epoch; it contains typical Mousterian scrapers and a cleaver. The fauna includes one species of rhinoceros, *Bos Primigenius*, reindeer, and a rodent belonging to the genus *Lepus*.

In 1921 the author saw the skull found by Rivière at Le Moustier. It is that of a youth and does not possess a single one of the characters that go to make the Mousterian type. It is certainly post-Paleolithic and is mentioned here by way of leave-taking.

Ipswich.—In October, 1911, workmen in the clay pit of Bolton and Laughlin at Ipswich (Suffolk) found a human skeleton. Reid Moir was notified and on his arrival found that a portion of the skull with its encephalic cast of boulder clay had been recovered. Mr. Moir removed the remainder of the skeleton in the presence of witnesses. In order to preserve the fragile bones, the containing beds were removed with them. Later, three geologists were called to Ipswich to examine the section.

A sheet of hard, chalky boulder clay of varying thickness is spread over East Anglia, overlying stratified midglacial sands. Between these deposits and at depth of only 1.4 meters (4.5 feet) the skeleton was found. Was it interstratified? This question will probably never be answered to the satisfaction of all. According to Moir, a "most careful examination of the section before the disinterment took place showed clearly that no signs of any previous digging were visible, the clay above the skeleton appearing to be in every way the same as that which extended for some distance on each side of it." The presence of a calcareous band immediately underneath the skeleton was noted, as well as the fact that it "extended more or less continuously on either side of the spot where the remains were found"; and it is pointed out by Moir that if a grave had been dug through the boulder clay, rain water percolating through the loose grave-filling would have dissolved away the calcareous deposit. One of the best bits of evidence is that the skeleton was partly embedded in glacial sand and partly in boulder clay; "this sand showed clearly lines of stratification and was conformable with that underlying it."

On the other hand, Slater, one of the three geologists called to view the place, but not until after the bones had been removed to London, looks upon the site as highly unsatisfactory. Considering the loss by infiltration, he would not expect to find distinct signs of a grave after a lapse of some thousands of years. The position on the side of a valley points to the possibility of hill wash or redeposited boulder clay.

The right side of the skeleton in contact with glacial sands was much better preserved than the left. The latter, being imbedded in the boulder clay, was most subjected to the destructive effects of roots as well as the action of the clay itself. The roots even penetrated the glacial sands, and their effects on the skull and pelvis were marked. The corroding effects of the boulder clay (sandy, chalky loam) played havoc with the soft, spongy portions of the skeleton, which are now represented by dense clay with here and there fragments of bone. The only complete bones recovered were those of the right hand.

The skeleton is that of a man about 1.8 meters (5 feet 10 inches) in height and 40 to 50 years of age. In addition to the complete brain cast (of boulder clay), there remains a "fragment of the frontal bone sufficient to show the characters of the forehead, parts of both temporal bones, with the joints of the mandible, and fragments of the parietal and occipital bones." Nine of the teeth were recovered; these differ in no way from the teeth of Neolithic man. Judging from the skull fragments and the brain cast, Keith concludes that the head did not differ essentially from that of modern Europeans except that the maximum width of the skull is situated rather far back, recalling in this respect alone the Neandertal race. With the exception of the lower leg bones (tibia and fibula) and the upper arm bone or humerus, the limb bones are of the modern European type. The tibia lacks the sharp anterior crest or shin of modern man; in this it suggests the Neandertal type, but not in respect to size and general shape.

Moir's admission in 1916 that he was wrong in attributing great antiquity to the Ipswich skeleton will no doubt serve to place this specimen outside the pale of authentic fossil human documents.

Oldoway.—In 1914 Hans Reck reported the discovery of ancient human skeletal remains in the northern part of German East Africa. The Oldoway gorge, situated on the eastern margin of the Serengeti steppe, lays bare a series of tufaceous layers that had been deposited in a fresh-water lake. Five deposits can be distinguished stratigraphically as well as paleontologically. In the lowest deposit fossil remains are rare, the chief specimen being a rhinoceros skeleton. The second deposit is rich in fossil mammalian remains, including the human skeleton. Remains of two types of fossil elephant, both different from the living *Elephas africanus,* were especially abundant; the skull of a hippopotamus was also found in deposit number two. Bones of the antelope appear for the first time in the third deposit, which also contains bones of the elephant. Elephant remains are dominant in the fourth deposit; fish bones are also abundant. The fifth and latest of the deposits is the richest of all in fossils. It is characterized by an antelope and gazelle fauna similar to that now living on the Serengeti steppe. In this deposit Reck found no elephant remains.

The change in fauna represented by the series corresponds to a change in climate. The climate of the upper horizon was similar to that of to-day; while the elephant, rhinoceros, hippopotamus, crocodile, and fish of the lower horizons bespeak a damp, woodland climate that was probably synchronous with the Würm Glacial Epoch in Europe.

The human skeleton, as has been said, came from the next to the lowest horizon (No. 2). It is not only in a good state of preservation, but is practically complete. The skeleton was found some 3 or 4 meters (10 to 13 feet) below the rim of the Oldoway gorge, which here is about 40 meters (164 feet) deep. The skeleton bore the same relation to the stratified bed as did the other mammalian remains, and was dug out of the hard clay tufa with hammer and chisel just as these were. In other words, the conditions of the find were such as to exclude the possibility of an interment. The human bones are therefore as old as the deposit (No. 2).

An attempt to determine the age of the human skeleton with any degree of accuracy must of course wait upon a further study of the geologic and paleontologic data as well as on a more thoroughgoing somatologic study of the skeleton itself. Reck is, however, already convinced that it antedates the so-called alluvial or recent period. The thickness of the deposits indicates a considerable lapse of time, especially when one recalls that at least two of the superposed deposits were laid down before the faulting occurred, and with it the drying up of the lake. The change in fauna from rhinocerous, hippopotamus, and two types of elephant both different from the living African elephant, to a gazelle and antelope fauna is likewise proof of considerable antiquity. Judging from the photograph of the skeleton still *in situ,* the man of Oldoway gorge did not belong to the Neandertal, but rather to the Aurignacian type. In the absence, however, of industrial remains and even photographs in detail, any pronouncement as to racial affinities with known European Quaternary human remains would be merely a guess. The Oldoway skeleton is said to exhibit all the features of a typical negro, and with the negroid skeletons from Grimaldi it may eventually serve to prove the existence of a negro type in the Pleistocene Epoch.

Boskop.—The finding of a new species of any animal is an important matter; the discovery of a new species of *Homo* is a scientific event of the first importance. Has this feat been accomplished in the case of the Boskop skull, found about 1914 in a surface laterite deposit of the Transvaal, South Africa? The skull, found by a farmer while digging a trench, is fragmentary, consisting of the greater part of the frontal and parietals with a small portion of the occipital. Later excavation at the same spot yielded a portion of the right temporal, most of the left horizontal ramus of the mandible, and some fragments of limb bones.

The depth at which the bones are supposed to have occurred is about 1.4 meters (4.5 feet). They all belonged presumably to one individual, and in view of their number might represent a burial, hence are not necessarily so old as the deposit in which they were found. There were no associated fossil remains or artifacts, so that the age of these fragments and their title to rank as the type of a new species of man must rest alone on their physical characters. What are these characters? In Broom's opinion, they are out of the ordinary because of their great size: skull length 210 millimeters (8.2 inches), breadth 160 millimeters (6.2 inches), height 148 millimeters (5.8 inches), capacity, 1,980 cubic centimeters, and powerful lower jaw with incisors and canines much larger than those in modern man—hence a new species of man for which he proposes the name *Homo capensis.*

If the above were measurements in the true sense, there might be some reason for his conclusion, but they are simply estimates. The skull was not complete enough to obtain the antero-posterior diameter; even if it had been, the unusual figure of 210 millimeters might still be due to a post-mortem spreading of the sagittal arch. Likewise a slight spreading of the transverse arch would lead to erroneous conclusions as to the breadth of the skull. That such is the case is apparent from the reproduction of the norma occipitalis published in *Nature,* August 5, 1915, and later by Haughton.[6] Elliot Smith, to whom an intracranial cast was sent, thought it to be "relatively remarkably flat" in view of the great breadth of the skull. Would it not be safer to assume a post-mortem flattening of the skull cap rather than that a new human type with enormous head had been discovered?

As to the lower-jaw fragment, Haughton called it "small and akin in characters to that of the Bantu or Bushman type." Judging from the half-tone illustration accompanying the paper by Broom, reproducing the fragment of the mandible natural size, the Boskop lower jaw does not differ from the average human lower

---

[6] Transactions of the Royal Society of South Africa, Vol. VI, Part 1, 1917.

jaw. Broom is correct in calling the only tooth present the second molar, but it is no larger than any other well developed second lower molar. He gives the greatest diameter of the canine socket as 9 millimeters (0.35 inch), and questions whether "this large socket has been formed by the root of a tooth or very largely by pyorrhoea." The latter of course might be the case, but it is not necessary to invoke a pathological condition in order to account for the socket's amplitude. In a collection of Inca skulls, for example, the author had no difficulty in finding lower canines with a maximum diameter of 9 millimeters. No importance need be attached to the fact that the canine was twisted on its axis and hence not set with its greatest diameter transverse to the jaw. The distance from the anterior margin of the canine socket to the plane of the symphysis proves likewise that the Boskop incisors were not of unusual size.

Thus the characters on which Broom relied to build up a new species of *Homo* do not seem to have any basis in fact. Would it not be better, therefore, to withhold the use of the term *Homo capensis* until a less "annoyingly imperfect" specimen shall have been found in South Africa?

Höhlefels.—This station (in Württemberg) belongs to the Upper Paleolithic. Scattered among the cultural remains were isolated human skeletal fragments—part of the right side of a cranium consisting of parietal, temporal, and half the occipital, apparently of the Cro-Magnon type; the right ramus and chin of four lower jaws, somewhat more primitive than modern European lower jaws and partaking of certain characters seen in both the Neandertal and Aurignacian races. According to Klaatsch, we have here a mixed type.

Bibliography *

ANTHONY, R. (with M. Boule), "L'encephale de l'homme fossile de la Chapelle-aux-Saints," *Anthr.*, xxii, 129–196 (1911).
—— "Le cerveau des hommes fossiles," *BMSA*, 7th ser., iv, 54–68 (1923).

*Supplementing the bibliography under Paleolithic sites in Appendix I.

AICHEL, O., "Die Beurteilung des rezenten und praehistorischen Menschen nach der Zahnform," *Zeitschr. f. Morphol. und Anthr.,* xx, 457 (1917).

ARCELIN, F. (with L. Mayet), *RA*, xxxiv, 38–66 (1924).

BERRY, Edward W., "The Age of *Pithecanthropus erectus*," *Science,* N. S., xxxvii, 418–420 (1913).

BLANCKENHORN, M., "Vorlage eines fossilen Menschenzahns von der Selenka Trinil Expedition auf Java," *Zeitschr. f. Ethnol.,* xlii, 337–354 (1910).

BOULE, M., "L'homme fossile de la Chapelle-aux-Saints," *Ann. de paléont.,* vi, 111–172 (1911); vii, 21–192 (1912); viii, 1–170 (1913).

BROCA, Paul, "Sur les crânes et ossements des Eyzies," *BSA*, 2d ser., iii, 350–392, 432–446, 454–514 (1868); also Broca's *Mémoirs,* Vol. II.

BROOM, R., "Evidence Afforded by the Boskop Skull of a New Species of Primitive Man (*Homo capensis*)," *Anthr. Papers Amer. Mus. Nat. Hist.,* xxiii, Part 2, 67–79 (1918).

COOK, W. H. (and Arthur KEITH), "On the Discovery of a Human Skeleton in a Brick-Earth Deposit in the Valley of the River Medway at Halling, Kent," *JAI*, xliv, 212–240 (1914).

DUBOIS, Eugene, *Pithecanthropus erectus, eine menschenähnliche Ueber-gangsform aus Java,* 4to, 39 pp., 2 pls. (Batavia, 1894).

—— "The Proto-Australian Fossil Man of Wadjak, Java," *Proc. Akad. van wetenschappen,* xxiii, No. 7 (Amsterdam, 1921).

——, "On the Cranial Form of *Homo neandertalensis* and of *Pithecanthropus erectus* Determined by Mechanical Factors," *Proc. Akad. van wetenschappen,* xxiv, Nos. 6 and 7 (Amsterdam, 1922).

FILHOL, H., "Note sur une mâchoire humaine trouvée dans la caverne de Malarnaud près Montseron (Ariège)," *Bull. Soc. philomathique de Paris,* 2nd ser., ii, 69–82 (1889).

FRAIPONT, Julien, and LOHEST, Max, "La race de Neanderthal ou de Canstadt en Belgique," *ABB*, vii, 587–757 (1887).

FRIEDENTHAL, H., "Ueber einen experimentelen Nachweis von Blutverwandtschaft," *Arch. Anat. Physiol., Phys. Abt.,* 494–508 (1900); 1–24 (1905).

GORJANOVIĆ–KRAMBERGER, K., *Der diluviale Mensch von Krapina in Kroatien,* 4to, 277 pp., 14 pls. (Wiesbaden, 1906).

GREGORY, Wm. K., "Studies on the Evolution of Primates," *Bull. Amer. Mus. Nat. Hist.,* xxxv, 239–355 (1916).

—— (with Milo Hellman), "Notes on the Type of *Hesperopithecus haroldcookii* Osborn," *Amer. Mus. Novitates,* No. 53, 16 pp., 6 figs., January 6, 1923,

GREGORY, Wm. K. (with Milo Hellman), "Further Notes on the Molars of *Hesperopithecus* and of *Pithecanthropus*" (with an Appendix: "Notes on the Casts of *Pithecanthropus* Molars," by Gerrit S. Miller, Jr.), *Bull. Amer. Mus. of Nat. Hist.*, xlviii, art. xiii, pp. 509–530, December 4, 1923.

HERVÉ, Georges, "La race des troglodytes magdaleniens," *REA*, iii, 173–188 (1893).

HRDLIČKA, A., "The Most Ancient Skeletal Remains of Man," 2d edit., *Smithsonian publ.* No. 2300 (Washington, 1916).

—— "The Piltdown Jaw," *Amer. Journ. Phys. Anthr.*, v, 337–347 (1922).

—— "New Data on the Teeth of Early Man and Certain Fossil European Apes," *Amer. Journ. Phys. Anthr.*, vii, 109–132 (1924).

KLAATSCH, H., *Homo mousteriensis hauseri*, AA, N. F., vii, 287–297 (1909).

—— "Die Aurignac-Rasse und ihre Stellung im Stammbaum der Menschheit," *ZE*, xlii, 513–577 (1910).

—— (with O. Hauser), *Homo aurignacensis hauseri*, PZ, i, 272–338 (1910).

MAHOUDEAU, Pierre G., "Le Pithecanthrope de Java," *RA*, xxii, 453–472 (1912).

MAKOWSKY, Alex., "Der diluviale Mensch im Löess von Brünn mit Funden aus der Mammuthzeit," *MAGW*, xxii, 73–84 (1892).

MANOUVRIER, L., "Discussion du *Pithecanthropus erectus* comme précurseur présumé de l'homme," *BSA*, vi, 12–46, 555–651 (1895); vii, 396–460 (1896).

MARTIN, Henri, *L'homme fossile de La Quina*, monograph published in 1923 by Doin, Paris.

MAŠKA, K., "Fund des Unterkiefers in der Schipka-Höhle," *VBGA*, xviii, 341–350 (1886).

MATTHEW, W. D., "Climate and Evolution," *Annals N. Y. Acad. of Sci.*, xxiv, 171–318 (1914).

MILLER, Gerrit S., Jr., "The Jaw of Piltdown Man," *Smithsonian Misc. Colls.*, lxv, No. 12, 31 pp., 5 pls. (1915).

—— "The Piltdown Jaw," *Amer. Journ. Phys. Anthr.*, i, 25–52 (1918).

MORTON, Dudley J., "Evolution of the Human Foot," *Amer. Journ. Phys. Anthr.*, v, 305–336 (1922); vii, 1–52 (1924).

—— "Evolution of the Longitudinal Arch of the Human Foot," *Journ. Bone and Joint Surgery*, vi, 56–89 (1924).

OBERMAIER, H., "Les restes humaines quaternaire dans l'Europe centrale," *Anthr.*, xvi, 385–410 (1905); xvii, 55–80 (1906).

OPPENHEIM, Stephanie, "Zur Typologie des Primatencraniums," *Zeitschr. für Morphol. und Anthropol.*, xiv, 203 pp., 14 pls., Stuttgart (1911).

Osborn, Henry F., "*Hesperopithecus*, the First Anthropoid Primate Found in America," *Amer. Museum Novitates*, No. 37, pp. 1–5, 3 figs. (1922).

Pearson, Karl, "Mathematical Contributions to the Theory of Evolution: V, Reconstruction of the Stature of Prehistoric Races," *PT*, Ser. A, cxcii, 169–244 (1899).

Pilgrim, E., "New Siwalik Primates and Their Bearing on Questions of the Evolution of Man and the Anthropoidea," *Records Geol. Sur. of India*, xlv, Pt. I, 1–74, 4 pls. (1915).

Schoetensack, Otto, *Der Unterkiefer des Homo heidelbergensis aus dem Sanden von Mauer bei Heidelberg*, 4to, 67 pp., 13 pls. (Leipzig, 1908).

Schwalbe, G., *Die Vorgeschichte des Menschen*, 8vo, 52 pp. (Braunschweig, 1904).

Selenka, L. (and M. Blanckenhorn), *Die Pithecanthropus-Schichten auf Java. Geologische und palaeontologische Ergebnisse der Trinil-Expedition*, etc., 4to, 310 pp., 32 pls. (Leipzig, 1911).

Smith, G. Elliot, Address as President of Section H, Brit. Assoc. for the Adv. of Science., Rept. of meeting at Dundee, pp. 578–598 (1912).

—— *Proc. Roy. Soc.*, lxxxvii, Ser. B, meeting of Feb. 19, (1914).

Sollas, W. J., "On the Cranial and Facial Characters of the Neandertal Race" *PT*, Ser. B, cxcix, 281–339, 1 pl., (1907).

—— "Paviland cave: an Aurignacian station in Wales," *JAI*, xliii, 326–374 (1913).

Spitzka, E. A., "A Study of the Brains of Six Eminent Scholars belonging to the American Anthropometric Society," *Trans. Amer. Philos. Soc.*, N. S., xxi, Part III, 175–308 (Phila., 1907).

Stolyhwo, Kazimierz, "*Homo primigenius* appartient-il à une espèce distincte de *Homo sapiens*," *Anthr.*, xix, 191–216 (1908).

Teilhard de Chardin, Pierre, "Sur quelques primates des phosphorites du Quercy," *Annales de Paléontol.*, x, 1–20 (1916–1921).

—— "La présence d'un Tarsier dans les phosphorites du Quercy et sur l'origine tarsienne de l'homme," *Anthr.*, xxxi, 329–330 (1921).

Testut, L., "Recherches anthropologiques sur le squelette quaternaire de Chancelade," *Bull. Soc. d'Anthr. de Lyon*, viii, 121 pp., 14 pls. (1889).

Verneau, R., "La race de Spy ou de Neanderthal," *REA*, xvi, 388–400 (1906).

—— *Les grottes de Grimaldi* (*Baoussé-Roussé*), II, Fasc. 1, 4to (Monaco, 1906).

Verworn, Max (with Bonnet and Steinmann), "Diluviale Menschenfunde in Obercassel bei Bonn," *Die Naturwissenschaften*, ii, 645–650 (1914).

—— (with Bonnet and Steinmann), "Der diluviale Menschenfund von Obercassel bei Bonn," 4to, 193 pp., 42 text figs., 28 pls. (Wiesbaden, 1919).

VIRCHOW, Hans, "*Die menschlichen Skeletreste aus dem Kämpfe'schen Bruch im Travertin Ehringsdorf bei Weimar*, 141 pp., 8 pls. (Jena, 1920).

VOLKOV, Th., "Variations squelettiques du pied," *BMSA*, 5th ser., iv, 632–708 (1903); v, 1–50, 201–331 (1904).

WERTH, E., "*Der Fossile Mensch*," Part I, 366 pp., 217 text figs. (Berlin, 1921).

WEIDENREICH, Franz, "Evolution of the Human Foot," *Amer. Journ. Phys. Anthr.*, vi, 1–10 (1923).

WOOD-JONES, F., *The Problem of Man's Ancestry* (London, 1918).

WOODWARD, A. S. (with C. Dawson), "On the Discovery of a Paleolithic Human Skull and Mandible, etc., at Piltdown, Fletching (Sussex)," *QJGS*, lxix, 117–151, pls. 15–21 (1913) (appendix by G. Elliot Smith).

—— *A Guide to the Fossil Remains of Man, etc., in the British Museum (Nat. Hist.)* (London, 1922).

CHAPTER IX

Having now reached the end of our survey of the Old Stone Age, we may well pause to summarize the main features of this dawn era before entering the transitional period which foreshadowed the increasingly rapid and complex development of human culture during the later ages of prehistory.

With the exception of two isolated discoveries, our present knowledge of the Old Stone Age is based on finds that have been recorded during the past one hundred years. With progress in this field of research there has come the realization of the unity of prehistory. There are no longer hiatuses between the Old Stone Age and the New Stone Age, or between the New Stone Age and the Age of Metals. There are still gaps in our knowledge; but these are gradually being filled so that the prehistoric fabric is seen as a homogeneous whole in proportion as new light is shed upon it.

The Old Stone Age being farthest removed from us, the evidence bearing upon it is less distinct and more fragmentary than that bearing upon the later prehistoric ages. The mass of evidence that has been accumulated is due to the immeasurably long period covered and to the aid furnished by such cognate sciences as geology, paleontology, zoölogy, and comparative anatomy. The geologic sciences have been indispensable in determining the age of the deposits that have yielded the oldest relics of man, while zoölogy and comparative anatomy have made it possible to fix approximately the place man occupies in nature. These are two of the cardinal points in the exploration of the prehistoric world; the other two are a comparative study of the material and social cultures of primitive living races.

Prehistory is not an exact science in the sense that astronomy is. Much of the record is yet to be revealed, and still more has been irretrievably lost. Moreover, the Old Stone Age chapter of prehistory lies wholly within the realm of the geologic past, and geologic time can be measured only approximately. Hence, Old Stone Age chronology is relative, not absolute, and this applies both to its beginning and to its close. We have to deal with an age without definitely known boundaries (as was the earth in the time of Columbus), but we have explored it sufficiently to know that it was enormously long.

To the average mind, much more importance attaches to the beginning than to the close of the Old Stone Age. When did it begin? This question can be answered by saying that the Old Stone Age is as old as man himself, which leads to the query, How old is man? This can only be answered by first determining what constitutes man, and here one comes to the difficulty that has always been a stumbling-block to progress in the study of human origins—the difficulty inherent in passing judgment on one's own case. An answer which should approximately meet the requirements would be that the human prototype passed the line into man's estate when he became man both physically and mentally (or culturally).

At least three physical factors are requisite—the hand, a brain that is fairly well balanced on a spinal column normally approximately erect, and stereoscopic vision. These three factors cannot be divorced; rather do they form a physical complex, triplets born at one and the same time. Given this physical complex, a culture that we may call human would as surely follow as does the day the night. Tools without a hand are unthinkable; but given a hand controlled by a brain unfettered in its chances of development, tools will in due course be selected and even invented.

The record, fragmentary though it is, points to a conjunction of the physical and cultural requirements necessary to constitute nascent *Homo* somewhere in the late Tertiary Epoch; this conjunction marks the beginning of the Old Stone Age. A period when man without a cultural background began to create one for himself may properly be spoken of as an Age of Stone, because stone tools, being practically imperishable, have been preserved to our day. Even had there been no stone, or if nascent man had neglected to use it, cultural evolution would have begun just the same, for its day had come; but evidence of its existence would be lacking because of the perishable nature of the materials employed.

The human era is infinitesimally short when compared with the age of the earth, or even with the era of life on the earth. It is long only when compared with that portion which is measured by historic records. The moment one leaves the historic realm and attempts to penetrate into and explore the prehistoric realm, he must plant one foot firmly on the rock of stratigraphy and the other on the solid basis of organic and cultural evolution.

A prehistoric relic found *in situ* in an undisturbed deposit is at least as old as the deposit itself, and here is where the geologist and paleontologist can help the prehistorian. A prehistoric relic found beneath an undisturbed deposit is older than the deposit. Among the first things to attract the attention of the student of the Pleistocene Epoch are the series of glacial and interglacial deposits and the succession of warm and cold faunas. A knowledge of the Ice Age, therefore, is a prerequisite to the proper dating of associated human skeletal and cultural remains.

So much for the stratigraphic record in deposits formed wholly (or almost wholly) through natural agencies, such as boulder clay, glacial till, loess, travertine, forest bed, peat, erosion and fill, etc. There is another important class of evidence found in deposits resulting for the most part from human habitation, for example, the floor deposits of caves and rock shelters and the accumulations under lake villages. These sites, more or less protected from natural agencies, reveal a sequence second in importance only to the stratigraphic record proper. The evidence from these two sources is so fundamental and indispensable that the author has deemed it worth while to list in Appendix I every known station that has more than one relic-bearing horizon. The number and character of the stations in this list afford abundant proof of the solid foundation on which prehistory now rests. In this list it is always stated whether the deposits are due primarily to natural agencies (stratigraphy properly so called) or primarily to human agency (culture sequence).

Only in rare instances does one find the two kinds of deposits in the same site. Such happy combinations make possible a direct correlation of the evidence from both sources; but they can hardly be called absolutely necessary because of the consistently harmonious character of the evidence afforded by each of the two classes of sites in question. There are other points of contact between the two groups of sites, such as faunal remains, both human and animal, and the nature of the cultural remains. The evidence from these sites shows not only that man inhabited Europe throughout the Pleistocene, but also that flint tools were chipped and utilized by the precursor of man as far back as the Upper Tertiary.

At no time during the Ice Age did man have to abandon Europe. The successive glacial and interglacial eras served rather as an aid than a hindrance to his organic and cultural evolution; they served to make him more adaptable to changing climatic conditions and eventually fitted him to survive in all latitudes and altitudes. The authors of Pre-Chellean culture lived through the first two glacial epochs and the intervening interglacial epoch (Günz-Mindel). The long Mindel-Riss Interglacial Epoch witnessed a distinct advance over Pre-Chellean culture to the stage known as Chellean and to the first phase of the succeeding Acheulian stage. The Acheulians lived on through the Riss Glacial to be followed during the Third, or Riss-Würm, Interglacial by the Warm Mousterians, whose skeletal and cultural remains are so well typified at Ehringsdorf and Taubach. During the Fourth, or Würm, Glacial there was a distinct quickening of cultural progress. The Cold Mousterians (Neandertal race) survived the Würm advance to fall before the Aurignacians, far superior both physically and culturally. The later phases of the Würm Glacial are synchronous with the Solutrean and Magdalenian Epochs. With these phases there disappeared from western Europe the last of the races of the Old Stone Age and a culture possessing many interesting and unique features, some of which are linked up with the dawn of art.

The oldest known cultural relics are the utilized flints commonly known as eoliths, but in all probability the earliest races of man made use also of bone and wooden tools and perhaps shells. The oldest known bone fashioned by man, presumably for some definite purpose, is the piece of leg bone of *Elephas* found at Piltdown (Sussex). Wood is still more perishable, yet under suitable conditions it may be preserved for an indefinitely long period. A pointed wooden implement, perhaps as old as the one of bone from Piltdown, was found in *Elephas antiquus* beds (early Pleistocene) at Claxton-on-Sea. These are but rare exceptions which prove the rule and justify the use of the term Old Stone Age.

The Age as a whole admits of division into two periods, the Eolithic and the Paleolithic, corresponding to a portion of the Tertiary and the Quaternary respectively. The Paleolithic Period is subdivided into seven epochs: the Pre-Chellean, Chellean, and Acheulian (Lower Paleolithic); Mousterian (Middle Paleolithic); and the Aurignacian, Solutrean, and Magdalenian (Upper Paleolithic). The culture of the Lower Paleolithic Period is best represented in France, southern England, Belgium, Spain, and Portugal; it occurs, but to a lesser extent, in Germany, Italy, Poland, and Monaco. Outside of Europe we find it in Algeria, Egypt, Sahara, Congo, and South Africa; also in various parts of Asia—notably in Syria, Mesopotamia, Indo-China, and Japan. These countries all lie in tropical to middle latitudes.

The chipped flints of the Upper Tertiary are associated with a warm Eurasiatic fauna—*Elephas meridionalis, E. antiquus, Mastodon arvernensis,* and a large horse (*Equus stenonsis*). Pre-Chellean man came in with a forest to steppe fauna, which gradually changed to tundra, including the woolly elephant, musk ox, and reindeer as the Günz Glacial reached its maximum; he lived on through the Günz-Mindel Interglacial to witness a complete faunal change to such types as the hippopotamus, *Rhinoceros etruscus,* the saber-toothed tiger (*Machaerodus*), and the two southern elephants (*E. meridionalis* and *E. antiquus*); and he even survived another glacial epoch, the Mindel, with a fauna not unlike that of the First Glacial Epoch. Chellean and Lower Acheulian man was contemporaneous with the second warm fauna, composed of hippopotamus, *Rhinoceros merckii, Elephas antiquus,* the red deer, and such shells as *Belgrandia marginata* and *Unio littoralis.* The red deer points to the beginning of a temperate climate; before the close of the Acheulian Epoch a cold fauna, represented by the mammoth (*Elephas primigenius*) and the woolly rhinoceros (*R. tichorhinus*), reappeared.

The first part of the Middle Paleolithic Period was synchronous with the last warm fauna known to Europe—*Elephas antiquus* and *E. trogontherii, Rhinoceros merckii,* and hippopotamus. The latter part of this period, the Cold Mousterian Epoch, witnessed the return of a cold fauna once more, for one finds the remains of Neandertal man associated with the mammoth, woolly rhinoceros, reindeer, horse, etc. This cold fauna persisted throughout the Upper Paleolithic Period and thus was associated with the Cro-Magnons and Magdalenians, who immortalized it in their art.

The physical record of man's existence during the Old Stone Age is even less delible than the cultural, because the only parts of man's physique which stand any chance of being preserved are his bones, and these are less durable than stone. Passing over the fossil apes, we come to *Pithecanthropus erectus,* which lived during an early phase of the Old Stone Age, and which itself might well have been a tool user (although no tools were found in the same deposits). Recent studies of *Pithecanthropus* based largely on endocranial characters lead one to the conclusion that its kinship with *Homo* is very close.

The apelike lower jaw found at Piltdown is probably that of an early type of *Homo,* to which the name *Eoanthropus* might well apply; besides, it was found associated with flints that had apparently been utilized. Whether it actually belongs with the Piltdown cranium is a question largely of academic importance; the latter certainly represents *Homo* both physically and culturally. Of equal or perhaps greater antiquity is the human lower jaw from Mauer, near Heidelberg, which physically measures up to that required for *Homo,* although it was not directly associated with cultural remains.

Coming to the long Mousterian Epoch corresponding to the Middle Paleolithic Period, one finds the earlier skeletal remains associated with a warm fauna, and the later with a cold fauna. During this long period, partly glacial and partly interglacial, the human type (as well as the culture) changed but little; it is generally referred to as *Homo neandertalensis,* best known by the remains associated with a cold fauna. The best preserved remains of Cold Mousterian, or Neandertal, man are those from La Chapelle-aux-Saints, Neandertal, Spy, La Ferrassie, La Quina, and Gibraltar. About a dozen other stations have yielded skeletal remains of Cold Mousterian Man. The record in regard to interglacial man of practically the same type, but living at an earlier period, is much less complete, being limited to parts of at least two individuals from Ehringsdorf, one from Taubach, and probably some ten from Krapina.

Enough finds of human skeletal remains have already been reported to show that man of the Cold Mousterian Epoch roamed over practically the whole of western Europe—Germany, Czechoslovakia, Belgium, France, Island of Jersey, southern England, and Spain. His cultural remains have been found over a much more extended area in Europe (everywhere in middle latitudes except Bulgaria and Rumania), as well as in parts of Africa and Asia. Recently the skeletal remains of an individual with certain very pronounced Mousterian characters were discovered at Broken Hill (Rhodesia); there is, however, as yet no way of determining its age. The associated artifacts are nondescript in character, the animal remains are closely allied with species still living in that region.

The Upper Paleolithic Period is marked by a very pronounced change in both cultural and physical types, also by the origin and evolution of Paleolithic art. The passage from the Middle to the Upper Paleolithic Period represents what might be called a mutation rather than a gradual unfolding of that which had gone before. The climate was glacial to steppe, and the fauna furnished not only food, but also warm clothing for the hunter population as well as raw materials useful in tool making and in the fine arts—bone, ivory, and reindeer horn. Fine needles with delicate eyeholes were made from bone and ivory. Reindeer horn was the favorite material from which harpoons were shaped. The first harpoons appeared about the middle of the Magdalenian; these usually had but a single row of barbs, relatively small and numerous. In the next stage, harpoons with a single row of lateral barbs predominate, but the barbs are longer and there is a protuberance near the base of the shaft. During the closing phase of the Magdalenian, harpoons with a double row of prominent barbs take the lead. Javelin points of bone, ivory, and reindeer horn were much in evidence; dart (or javelin) throwers of similar materials, sometimes bearing beautifully carved or incised ornaments, were used to increase the velocity of the projectile.

The cleaver and scraper of the Lower and Middle Paleolithic Periods were superseded by a stone industry based on the evolution of the bladelike flint flake, made possible by improved methods in the preparation of the flint nucleus. In addition, there appear during the Solutrean Epoch two new types of flint implement evincing a high degree of skill in workmanship, the laurel-leaf point and the willow-leaf point produced by means of pressure flaking.

Art of the Old Stone Age was born, realized its full fruition, and finally decayed during the Upper Paleolithic Period. Its inception was closely linked up with the religious instinct and with the belief in the efficacy of magic; it flourished until better means of accomplishing the desired results had been devised, and finally became dormant through lack of a vital stimulus. Paleolithic art was primarily realistic; its models existed in nature rather than in phantasy or the desire to please the eye.

The oldest examples of art include animal figures sculptured in the round. The execution of figures in relief or on a plane surface requires a greater stretch of the imagination and a more complex degree of artistic skill, because the artist much first decide upon a given aspect for representation—side view, three-quarter, front, or rear. In actual practice the Paleolithic artist found that the profile of a quadruped offered the fewest difficulties, and very seldom did he undertake to render any other aspect. The human form was for him a problem apart. So far as the human head was concerned, the rule that applied to quadrupeds held good; but the front view of the rest of the figure must have been more satisfactory than the profile, since a majority of engraved and relief figures of the human form show the front view of the body with the head turned to one side.

The use of color was the third step in the evolution of Paleolithic art. Simple contour drawings in red or black may be as old as the earliest engravings; shading in one or more colors was a later development. The mural frescoes in polychrome are all obviously the work of Magdalenian artists, who also left behind them some interesting examples of stylistic art, including motives derived from animate forms.

In the field of Paleolithic art fauna plays the dominant rôle; among faunal representations, mammals (including man) far outnumber all others. Birds and fishes are a poor third, while invertebrate figures have thus far been reported from only five localities. Figures representing plants are likewise very rare. If one take as a guide the number of stations where representations of a given animal have been found, the horse family makes the best showing, occurring as it does in seventy-eight stations distributed over England, Germany, Switzerland, France, Spain, and Italy. The artist's work is so true to nature that one can distinguish three varieties of the horse and one of the ass.

After *Equus,* the following mammals (including man) occur in the order given: *Homo,* sixty-three stations; red deer (*Cervus elaphus*), fifty; Bison, forty-six; wild goat, thirty-six; Bovidae (chiefly *Bos primigenius*), thirty-five; reindeer, twenty-eight; Cervidae, twenty-seven; *Ursus,* thirteen; *Felis,* eleven; *Rhinoceros,* nine; chamois, seven; Canidae (not distinctive enough to admit of finer classification), six; seal, six; Capridae, saiga antelope, and wolf, five each; musk ox and wild boar, four each; elk, fox, and hyena, three each; glutton, two; and badger, hare, lynx, otter, Rodentia, roebuck, and wild sheep, one each.

Representations of the bird family have been found in nineteen stations: in France, twelve; Spain, four; and Germany, Italy, and Russia, one each. They include the crane, duck, goose, grouse, owl, penguin, partridge, and swan—practically all edible forms.

The varieties of fish that figure in cave art include the carp (?), flounder, pike, plaice, salmon, Spanish mackerel, and trout. Examples have been found in thirty-one stations: in France, twenty-five; Spain, three; Belgium, Czechoslovakia, and Poland, one each. Figures that might have been intended for the eel, or the snake, have been reported from five stations in France.

Invertebrates and plant life played a very insignificant rôle in cave art. In the Daleau collection there is an ivory facsimile of a *Cypraea* shell from Pair-non-Pair (Gironde); Hamal-Nandrin has recently found an ivory insect (Coleopter) in the Grotte du Coléoptère (Luxembourg); other Coleopters have been reported from Arcy (Yonne); and the rock shelters of Cap-Blanc and Laugerie-Basse (Dordogne). Plantlike representations are almost as rare, having been reported from only seven French stations.

The inanimate world did not appeal to the Paleolithic artist. Stylistic figures and decorative motives are not very plentiful; moreover, most of them, including alphabetiform signs, chevrons, frets, spirals, volutes, and wave ornaments, can be traced to their original sources in the animate world. Exceptions to this rule are practically confined to claviforms, darts, and tectiforms.

The cave artist had his own canons, among which the matter of scale had its place. In proof of this many examples might be cited—the male and female of the same species, the mother and its young, a herd of a given kind, the hunter and the animal hunted, etc. Figures in accidental juxtaposition, executed at different times and by different artists, would of course not conform to the same scale. There are also cases where, for the sake of emphasis, the rule was ignored.

Cave art can be dated where a definite time relation can be traced connecting it with datable objects: for example, when it occurs *in situ* in deposits of known age or, if mural art, when it is found to have been covered by deposits the age of which can be determined. Again, where there is a superposition of engraved and painted mural figures, one can tell which is the older and which the younger. There remain a large number of examples the age of which can only be determined approximately by comparison with art objects bearing a definite relation to the established time scale.

The Upper Paleolithic Period is divided into three epochs: the Aurignacian, Solutrean, and Magdalenian. The Aurignacian and Solutrean agree with the Cold Mousterian in being absent from Switzerland, a fact which throws light on the relation of these three epochs to the last glacial epoch (Würm). The Cold Mousterians, Aurignacians, and Solutreans who happened to live during a glacial epoch, were all barred from Switzerland because of its elevation. Before the close of the Magdalenian Epoch, the Würm glaciation had spent its force to a degree sufficient to allow an invasion of the lower altitudes by the Magdalenians, who left their traces in some thirteen Swiss stations. Outside of Switzerland the geographic distribution of Aurignacian culture coincides with the Magdalenian to a marked degree, although the latter reaches somewhat higher latitudes, as might be expected.

A primitive culture like the Magdalenian, stretching as it does all the way from Spain to Irkutsk (and probably to China), would be subject to regional variation. In the Yenisei valley it has a Siberian facies, harking back to Mousterian prototypes; in northern Africa there is a Getulian facies; while in Italy there is a Magdalenian equivalent in which the reindeer plays no part. In comparison with Aurignacian and Magdalenian, Solutrean culture is limited both in geographic distribution and in the number of stations where it occurs.

The Old Stone Age represents a long period of human struggle for mastery over self and environment. The period was long because there was at first no background of inheritance on which to draw for help. Man had first to learn to stand erect then to stand alone until in due time there was accumulated a mass of racial experience on which he could rely and against which he could support himself when hard pressed. Then, as now, the big battles were fought and won by the gifted few who shaped the course of progress, while the rate of progress depended on the ability of the many to profit by the achievements of the few.

One of the biggest achievements of all time, the taming of fire, must be credited to the Old Stone Age. Another notable achievement bearing a certain relation to the fire-using habit was that of supplementing natural means of conserving body heat by the use of clothing made of the skins of animals. Provided with fire and clothing, it became possible for man to penetrate into more northern latitudes and to ascend to higher altitudes; these served as a double door through which man could pass to all parts of the world. When to these two great achievements there is added what man of the Old Stone Age accomplished in the field of art, one is amazed at the results. His was the prototype of all pioneer work, and the legacy he left to the succeeding races was immeasurably great because he began as a bankrupt so far as culture was concerned. Fire, clothing, and the fine arts, if these three cornerstones of present-day civilization were removed, the structure would be, to say the least, in dire need of repair.

Certain elements in the prehistoric culture complex can be treated to best advantage if followed vertically upward through succeeding epochs. Such a treatment makes it possible to pick up a given cultural thread and follow it as long as there is no break. There are threads that bind age to age, period to period, and epoch to epoch. These are the time-binding elements of the complex, running through and overlapping on the periods. They bind the Old Stone Age to the New to such a degree as to form a single complex for the Age of Stone as a whole. A fuller treatment of this combined complex is reserved for Chapter XII.